ANIMAL SCIENCE, ISSUES AND PROFESSIONS

MONKEYS

BRAIN DEVELOPMENT, SOCIAL AND HORMONAL MECHANISMS AND ZOONOTIC DISEASES

ANIMAL SCIENCE, ISSUES AND PROFESSIONS

Additional books in this series can be found on Nova's website
under the Series tab.

Additional e-books in this series can be found on Nova's website
under the e-book tab.

ANIMAL SCIENCE, ISSUES AND PROFESSIONS

MONKEYS

BRAIN DEVELOPMENT, SOCIAL AND HORMONAL MECHANISMS AND ZOONOTIC DISEASES

HUGO A. BARRERA SALDAÑA
EDITOR

nova publishers
New York

Library of Congress Cataloging-in-Publication Data

Monkeys : brain development, social & hormonal mechanisms and zoonotic diseases / editor, Hugo A. Barrera Saldaqa (Universidad Autsnoma de Nuevo Lesn, Pedro de Alba Ciudad Universitaria, Nuevo Lesn, Mexico).
pages cm. -- (Animal science : issues and professions)
Includes index.
ISBN: 978-1-63117-851-1 (hardcover)
1. Monkeys--Nervous system. 2. Monkeys--Psychology. 3. Monkeys--Diseases. I. Barrera Saldaqa, Hugo A.
QL737.P9.M597 2014
599.8--dc23

2014013783

Published by Nova Science Publishers, Inc. † New York

CONTENTS

PREFACE

Monkeys are superior experimental models than those traditionally used in scientific laboratories to address questions on complex research issues aimed at dealing with some unique features of our human nature. Given their close evolutionary proximity to our species, most research tools designed for human studies easily help us to dig deeper for clues to support hypotheses that otherwise are testable with limitations in simpler models, such as non-primate mammals.

If as said by the character in the movie Shrek, humans are like onions—made of layers, monkeys are made perhaps of the same layers except for the outermost. Thus we share most of our genetics, physiology and pathology, in addition to our behavior and social conduct. Moreover, in some areas in these two external layers with which we are different from monkeys, thickness is uneven and thus this more or less separates us from them, depending on the area in question.

This book is a collection of fascinating contributions from research experts around the world studying, in monkeys, their brain, behavior, metabolic control, and diseases that pose potential treats to our species.

IN THE BRAIN DEVELOPMENT SUBJECT

Ibañez-Contreras and coworkers and Hernández-Godinez and coworkers describe that since their early appearance during the late Paleocene primates have developed and perfected most of their basic morphologic characters, such as prehensile extremities and other characteristics such as the frontalization of both ocular orbits, which constitute mere adaptations for their effective displacement in an arboreal environment but without limiting them to exercise a strictly arboreal life. The developmental maturation of the primate brain including sound cranial morphologic changes, have reached the maximum extent of transformation within the Mammalian class and throughout the Animal Kingdom. Anatomically speaking, although the brain of primates preserves the same structure seen in other mammals their larger size and inner complexity allows them to gather a broader knowledge of their close surrounding world, which in turn generates more elaborated and original responses to their immediate environment. In the more advanced primates their associative areas acquire larger predominance, since designed as such these are responsible for the integration of all activities in complex segments of their brains, while acting also as basic units of memory storage, therefore allowing them to exercise a coordinating role among the primary perception of stimuli and voluntary responses. Given the position that NHP occupy within the evolutionary tree and for being the closest phylogenetic species to humans, the rhesus monkeys (*Macaca mulatta*) constitute an ideal animal model for the study of fundamental neural mechanisms occurring throughout the developmental and maturation phases of life from neonatal to senile stages.

In the Social Aspects

In the chapter by Amici, the most recent findings on social cognition in New World monkeys are reviewed to understand the evolutionary factors that most likely shaped the distribution of social cognitive skills. The author discusses the different paradigms that have been used to test social cogniton, including the ability of New World monkeys to recognize other individuals, to understand their behaviour and direct and third-party relationships. The author reports the results of studies on New World monkeys' ability to reciprocate cooperate and interact with other group members. Finally, the author provides information on monkeys' gestural and vocal communication, gaze following skills, tool use, culture and theory of mind.

In the chapter by Amici & Braeuer, the authors review the most important findings on physical cognition in New World monkeys. The authors discuss the different paradigms that have been used to test physical cognitive skills in captive and partially wild monkeys. Moreover, they originally put together the results of the reported studies to obtain a more general picture of the distribution of physical cognitive skills across monkey species, in order to identify the evolutionary factors that might have shaped their specific cognitive skills.

In the Zoonotic Field

Chaves-Ramírez and Gutiérrez-Espeleta reminded us that non-human primates are an example of animal populations representing many wild reservoirs for zoonotic diseases. Habitat loss and fragmentation, environmental degradation, climate change and other anthropogenic factors can modify the dynamics of these populations.

The impact of infectious diseases on wild animal populations has been recognized as an important factor influencing the distribution and abundance of species. It is considered a critical component in conservation biology. As the natural habitat of non-human primates is reduced and fragmented by human activities, the populations are concentrated to small confined areas, favoring opportunities for transmission of diseases.

The occurrence of zoonoses and public health problems obey well-established causes and potentially preventable. Understanding the implications of the conservation of wild populations is essential to ensure their current and future survival.

Investigations of non-human primates requires a coordinated set of actions, where local specialists, fully identified with local realities and priorities, work together in a multidisciplinary task to develop solutions for the short, medium and long term conservation.

Shlabritz-Loutsevitch and coworkers give an account of the role of natural occurring infections in NHP used for experimental research. Among the Non-human primates (NHP) preferred by experimental medicine Baboons (*Papio spp.)* have been particularly valuable for reproductive research. They are a well characterized model of premature labor, endometriosis and pregnancy perturbations. And since the role of the naturally occurring infectious pathologies in the reproductive research in this species has received less attention, Natalia Schlabritz-Loutsevitch and coworkers call our attention to the essential participation of microbes in host functions in reproduction, metabolism, and even defense. They recall us that changes in fecal microbiome have been recently recognized as an essential part of both - inflammation and pregnancy. Thus, their chapter describes the most common infectious pathologies in NHPs and its influence on the reproductive research.

In the Hormonal Domain

Wallis and Wallis deal in their chapter with evolutionary aspects of growth hormone (GH) and prolactin genes in New World monkeys (NWM). The authors describe that in some mammalian groups duplications of the GH or prolactin gene have given rise to complex gene families expressed primarily in the placenta. In rodents and ruminants such duplications have involved the prolactin gene branch, while in primates this evolutionary process acted on the GH gene branch.

In most non-primate mammals pituitary GH is encoded by a single gene. This is true also for most prosimians, but in higher primates the GH locus comprises a cluster of GH-related genes, including the gene encoding pituitary GH at the 5' end, followed by a number of genes (and/or pseudogenes) at least some of which are expressed in the placenta. Such GH gene clusters arose independently in Old World Monkeys/apes and in New World Monkeys (NWM), as a consequence of multiple, tandem gene duplications followed by divergent evolution.

The authors continue by describing that following the independent duplications that gave rise to GH clusters on each of these lineages the rate of evolution of pituitary GH decreased, but the evolution of the other GH-like proteins remained high, apparently reflecting adaptive evolution. Although the biological function(s) of the GH-like proteins in NWM remain largely unknown, they go on to explain in detail the nature of their genes and encoded proteins. They close their chapter revealing that in contrast to the complexity shown by genes and proteins in higher primates, the related hormone prolactin is encoded by a single gene in all primates and shows only modest variation between NWM species.

On the other hand, Perez-Amaya and coworkers describe for us the rest of this amazing evolutionary story of GH and GH-related genes and their encoded proteins in primates, concentrating their chapter on the processes that took place in Old World Monkeys (OWM) and Great Apes, including man.

After divergence from NWM evolution continued acting on the ancestral pituitary GH gene in OWM to generate multiple copies of it, with the resulting gene cluster following two opposite evolutionary paths: Pituitary GH gene on one hand evolving slowly to accumulate just minor changes and on the other hand placental GH and placental lactogens (PLs, also known as Chorionic Somatomammotrophin Hormones or CSHs) experiencing accelerated evolutionary rates and structural and supposedly functional changes.

At the same time that the authors describe the gene composition and order in the multigene GH loci in species representative of OWM and great apes, they address the issue of regulation of gene expression. Control networks evolved to restrict expression of the ancestral GH gene mainly to the pituitary gland and of the remaining more recently acquired evolutionarily-speaking two types of genes (placental GH and PLs) primarily to the placenta.

Taking together these two chapters of the endocrine section of the book, they describe the unique forces that over 60 millions of years ended shaping the multigene GH family of the human genome, with the functions of its encoded hormones although still being elusive, maybe played a key role in the survival of our ancestors and thus of our species; this being a new challenge for the research in the years to come in the molecular biology and evolution of this fascinating endocrine gene family.

Finally, Tejero and coworkers highlight the these great similarities particularly on metabolic aspects is thoroughly documented by the authors that closes the book describing the baboon as an experimental animal model being used to discover the genetic and hormonal alterations behind the appearance of those chronic diseases that have become a pandemic not only in western societies, but also in the emerging economies that along their macro and micro economic successes, are changing their life styles, feeding habits and food type preferences.

ABOUT THE AUTHOR

Prof. Dr. Hugo A. Barrera-Saldaña
School of Medicine of the Autonomous University of Nuevo Leon (UANL).
Monterrey, Nuevo León, México.

Tel: +52 (81) 8329-4173
habarrera@gmail.com

1. Self-taught Biochemist (B.S. Magna cum laude, Biology plus many extracurricular courses in Genetics, Chemistry and Microbiology), School of Biological Sciences, UANL.
2. PhD, Biomedical Sciences with specialization in Molecular Biology, The University of Texas Health Science Center at Houston.
3. Postdoctoral Fellow, Louis Pasteur University, Strasbourg, France.
4. Professor, Department of Biochemistry and Molecular Medicine, School of Medicine, UANL.
5. Science and Technology Commercialization Specialist, The University of Texas at Austin.
6. Member, National Researchers´ System since its inception (currently at highest level: III).
7. Founder, Genetic Engineering and Gene Expression (ULIEG), Molecular Diagnostics (UDEM) and Medical Biotechnology (UBM) Units, Department of Biochemistry and Molecular Medicine, UANL School of Medicine (first genetic screening and gene therapy trials centers).
8. Creator, Graduate Programs of Genetic Engineering and Molecular Biology at UANL and of Biotechnological Genomics at National Polytechnic Institute (IPN).
9. Founder and first Director, Biotechnological Genomics Center (first genomics institute in Latin-American) of IPN and Cofounder, Technology Transfer Center of UANL and of this university´s Biotechnological Genomics college program.

10. Author of more than 200 peer-reviewed publications with more than 2,000 citations, two international books on biomedicine and patents in biotechnology, and several technology transfers to the Mexican Life Sciences Industry.
11. Supervisor or Co-Supervisor to over 100 graduate students.
12. Recipient of numerous research awards, including 18 from the UANL and more than 20 national awards including the National Food (1987), the "Dr. Jorge Rosenkranz" Medical Research (1989, 2005, and 2012) and several CANIFARMA awards.
13. Member, Mexican Academies of Sciences and Medicine, and several other prestigious Scientific Societies, as well as lecturer of universities throughout Mexico and abroad.
14. Contributor to the modernization of genetics units in Mexico, Colombia, Venezuela, and Peru.
15. In Mexico recognized as a pioneer in the molecular diagnoses of various diseases, human molecular biology research, and gene therapy (directed the first clinical protocol in Latin America).
16. 1988 world record, together with researchers from Genentech, Inc. and the Universities of Texas and Washington, sequenced the largest human gene manually. *Science* considered this evidence for the feasibility of the Human Genome Project).
17. World recognition for work on evolution, dysfunction and biotech use of genes of growth hormones. Coauthor of Nature publication of landscape of genomic alterations in cervical cancer.
18. *Nature Medicine* published his career profile in the July 2003 issue and Distinguished Alumnus, The University of Texas Health Science Center Houston Graduate School of Biomedical Sciences.
19. Chairman, National Scientific Committee of BioMonterrey 2006; Review Committees of CONACyT and SNI; Health Section of the Universal Forum of Cultures, National Advisory Council on GM foods, and CONACyT´s National Network of Pharmaceutical Innovation.
20. Secretary, Science and Technology; Coordinator, Laboratory for Genomics and Bioinformatics and Chief, Medical Biotechnology Unit, all of the UANL´s School of Medicine.
21. Founder, Consulting Firm Innbiogem, S.C. and Diagnosis Laboratory Vitagénesis, SA; both incubated in UANL´s

Center for Enterprises Incubation.

22. Declared distinguished citizen of his hometown (Miguel Aleman, Tamaulipas) and awarded the Medal of Academic and Science Merit at the fiftieth anniversary of its emancipation. Also, Medal of Merit "Dr. Carlos Canseco" 2012, for outstanding achievements in Scientific Research, granted by the Mayor of the city of Monterrey. Recognized as the most distinguished fellow of the Mexican Academy of Sciences, Northeast section, 2011.
23. President, Association of Human Genetics (Mexico), 2003-2004
24. UNESCO declared his lab a training-site for the Latin American Human Genome Program.

BRAIN DEVELOPMENT

In: Monkeys
Editor: Hugo A. Barrera Saldaña
ISBN: 978-1-63117-851-1

Chapter 1

BRAINSTEM'S AUDITORY EVOKED POTENTIALS AND INTERVAL VALUES IN RHESUS MONKEYS (*MACACA MULATTA*) AT DIFFERENT INTENSITIES (70-30 DB)

Alejandra Ibáñez-Contreras, [1,2,3,4,*]
Salvador Abraham Solís-Chávez, [1,2,3,]
Eduardo Tena-Betancourt [5]
and Braulio Hernández-Godínez [1,2,3,4]
[1]Applied Research In Experimental Biomedicine S.A. de C.V. (APREXBIO). Laboratorio de Primatología. Ciudad de México México D.F.
[2]BIOINVERT ®, Biología Integral para Vertebrados, Unidad de Experimentación Animal, Estado de México, México
[3]Centro de Investigación, Proyecto CAMINA, A.C., Unidad de Primates no Humanos. Ciudad de México, México D.F.
[4]Posgrado en Ciencias Biológicas y de la Salud. Universidad Autónoma Metropolitana- Iztapalapa.
[5]Departamento de Etología Fauna Silvestre y Animales de Laboratorio, Facultad de Medicina Veterinaria y Zootecnia, UNAM, México D.F.

ABSTRACT

Audition plays a critical role in the development of elaborated behaviors strongly dependent on the phylogenetic scale or evolutionary tree. Recordings of Brainstem Auditory Evoked Potentials (BAEPs) tend to show definite differences of response which are deeply influenced by the parameters of stimuli, environmental conditions of registration and characteristics of the species; such facts make the establishment of baseline data-values for each laboratory essential. The aim of this study

[*] Corresponding Author: ibanez.alejandra@hotmail.com.

was to study auditory evoked responses in the *Macaca mulatta* species through the brainstem's auditory evoked potentials (BAEP), in order to obtain normative data on latency values and electrophysiological responses of the different components of BAEPs in rhesus monkeys using different intensities (70-30 dB). This study was carried out using 14 Rhesus macaques (*Macaca mulatta*) averaging 7 years-old and allotted into two groups of 7 males and 7 females weighing approximately 7 kg and 5 kg respectively. BAEP`s were recorded through the stimulation of the ear using rarefaction *"clicks"* at 70, 50 and 30 dB of intensity. The cerebral electric activity was collected using gold disk electrodes placed at the Cz (+), A1, A2 (-) and Fz derivations as ground according to the 10/20 international system. Four waves were observed in males and females in response to the different intensities applied. Since no significant differences were found in the Student t-test for the separated afferences, afferents were unified, thus evaluating absolute latencies of the four waves and the intervals I-II, II-IV and I-IV in the 14 subjects at different intensities.

Keywords: Latencies, brainstem's auditory evoked potentials (BAEP), interpeak latencies

INTRODUCTION

The auditory system in mammals includes the cochlea, the cochlear nerve and the central auditory pathway from the cochlear nuclei into the brainstem towards the cortex of the temporal lobe. The sense of hearing itself plays an important role in the development of more elaborated behaviors, insofar as it ascends according to the status assigned in the phylogenetic tree (Ramsier and Dominy 2012; Cunningham and Acker 2000). Based on the aforementioned concept the use of non-human primates (NHP) is widely justified since they constitute an excellent model for the study of basic neural mechanisms related to auditory function (Ibáñez-Contreras et al., 2012, 2011; Fowler et al., 2010; Populin, 2006; Tower P 3rd et al., 2000), a true advantage supported by the fact that within the evolutionary tree primates are the most phylogenetically advanced and similar to humans (Fowler et al., 2010; Wilson and Sarich 1969). Among NHP one of the most frequently used species in biomedical research during the last 50 years is the rhesus monkey (*Macaca mulatta*), once described as the *basic primate*, given its closeness to humans anatomically and physiologically (Ibáñez-Contreras et al., 2012, 2011; Isa et al., 2009; Kemnitz et al., 2000; Schapiro 1996; Rowe 1996). The lifetime of these models has been estimated at approximately 30 years in captivity (Ibáñez-Contreras et al., 2012, 2011; Toledano et al., 2011; Fowler et al., 2010), which is a fact that complements and eases the design of longitudinal studies for the understanding of lifetime processes such as development, maturation and ageing of the auditory pathway.

It is extensively known that one of the main functions of the sense of hearing is the detection of environmental sounds to allow inter-specie communication; nevertheless, almost all vertebrate animals are born immature and such state does not allow them to compete with other members of their same kind or against animals belonging to other species. Therefore, in early life these models develop behavioral patterns only suitable for such living transition, which depends fundamentally on maternal protection for survival and con-specifics interaction through vocalizations and visual signals and among others facial expressions (Kirwan et al., 2013; Ramsier and Dominy 2012; Ibáñez-Contreras et al., 2011).

In vertebrates the ear transforms acoustic waves obtained from the environment into nerve impulses, and then are efficiently transmitted to the central nervous system (CNS) (Swenson and Reece 1999). The inner ear is responsible for the perception of sound-waves and its natural sensory functional unit consists of different types of nerve cells; some are constituted by hair cells and others are support cellular lining arranged in a mosaic fashion. In this functional unit sound waves are transformed into nerve impulses sent to the brain, a final function that makes possible the hearing. The mechanosensory information is transmitted through the hair cells that in turn activate bipolar sensory afferent neurons leading thus the information to the brainstem (Sapède and Pujades 2010; Kardong 1999; Climent et al., 1998).

The central auditory pathway is by far more complex than the somatosensory or visual pathways; however, macaques possess a limited neocortex that including the auditory zone comprises only 3% of its total extent, where more than 50% of the cortex is devoted to processing visual information; the remaining 11% of the crust is of somatosensory nature. According to such concept, in order to determine the specific location of sound a highly complex neural network exists in mammals, which provided with numerous connections (mostly inhibitory) lying between the two main central pathways enables it to identify the tiniest differences in sound intensity and timing, occurring during normal binaural hearing (Fitzgerald et al., 2012; López-Antúnez 2003; Troy et al., 2001; Kandel et al., 2001).

Concerning the use of NHP for the study of auditory physiology most researchers have employed the rhesus macaques to assess the functioning of specific anatomic segments of the auditory cortex. Such studies had enabled them to learn about the functioning and eliciting mechanisms related to the neural auditory generators, and their response in the evaluation of the hearing under dietetic calorie restriction; unfortunately only few authors have published work dealing with the study of the complex and detailed mechanisms of maturation of the auditory pathway in these species (Ibáñez-Contreras et al., 2012, 2011; Fowler et al., 2010; Torre et al., 2004; Torre et al., 2000; Torre and Fowler 2000; Laughlin et al., 1999; Doyle and Webster 1991; Doyle and Rood 1979).

EVOLUTION OF THE HEARING

As mentioned, hearing plays an important role in the development of more elaborated behavior while ascending into the phylogenetic scale, (Cunningham and Acker 2000) although anatomical differences are observed in different animal species that develop and share the same basic structures and operational mode of the neocortex (Solís-Chávez et al., 2013; O'Neill et al., 2012; Kardog 1999).

During the evolution of terrestrial vertebrates major changes in the peripheral and central structures of their CNS occurred, mainly to match impedance during the transmission of sound waves amid an outer aerial environment into an enclosed liquid medium; it is thought that such phenomena occurred accompanied by evolutionary change ending in the formation of the small ossicles, which constitute the middle ear and a more efficient membrane (Ladich 2000; Lowenstein and Roberts 1951).

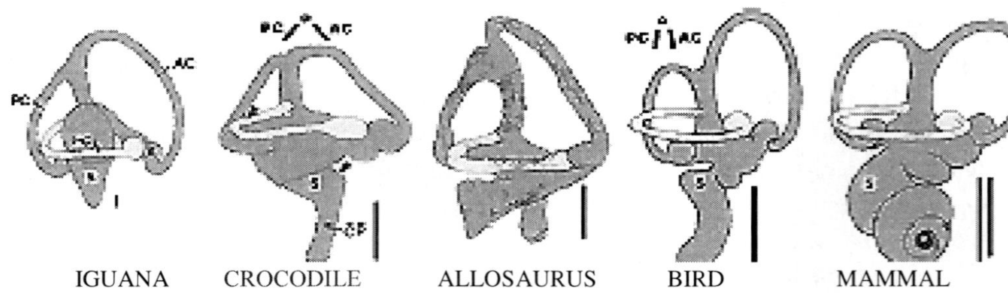

Figure 1. Major Anatomic Structures of the Inner Ear of Different Species (Kardog 1999).

It has been described that the first marine multicellular invertebrates such as the well-known jellyfish have a nervous system without brain or ganglia, enabling limited reflex reactions necessary for survival. Instead of brain jellyfish possess an elementary nervous system formed by numerous radial annular fibers arranged in an upper hood, where a fringe of closely-spaced delicate tentacles provides regularly spaced notches containing sensitive organs. Such segment comprises an eye-spot or ocellus containing floating calcareous granules resembling otoliths, which act as a balance organ that by inertial effect place the gravidity center on its own position drifting away vertically in the water (Satterlie 2011; Kardong 1999; Tavolga 1977).

The first kind of ear and hearing system occurred in the Class Cyclostomi and some primitive Ostracoderm which possessed three semicircular canals; later on, biologists observed that amphibians evolved accessory ear structures and developed semicircular canals complemented by three otolithic end organs, namely the utricle, the saccule and the lagena. These last structures are responsible for the perception of sound. Each chamber has an otolith called lapillus in the utricle and asteriscus in the lagena (Hildebrand and Goslow 2001).

It has been long recognized that the hearing sense in mammals and birds is much more advanced that in any other animal species; this occurred as a result of the complex evolutionary mechanism of survival developed by these organisms. In contrast, it is considered that hearing among vertebrates was developed as an adaptive mechanism for vibratory perception, initially processed by mechanosensory organs situated off the lateral line in fish which is a cluster of pear-shaped sensory cells, distributed along the body in sub-epidermal channels (Manley 2000).

The cochlea was developed from an adaptation of the saccule and improved further in a progressive fashion. In fish, there is a distal area of the saccule called lagena whose epithelium differentiates into a specialized sensory portion called the maculae of the lagena. In amphibians, such segments are distinct and appear as independent shaped- rounded units emerging from the back of the saccule containing specialized sensory epithelium called papila basilaris. In all these animals the sensory neuro epithelium remains a particular portion always situated at the bottom of the lagena (Gil-Carcedo et al., 2004; Manley 2000; Gans and Wever 1975).

At the cerebral level the predominant sensory functions in reptiles are the vision and hearing. The brain of reptiles is a very small organ that does not exceed 1% of the total body mass (Mader et al., 2006). From these characteristics it is thought that due to its size and position within the flow of nervous information, the dorsal ventricular ridge (DVR) could be one of the main areas for higher association in reptiles and birds (Kardong 1999), a theory that rests in the fact that the auditory, visual and somatosensory pathways generate efferent nerve projections from the same DVR nucleus. Such argument has generated the hypothesis that the dorsal ventricular ridge DVR of lizards, the dorsal palio of amphibians and the mammalian isocortex share an analogous anatomical origin (Striedter 1997; Northeutt and Kaas 1995; Bluter 1994; Karten 1967). On the other hand, most reptiles have a large tympanum that tilts along the surface of the head or is protected by the skin. The Eustachian tube is wide open towards the pharynx. The ossicles or stapes comprise a thin structure which has a cartilaginous portion provided with arms at its distal end, while the lagena stretches only in mammals and reptiles (Hildebrand and Goslow 2001; Gans and Wever 1976). In the case of birds, such creatures possess an external ear containing a very small ear canal channeling sound efficiently into the tympanic membrane. In the middle ear there is has a single bone stretched across that it is called the columella, supported by an osseous frame of the square and scamous bones (Gil-Carcedo et al., 2004).

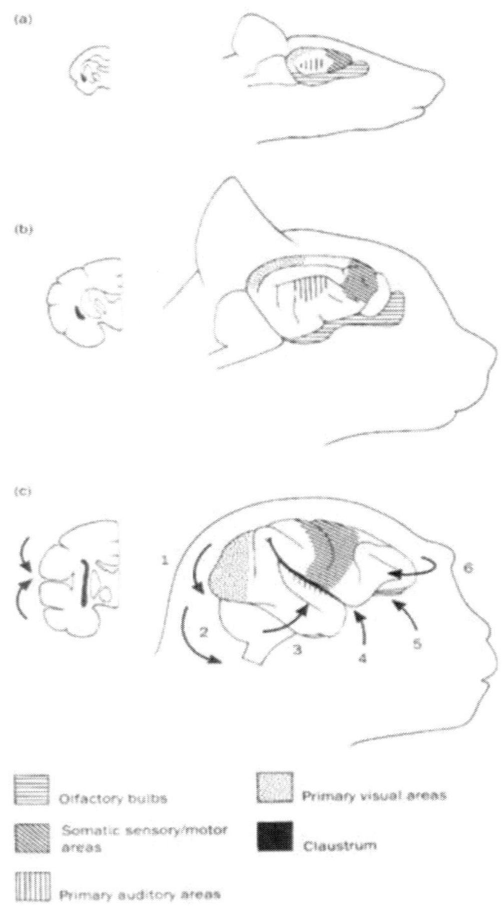

Figure 2. Schematic representation of the brain in different species demonstrating specific sensory input sites and specialized pathways contributing to the developmental morphology of primate brains. Rats (a), Cats (b) Monkey (c). (Jones 1999).

The stapedial bone is a long and thin complex that protrudes into the tympanic membrane, while the median tympanic membrane is occupied by the supraestapedial process which continues descending along the infrastapedial segment. The final development of the inner ear of birds shows homologies with crocodiles; however, these structures show considerable morphologic differences considering that the lagena in the case of poultry is found completely individualized of the saccule (Gil-Carcedo et al., 2004; Moiseff and Konish 1983).

ANATOMY AND PHYSIOLOGY OF EAR IN HUMANS AND NON-PRIMATES

During the hearing process, both human and NHP transforms very efficiently environmental acoustic information to form nerve impulses transmitted to the CNS (Swenson and Reece 1999). This process is done through the collection of stimuli by specialized sensory receptors, which codified the energy of sound waves and transform them into nerve impulses, transmitted to the CNS via the afferent fibers (Kardong, 1999). Consequently, the ear performs a double function: initially collects stimuli in the form of sound waves entering the auditory canal and strike the eardrum, causing it to vibrate and transforming stimuli into nerve impulses, which produce auditory sensations at the level of the auditory cerebral cortex; at the same time such impulses will be the source for auditory, somatic, and vegetative reflexes. On the other hand, the ear is capable to collect stimuli arising from the position and movements of the head and turn them into nerve impulses that constitute the afferent pathway related to balance and postural stability in animals (Climent et al., 1998).

Outer Ear

The outer ear of primates shows considerable morphological diversity; the folds surrounding the ear canal vary in their dimensions according to the taxonomic group to which they belong. Auricular appendices of primates then are very variable (Coleman and Ross 2004).

The outer ear consists of the pinna, a cartilaginous structure whose function is to capture sound vibrations and redirect them inside of the ear through the tympanic membrane by the external ear canal (Swenson and Reece 1999). Largely the pinna executes a triple action, since protects the external acoustic meatus, locate the sound source, and collects sound waves. Several animal species are able to move their pinna at will to orientate themselves towards a sound (Guerrier and Uzziel 1985). In NHP the pinna decreases substantially its sized and suffers striking changes in shape (Figure 3). The free ridge of the pinna is coiled on itself forming the helix at the level of its upper portion (Gil- Carcedo et al., 2004). On the anterior surface a large number of sebaceous cells and hairs exists, which decreases towards the external auditory meatus (Banks 1986).

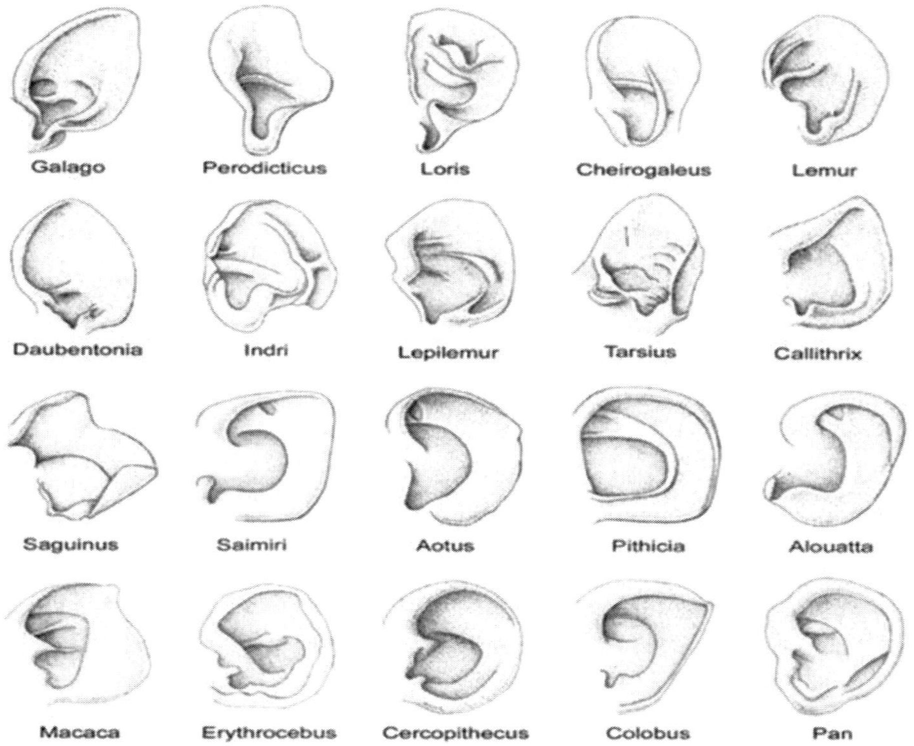

Figura 3. Scale morphology representation of the external ear of various representative primates (family and subfamily) (Hernández - Godinez et al., 2005).

The external acoustic meatus is a canal connecting the ear and the eardrum. It has a cartilaginous component surrounded by the auricular cartilage. The bony part runs within the external acoustic meatus and its length is variable (Climent et al., 1998). The superficial lining of the meatus is merely skin containing some hair, numerous sebaceous glands and modified tubular apocrine sweat glands. The secretion of these glands serves to protect the canal and keep the tympanic membrane moisten and flexible (Hartman et al., 1961). The external ear canal leads sound waves into the tympanic membrane and regulates the temperature of incoming air to assure they have similar levels on either side of the eardrum. Furthermore, this structure transforms spherical sound waves into flat waves and reinforces the resonance of frequencies between 2,000 and 4,000 Hz (Gil-Carcedo et al., 2004).

The outer ear protects the tympanic membrane against external environmental aggressions and also plays an important role in sound amplification. Such amplification is necessary and has been measured by placing against the eardrum miniature microphone probe tubes, which do not modify the internal configuration of the external ear canal. The outer ear determines an amplification of intensity of 10 to 15 dB over a range of frequencies between 1.5 and 7 kHz. Each of the two elements of the outer ear contribute to this amplification: the pinna with a peak of resonance of 5 kHz and the external ear canal with a peak of 2.5 kHz (Guerrier and Uzziel 1985).

Middle Ear

In terms of functionality, the essential function of the middle ear is to match acoustic impedance of the media-sound reaching the fluid contained in the inner ear; this transforming action allows a sound wave travelling in the air (outer ear) to become a sound wave travelling in fluid (Guerrier and Uzziel, 1985). It consists of the tympanum, which is elastic, semi-transparent and a little conical membrane that separates the external ear canal and the middle ear (Banks, 1986).

The eardrum consists of three layers; the outer layer, which continues with the ear canal skin; the middle layer composed of connective-tissue arranged by central circular and peripherally radial oriented collagen fibers, and the inner layer which is in continuity with the lining of the middle ear and the mucous membrane of the tympanic cavity (Swenson and Reece, 1999).

The middle ear forming a part of the middle ear cleft consists of three segments:

- *Epitympanum or attic:* this segment lies above the level of the eardrum. It contains the body of the incus (anvil) and the head of the malleus (hammer).
- *Mesotympanum or atrium:* lies at the level of the eardrum. It contains the malleus handle and neck, the long process of the incus, the stapes (stirrup) and other segments.
- *Hypotympanum recess:* this part of the middle ear space rests below the eardrum. It is usually empty but sometimes will have the jugular bulb within it.

The eardrum is highly sensitive to vibrate when sound waves are driven in by the external acoustic meatus; such vibration is retransmitted to the hammer-handle and consequently to the chain of ossicles in the ear. The ossicles of the middle ear are three small bones (hammer, anvil and stirrup) located in the tympanic cavity of mammals. They form a chain that extends from the tympanic membrane to the oval window, which is connected with the inner ear (Swenson and Reece, 1999). Furthermore, ossicles carry the vibrations across the middle ear and make the oval window vibrate. Such stimuli cause back-and-forth vibrations within the perilymph. (Climent et al., 1998). The hammer is intimately inserted between the layers of the tympanic membrane in the portion of the handle; from there the anterior/posterior malleolar ligaments run dividing the tympanic membrane in pars flaccid (PF) and pars tensa (PT). The hammer-head is articulated with the body of the incus (sinartrosis) and articulates through its long apophysis to the head of the stapes; similarly, the stapes articulates with the oval window via its footplate and stimulates endolymph contained in the membranous perilabyrinth during the hearing process (Hartman et al., 1961). The axis of rotation of the timpanic malleoincudal complex is fused together and runs around a horizontal rotational axis, represented by a line that would be the link between the posterior ligament of the incus and the anterior ligament of the hammer. This axis contains the center of gravity of the ossicular body and is also the axis of oscillation of the eardrum (Guerrier and Uzziel, 1985).

Functionally the middle ear acts as an adaptive segment for impedance, which occurs derived from two mechanisms of amplification:

1. Amplification of force resulting from the lever effect of the ossicular chain.
2. Amplification of pressure due to the difference between the surfaces of the tympanic membrane and the surface of the stapes.

The Eustachian tube connects the tympanic cavity with the pharynx and ensures that the air pressure in the tympanic cavity is equal to that on the outside of the body. The Eustachian tube opens actively by the action of the levator and tensor muscles of the palate during the swallowing, chewing and yawning. In order to allow the timpanoosicular system to vibrate in the tympanic box its pressure should be equalized to that existing in the external site. The function of the Eustachian tube is to provide an airway from the nasopharynx to the ear to equalize pressures on both sides of the tympanic membrane (Swenson and Reece, 1999).

The main functions of the middle ear are the transmission of sound waves to the inner ear; the mission of the timpano-ossicular chain is to transform air vibration into variations of pressure directed to the fluids in the inner ear. Hence, the transmission of pressure variation is possible by the role being played by the oval and round windows that vibrate in opposite phases, compensating the adaptation of the labyrinthine fluids (Guerrier and Uzziel, 1985). The motion at the base of the stapes generates waves towards the perilymph of the labyrinth making the basilar membrane (BM) to vibrate. When this membrane motions, movement is induced in the tectorial membrane (TM).

In mammals sound waves stimulates neuromasts (Organ of Corti) which are sensory organs for sound. In some species, the hair bundles or cilia appear directly in the fluid surrounding them. This structure known as hair cells is the common sensory receptor organ of the vestibular and auditory segments (Meza Ruiz, 1995). The unidirectional deflection of the cilium causes the excitatory increase in the flow and deflection towards its opposite direction diminishing ionic forces. Such fluctuating current reproduce within certain limits, the acoustic signals reaching the ear (Swenson and Reece, 1999).

Inner Ear

The inner ear consists of a membranous "labyrinth" encased in an osseous labyrinth. Its sensory structures are contained in the membranous labyrinth consisting of a vestibular and auditory part. The cochlea is the anterior part of the auditory labyrinth (Guerrier and Uzziel, 1985). The vestibular labyrinth encompasses semicircular canals and gravity receptors, which detect rotation. The acoustic portion of the labyrinth is the cochlea, a coiled tube filled with liquid and a central bone cortex coiled around the modiolus (Hartman et al., 1961).

The organ of Corti is a specialized neuro epithelium resting on the basilar membrane within scala media where the hearing sensory cells are contained. The pillar cells delimit and constitute the Tunnel of Corti containing cortilymph, a fluid filling the intercellular spaces of the organ of Corti covered by the tectorial membrane (Guerrier and Uzziel, 1985). Next to the modiolus there is the spiral ganglion of Corti, a collection of cell bodies where neurons are connected to hair cells and where the auditory nerve emerges (Swenson and Reece, 1999). The organ of Corti is lined up with sensory and support cells of epithelial nature:

a) The support cells are supported by the basilar membrane at its basal pole.
b) The auditory sensory cells form three or four external rows and one internal. These cells are cylindrical and their cilium protrudes above the reticular membrane formed from the support cells existing at the basilar membrane (Climent et al., 1998). The internal and external auditory sensory cells differ from its form and structure.

The cochlea of human and non-human primates (NHP) consists of a broad base and a spiraling channel running into a variable number of turns around a conical axis of bone tissue known as modiolus; around this, a bone called the blade of the modiolus separates the adjacent turns of the spiral channel (Climent et al., 1998; Lawrence, 1982). The cochlea is formed by three longitudinal chambers filled with fluids: the scala tympani and the scala vestibuli contain perilymph, while the scala media contains endolymph and is also called cochlear duct. The scala tympani merge and communicate with the vestibular in the helicotrema. These chambers are separated by two membranes: membrane of Reissner and the basilar membrane, a main structural element that separates the scala media from the scala tympani and determines the mechanical propagation properties of waves in the cochlear partition (Climent et al., 1998).

The innervation of the cochlea is composed of three types of fibers: the highly numerous afferent fibers that convey nervous impulses born at the lower pole of the ciliated cells to the nuclei of the brainstem; the efferent fibers that configure the medial olivocochlear bundle (MOC) pathway, and originates at the level of the superior olive complex that are distributed into the ciliated cells sending centrifugal impulses, and finally sympathetic fibers emerging from the superior and inferior cervical sympathetic ganglion. These constitute an adrenergic plexus at the level of the cochlea (Guerrier and Uzziel, 1985).

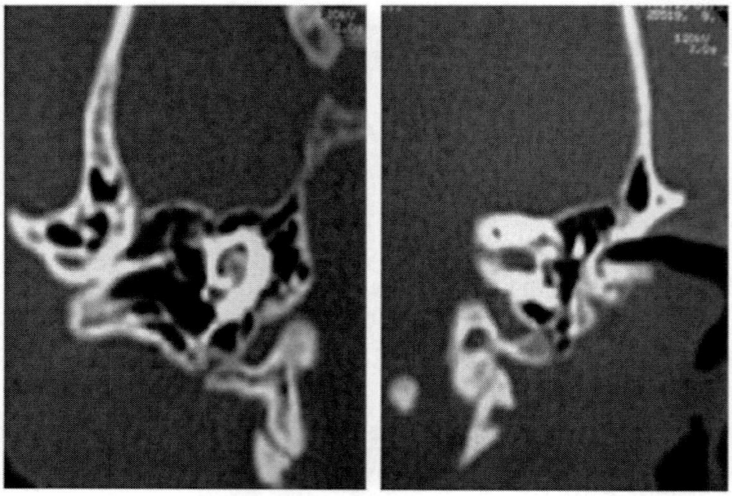

Figure 4. Cross-section CAT (computerized axial tomography) scan of a Rhesus monkey showing the normal arrangement and anatomical representation of the cochlea and the semicircular ducts.

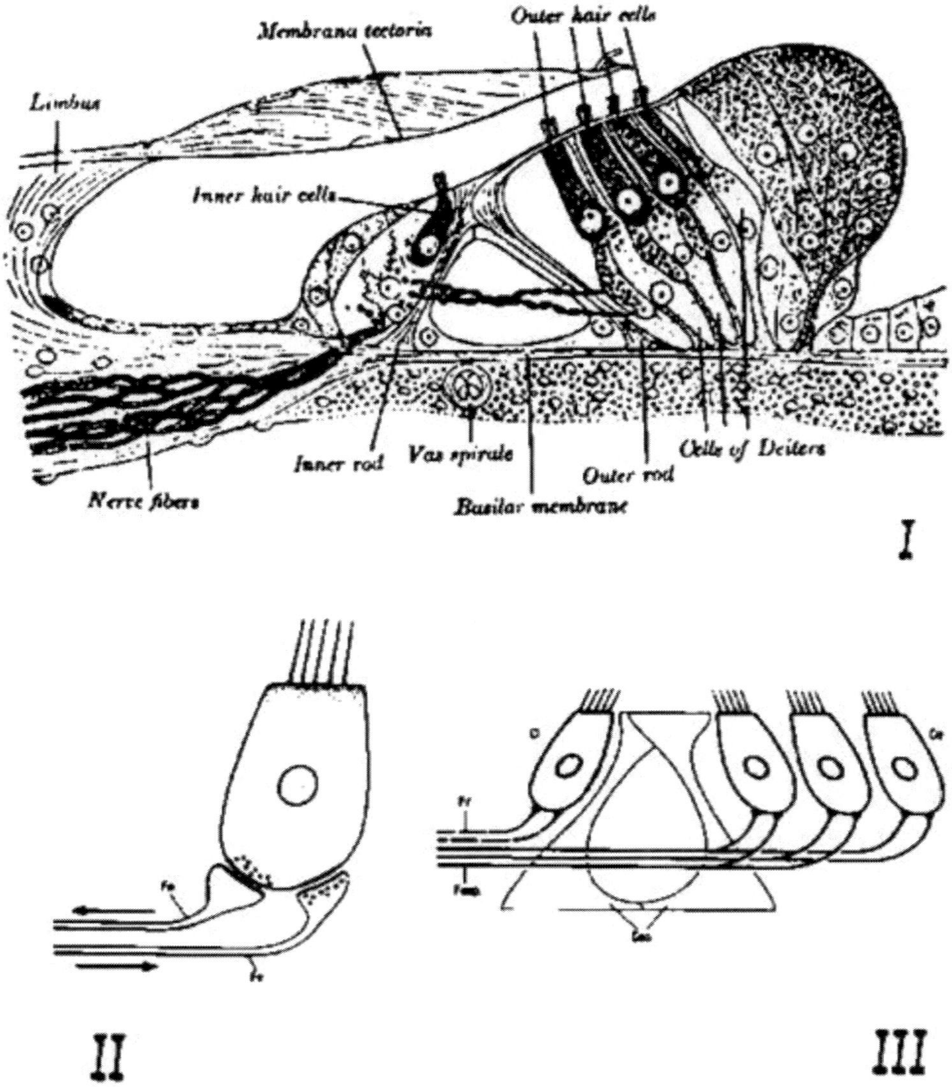

Figure 5. Cross-section of the cochlea illustrating the organ of Corti. Part I shows a representation of the structures of the organ of Corti. Bottom left (II) illustrates the relationship of a ciliated cell with its afferent (AF) and efferent fibers (EF) of the VIIIth nerve pair, and on the right (III) is shown the ending point of the radial (RF) and spiral fibers (Fesp), along with inner hair cells (IHC) and outer hair cells (OHC). (López-Antúnez 2003).

In cochlear physiology the first phase is basically stated as the mechanical events that take place in the cochlea when a sound wave enters the inner ear and motions fluids and membranes; a second phase consists of cochlear micromechanics and occurs during the displacement of the organ of Corti in relation to the tectorial membrane; a third phase occurs when the transduction or transformation of mechanical energy into bioelectric energy takes place and hence the flow of information from the ear to the brain occurs (Gil- Carcedo et at, 2004).

ORGANIZATION OF THE AUDITORY PATHWAY IN PRIMATES

The ascending auditory pathway ends at the cerebral cortex which has several different auditory areas located on the dorsal surface of the temporal lobe. The most important projection originates from the medial geniculate nucleus and extends into the primary auditory cortex (area A1, or 41 and 42 of Brodmann). Having a peculiar cytoskeleton structure this region contains a place information or characteristic tonotopic representation; neurons tuned with the low frequencies are located at the rostral end, while the caudal region corresponds to the high frequency-sensory cells. The primary auditory cortex is surrounded by separate regions involved in the elaboration of specific types of auditory information. It has been described that the tonotopic organization of the primary auditory cortex reflects the pitch rather than the frequency of the stimulus (López-Antúnez, 2003; Hackett et al., 2001; Kandel et al., 2001; Kaas et al., 1999).

Some mammals have at least nine of these regions mostly organized into tonotopic maps for detection of the frequency of stimuli (Kandel et al., 2001). The topographic organization of the various areas of hearing has been outlined by means of evoked potentials studies. The auditory cortex consists of three subdivisions: (AI), a central area surrounded by two areas, AII and PE (posterior ectosylvian gyrus). At the level of each of these areas there is a specific tonotopic representation of the different segments of the cochlea. In addition, each of these cortical areas receives information coming from both ears (Kaas et al., 1999).

In humans and primates the primary auditory cortex is deeply buried and located at the level of the transverse temporal Heschl gyri, which is located on the insular face of the first temporal convolutions hidden within the lateral sulcus (also called Sylvian fissure) (Hackett et al., 2001). The auditory cortex of primates has been studied by techniques of single-unit registration (Guerrier and Uzziel, 1985).

The basic terminology and architectural criteria used to define the auditory cortex of macaques are based on descriptions by Kaas and colleagues (Hackett et al., 1998; Kaas and Hackett, 2000). These researchers divided the auditory cortex of macaques into a core of primary or primary-like areas located on the lower bank of the lateral sulcus, which is a surrounding narrow belt of associated fields and a parabelt region laying just lateral to the belt on the superior temporal gyrus. Thus, two fields were concluded: the primary area, auditory area I (AI) and an adjoining rostral field named RF (R for rostral). Further work characterized the surrounding belt cortex into several fields, such as caudo-medial (CM) and lateral (L). All fields show clear evidence of tonotopic organization (Kaas et al., 1990), all also concluded the existence of a third primary-like field and named it RT for rostro-temporal, which in addition showed koniocortical features.

Neurophysiology of the Auditory System

The hair cells of the organ of Corti are columnar cells; each have a bundle of up to 200 specialized cili. As recipients of the auditory system the ciliated cells have three important functions; as hearing receptors they belong to the mechanoreceptors group and transduce sound waves into electrical activity in the nerve fibers (López-Antunez, 2003). Through numerous sensitive bundles of cilia transform mechanical impulses into electrical responses. In many cases hair cells make their mechanical or electrical contribution in response to the selectivity of frequency of the receptor organ (Kandel et al., 2001). The afferent neurons innervate the cochlear inner hair cells and communicates signals eliciting a response in the receptors to the information nuclei via the cochlear branch of the the VIIIth auditory nerve; their soma is located along the merge of the osseous spiral lamina to the modiolus, and together constitute the spiral ganglion containing nearly 30,000 spiral ganglion cells. The peripheral branches of the cochlear neurons are in contact with the basal segment of the ciliated cells whose cytoplasm contains synaptic vesicles at the nerve ends (López-Antunez 2003).

The peripheral efferent innervations comprise three main segments: the olivocochlear bundle (OCB), which originates at the superior olivary complex in the brainstem. This is situated contra and homolateral to the cochlea and the reticulo-cochlear bundle and originates at the reticular formation (Guerrier and Uzziel 1985). These efferent fibers originate from neurons at the dorsal bridge of the superior olivary complex and continue through the olivocochlear bundle to the organ of Corti (López-Antunez 2003).

Central Auditory System

The auditory pathway as well as other sensory systems consists of several stations that form an interconnected chain traveling upstream and trigger the respective nerve nuclei. The cochlear nuclei are the first relay station in the auditory system. A series of parallel information channels are generated from this site into the cerebral cortex and terminate directly or indirectly through the superior olivary complex and the lateral leminiscus, at the inferior colliculus. Through these stations the auditory signals transmitted from the cochlear nerve terminates reaching the auditory cortex.

VIIIth Craneal Nerve

The auditory nerve, also called vestibulocochlear is a sensory structure related to hearing and balance. It has two distinct components: the cochlear nerve that is responsible for hearing and the vestibular nerve which is related to equilibrium and carries out spatial orientation information to the brain (Kandel et al., 2001). The vestibulocochlear nerve consists mostly of bipolar neurons and is characterized by a cochlear branch emerging from the spiral organ or Corti ending at the bridge or annular bulge. The acoustic division has its cellular bodies in the spiral ganglion of the cochlea. The node is lined up with bipolar cells, whose peripheral projections propagate the auditory impulses from the specialized neuroepithelium of the inner ear (López-Antunez, 2003; Kandel et al., 2001; Guerrier and Uzziel, 1985).

Sound determines the motion of the basilar membrane that vibrates under stimuli. When the displacement of the basilar membrane takes place the stereocillia bends upwardly, which being somewhat fixed induce an stimulus to activate complex hair-like structures. Next, the stimuli are transmitted to the cochlear sensory fibers ending at the base of each hair cell. Each auditory afferent fiber and the hair cell connecting to them have a minimum threshold response at sound frequencies (Kandel et al., 2001).

Neurons project the information from the receptor cells into their centres via the cochlear branch of the VIIIth cranial pair. The information driven by the cochlear branch of the vestibulocochlear nerve reaches the cochlear nucleus, which constitute the first central synaptic level of the auditory pathway located at the junction of the bulb and the bridge (Durand-Rivera 1998). The highly organized combination of fibers of the auditory nerve is preserved at the level of the cochlear nucleus; fibers that have higher frequencies pass through the dorsomedial site while the lowest frequencies travel through the ventrolateral zone. In this way the cochlear nerve fibers react to all limits of audible sound (Kandel et al. 2001).

Cochlear Nuclei

The cochlear nuclei (CN) constitute the end-segment of the auditory nerve and consist of two heterogeneous collections of neurons in the mammalian brainstem. In mammals, the CN are divided into the dorsal cochlear nucleus (DCN) and the ventral cochlear nucleus (VCN), the latter is further divided into the posteroventral cochlear nucleus (PVCN) and the anteroventral cochlear nucleus (AVCN). After entering the CN the cochlear nerve axons bifurcate; each fiber sends an ascending and a descending branch crossing and innervating the posterior cochlear nucleus and ends at the dorsal cochlear nucleus. (Gil-Loyzaga and Poch, 2007; Sanchez and Merchant, 2007; Lopez – Antunez, 2003).

One of the main functions performed by the CN is to execute a primary analysis of the neuronal messages coming from the auditory system. In this sense, these nuclei are involved in the elaboration of complex messages carried to higher levels. The CN decode the intensity, analyze various temporary parameters such as duration, time of start and end of an auditory stimulus and maintain and/or transmit the frequency analysis performed by the specific peripheral receptor (Gil-Loyzaga and Poch, 2007). The information processed in the cochlear nuclei travels upstream through secondary fibers that make synapses at different levels: the superior olivary complex, the nucleus of the lateral lemniscus and the inferior colliculus (Sanchez and Merchant 2007; López-Antúnez 2003).

Superior Olivary Complex

The superior olivary complex (SOC) is the second great contact station of the central ascending and descending auditory pathways. It consists of a collection of brainstem nuclei located in the protuberance, in close relationship with the majority of the tracts that unite the cochlear nucleus with the inferior colliculus (Guerrier and Uzziel 1985). The SOC is divided into three primary nuclei: the Lateral Superior Olive (LSO), which is a brainstem nucleus, the Medial Superior Olive (MSO), the Lateral Superior Olive (LSO), and the medial nucleus of the trapezoid body. In the case of NHP the MSO is fully developed, while it is smaller in the LSO and the medial nucleus of the trapezoid body. The relative size of the three nuclei is related to the head size, the audible range frequency to which each species are sensitive, and the neural circuits involved in the detection of differences in time and intensity elicited for the localization of sound (Gil-Loyzaga and Poch 2007).

The LSO is a S-shaped nucleus. It consists of multipolar cells whose dendrites are developed in the transverse and vertical plane and receive bilateral afferences. The LSO neurons are activated by homolateral stimulation and inhibited by contralateral stimulation. The nuclei of the SOC contribute to define the spatial location of the sound source using binaural hearing. The axons of these neurons unite to the lateral lemniscus (Gil-Loyzaga and Poch 2007; Durand - Rivera 1998).

Nuclei of the Lateral Lemniscus

The lateral lemniscus is a tract of axons located on the side of the brainstem running in rostral direction; it is connected to neurons of first, second and third order emerging from the cochlear nucleus and the SOC along with the inferior colliculus of the midbrain (Durand-Rivera 1998).

The nuclei of the lateral lemniscus are a set of neural islets resting between the fibers of the lateral lemniscus, which functionally are divided into small nuclei namely ventral and dorsal. The ventral nuclei forms a monaural source sound reproduction system, while in the dorsal nucleus is binaural (Gil-Loyzaga and Poch 2007; Sanchez and Merchant 2007). The ventral nucleus of the lateral lemniscus (VNLL) receives mostly collateral projections from the contralateral dorsal cochlear nucleus. It also receives although to a lesser extent, the collateral projections of olivary complex.

The vast majority of the VNLL activates only by collateral stimulation. However, there are some binaural neurons receiving collateral inputs from binaural neurons of the olivary complex. It is believed that the neurons of these nuclei can detect and analyze variations in the temporary properties of sound, so they can be involved in decoding and perception of vocalization and language. It has also been shown also that neurons of the ventral complex are necessary to determine the duration of sound, an action executed by neurons of the inferior colliculus (Gil-Loyzaga and Poch 2007; Sanchez and Merchant 2007; Lopez - Amtunez 2003)

The dorsal nucleus of the lateral lemniscus (DNLL) receives afferences from the contralateral cochlear nucleus, although also receives binaural inputs from the homolateral MSO and the LSO on both sides. This cross-projection communicates via the commissural fibers of Probst. An essential function of the neurons in the dorsal nucleus of the lateral lemniscus is to improve and refine binaural processing previously initiated at the level of the Superior Olivary Complex, while helping establish a better tuning of neurons in the inferior colliculus for which tonotopic innervations exist. Finally, it is important to note that the dorsal nucleus also provides a timed-sustained inhibition to the inferior colliculus, allowing neurons to determine the source of the sound and not confuse them with multiple reflections or echoes (Gil-Loyzaga and Poch 2007; Sanchez and Merchant 2007; Kandel et al., 2001).

Inferior Colliculus

The inferior colliculus (IC) or posterior corpora quadrigemina is the principal midbrain nucleus linked to the auditory pathway and part of the midbrain roof (36). It consists of two nuclei: the central nucleus (CIC) and the dorsomedial nucleus surrounded by a cortex (pericentral nucleus and outer core). Most of the fibers of the lateral lemniscus terminate in the central cortex. The central nucleus of the IC receives a heterogeneous number of fibers from neurons of second, third and fourth-order (directly or crossed) from the lateral lemniscus. The input of the central nucleus of the IC comes from the cochlear nucleus of the complex of the superior olive or nuclei of the lateral lemniscus (Durand- Rivera 1998). The dorsomedial and pericentral nuclei receive excitatory afferencies mainly from the auditory cortex and constitute a relay for the descending auditory system. The outer core is innervated by efferent axons of the central core (Guerrier and Uzziel 1985). The neurons of the IC are projected into the geniculate body of the thalamus (GCB) being the last relay before reaching the cortex.

The IC and the GCB are not simple relays between the nuclei of the lower brainstem and the auditory cortex, but instead highly integration centers. One part of these two nuclei neurons is involved in the treatment of binaural information. The cells of the IC and the GCB may respond to independent stimulation of each ear (Guerrier and Uzziel 1985). Finally, the IC and the GCB constitute the multisensory integration center that effectively receives vestibular, somatic and visual afferencies at the level of some of their subdivisions. These neurons could participate in the multisensory integration of sounds in the auditory cortex required for the coordination of movement, allowing the displacement of the animal towards a sound source, while adding significance of behavior. (López-Antunez 2003; Kandel et al., 2001; Guerrier and Uzziel 1985).

NHP as Models for the Neurophysiologic Study of the Auditory Via

Studies related to hearing in animals have been executed under different approaches which have included Brainstem Auditory Evoked Potentials (BAEPs), a useful tool to determine auditory functionality, neonatal hearing loss associated to adverse risk factors during childbirth, auditory neuropathy and retro cochlear pathologies (Ibáñez-Contreras 2012, 2011; Fowler et al., 2010; Harland et al., 2006; Kretzmer et al., 2004; Henry 2004; Torre et al., 2004; Malmierca 2003; Fowler et al., 2001; Torre and Fowler 2000; Mc Fadden et al., 1999; Laughlin et al., 1999; Doyle and Webster 1991; Church and Shucard 1986), these assessments have been used also in multiple experiments with Nonhuman primates (NHP) to evaluate the clinical performance of cochlear implants, the effects of caloric restriction, etc., (Ibáñez-Contreras 2011, 2012; Fowler et al., 2010; Henry 2004; Malmierca 2003; Fowler et al., 2001; Mc Fadden et al., 1999; Kraus et al., 1985).

Nonhuman primates (NHP) constitute an excellent model for the study of fundamental neural mechanisms in auditory research & functionality (Ibáñez-Contreras 2012, 2011; Fowler et al., 2010; Populin 2006; Torre et al., 2000), such advantage derives from the position that primates occupy in the evolutionary tree since they are the most analogous animal species phylogenetically close to humans (Fowler et al., 2010; Wilson and Sarich 1969). In neurophysiology research the most frequently studied species of NHP include the *Cheirogaleidae* and *Cercophitecoideae* families (Toledano et al., 2011), while within the *Cercophitecoideae* species the most common species used in biomedical research are the Green monkeys (*Chlorocebus aethiops*), rhesus monkeys (*Macaca mulatta*), and the stump-tailed macaque (*Macaca arctoides*).

It is estimated that in captivity conditions the lifespan of these models is approximately 30-years. Among primates, senile individuals are those who have more than 20-year-old, and extremely aged or elderly those animals who reach 30 year-old or more (Ibáñez-Contreras et al., 2012, 2011; Toledano et al., 2011; Fowler y cols 2010), which facilitates and complements the design of longitudinal studies needed for the understanding of processes such as development, maturation and ageing of the auditory pathway.

In NHP the studies most commonly undertaken have been related to hearing and thus focused toward gaining novel knowledge in the discrimination of complex sounds, which have been evaluated through behavioral research. Therefore, using a wide range of intensities indicative of an analogous human sensibility notorious differences have been found. This is to say that monkeys and humans show sound physiological responses toward sound discrimination. However, only few authors have detailed the processes of maturation of the auditory pathway in these species (Fowler et al., 2010; Torre et al., 2004; Torre et al., 2000 a,b; Laughlin et al., 1999; Doyle and Webster 1991; Doyle and Rood 1979).

BRAINSTEM AUDITORY EVOKED POTENTIALS (BAEPS) IN NHP

Evoked potentials (EP´s) are defined as electrical signals generated by the nervous system of cerebral origin able to define the specific activity of sensory or motor pathways (Poblano, 2003; Alegre and Artieda, 2000; Shkurovich, 1997). They constitute a specific neuropsychological tool allowing the objective study of the functioning of sensory-neural pathways (auditory, visual, and somatosensory), after stimulation at specific sensory nerve pathways traveling to the cerebral cortex (Starr 1977; Jewett and Romano 1972). The information obtained through EP´s is used for the detection of normal/abnormal nerve conduction, as well as the specific position of deficiencies in a segment of the nervous canal, allowing also the diagnosis and characterization of disorders and the severity of an injury.

The Brainstem Auditory Evoked Potentials (BAEPs) are brain responses to auditory stimulation. The acoustic stimuli generate approximately 25 to 30 waves, which on the basis of its latency can be subdivided according to the analysis duration as follows: (Alegre and Artieda, 2000; Maurer et al., 1989):

- Short-latency auditory evoked potentials (SAEPs): occur within 0.5—15 ms
- Middle-latency auditory evoked potentials (MAEPs): occur within 11.0—50 ms
- Late-latency auditory evoked potentials (LAEPs): occur within 50.0—500 ms

In the case of (BAEPs), the peripheral receptor, the auditory pathway, the cortical projections and the cerebral cortex constitute a complex network of cells and nerve fibers, which form a intricate system that in response to stimuli generate a variety of waves of certain latencies and amplitudes (Maurer et al., 1982; Hoke 1979; Davis 1976; Picton et al., 1974). Usually, BAEPs studies carried out in NHP are of a short-latency type induced by brief stimuli and basally extracted from an electroencephalogram aided by a computerized instrument (Ibáñez-Contreras et al., 2011; Morales et al., 2002; Alegre et al., 2001; Recanzone 2000; Laughlin et al., 1999; Lasky et al., 1995; Møller and Burgess

1986).Typically, all recordings are characterized by at least four permanent positive waves in animals and five waves in humans. These waves reflect the electrical activity determined at specific nuclei of the auditory pathway (Ibáñez-Contreras et al., 2012, 2011).

There are diverse studies executed in different species of NHP with variable results reported in terms of the distribution sites of the neural generators of auditory pathways, which describe also different wave numbers and further difficulties in the standardization of BAEP recordings. Allen and Starr (1976) described the acquisition of five waveforms in rhesus monkeys under anesthesia; however, they considered only the first four waves equivalent to the five waves reported in humans and based on the latency values and anatomical location. On the other hand, Velasco et al., (1984) reported the presence of seven waveforms in rhesus monkeys describing it of analogous equivalency to those observed in humans based in their morphologic features; nevertheless, the latencies showed smaller values. Finally, Ibáñez-Contreras et al., (2011), Laughlin et al., (1999), Möller and Burgess (1986) indicated the recording of four well-defined and consistent waveforms, emphasizing that in *M. mulatta* species the more prominent and larger in amplitude ratio was the IV wave, which was also more stable and lasting longer after the intensity of the stimulus diminished until it disappeared. Hence, they concluded that wave IV would be equivalent to wave V in humans corresponding to the inferior colliculus, noting also the wavelength behavior in response to intensity and morphologic changes.

Based on the above the anatomical sites or neural generators of the auditory pathway in the *M. mulatta* species would be as follows:

Wave I: Auditory Nerve / Nuclei Cochlear
Wave II: Superior Olivary Complex.
Wave III: Lateral lemniscus.
Wave IV: Inferior Colliculus

The theoretical fundament of BAEP recordings relies on stimuli which will be transmitted via the neurosensory pathways, always occurring at the same speed and suffering a transient-constant delay at each of the synapses that constitute this neuronal pathway. It is necessary to apply a successive number of stimuli to isolate the evoked potential elicited by a sensory stimulus (noise) from the background activity of the brain. BAEPs show an excellent consistency among patient-to-patient; the habituation to response does not exist and they are not altered by sleep stages (Hecox et al., 1974). The majority of evoked potentials cannot be observed and assessed through an electroencephalogram, because it has very low amplitudes of 0.1 to 20 mVolts (Durand Rivera 1998).

Although the function of the auditory system is to analyze and recognize sounds produced in the environment, the animals under investigation must differentiate sounds produced in their own environment against artificial noises. Previously, it has been described that in order to know and understand the functioning of the auditory system is necessary to implement and experiment a variety of sounds. Several studies have been devised in order to subject animals to different auditory stimuli such as their own vocalizations, including also their exposure to different types of vocal communication interspecies (Hackett et al., 2001). It is clear that in the auditory system a decoding process exists for these kind of stimuli eliciting certain responses at the level of the CNS, where a historical storage of past experiences and learning stimuli exists in the subjects; such condition makes important to employ dissimilar auditory stimuli in order to break previous memory-records stored in the brain of the studied animals. This fact explains the need of using different sounds and intensity levels and record responses using a mixture of nerve responses to different sound stimuli (Poon and Brugge 1998)

Type of Sound Stimulus

The most commonly used type of stimulus for BAEP recordings is the click. A click is the sound generated by transmitting to the speaker or earphone an electrical square wave. There are two possible types of clicks, (the most commonly used) rarefaction and condensation depending on the traveling direction of the speaker membrane (inwards - rarefaction - or outward - condensation-), which in turn depends on polarity of the electric pulse generated by the displacement (Alegre and Artieda 2000).

Electrodes

The electrodes most frequently used in the clinical practice of BAEP recordings are the small disk skin surface electrodes. The silver chloride electrodes are properly attached to the scalp and other recording site using a collodion to achieve an impedance of < 5000 k ohms. (Shkurovich 1997) The disks serve as an interface between the skin and the registration system in the auditory pathway, so it is of utmost importance that such interface presents minimum impedance. Consequently, the majority of systems usually incorporate specific circuits to measure and keep impedance at less than 5 k ohms or even lower to 2 k ohms (Durand Rivera 1998; Maurer 1989).

Amplification

The amplification system serves to increase the amplitude of the electrical signal obtained by means of electrodes. Amplifiers in EPs are of the differential type and require three locations: A1 left earlobe / behind the left ear; - – A2 right earlobe / behind the right ear – Cz vertex (10-20 system). The ground electrode is usually placed at Fz. (Durand Rivera 1998).

Filtering

The main objective is to eliminate as much as possible part of the incoming noise without distorting the amplitude or morphology of EPs. The filters can be of two types: low-pass filtering that as its name indicate allows the passing of low frequencies and attenuate high-frequencies, and high-pass filtering allowing the passage of high- frequencies and attenuate low-frequencies (Durand Rivera 1998).

Methodological Aspects of BAEPs

In order to obtain the recordings of the auditory electrical activity each ear is explored separately.

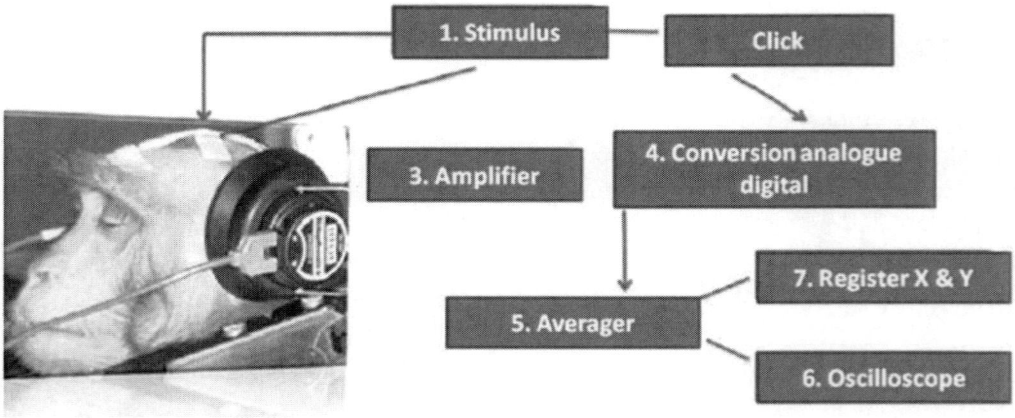

Figure 6. Basic Diagram of EPs Procedure: (1) Synchronized Equipment – (2) Electrodes register the baseline spontaneous electric activity of the brain, as well as the stimulus response- Such activity is then amplified (3) and rendered as an average through the PC analog-digital conversion (4, 5).

The stimulus employed is the click or rarefaction snap of 0.1 ms of duration applying an intensity established individually according to the auditory threshold of each patient, and considering the noise level existing in the laboratory. In the case of rhesus monkeys a descending stimulation series was executed in order to establish three different intensities (70dB, 50dB, 30dB); the stimulation was initiated at 70 dB and ended at 30 dB (Ibáñez - Contreras et al., 2011). In order to avoid that the sound might be perceived through the contralateral ear by bone transmission, we used a blank noise masking; in such cases a 20 dB stimulus below the employed stimulation is commonly used (Ibáñez-Contreras et al., 2011; Alegre and Artieda 2000).

The most common montage for registration of BAEPs in NHP consists of placing disk electrodes in A1 and A2 (–) and the vertex of the skull Cz (+), and finally Fp (frontal polar) as ground according to the 10/20 International System (Ibáñez-Contreras et al., 2012, 2011).

In the case of NHP it is recommended to shave and clean with mild soap and water the area where the disk electrodes would be placed in order to reduce impedance (Ibáñez-Contreras et al., 2011).

The filters most commonly used for early BAEPs recordings are situated at 3000 Hz. Lower values can modify the morphology and latency of waveforms, while higher values do not provide advantages and may induce the appearance of artifacts. The high-pass filter values commonly used were placed in a range between 10 and 100 Hz (Alegre and Artieda 2012; Ibáñez-Contreras et al., 2012, 2011; Durand - Rivera 1998).

Factors Affecting BAEP Recordings

There are a number of factors that can influence the electrically generated brain responses. As mentioned before, in recent years several works have been published aiming to enhance the current knowledge and standardize normal values of latencies observed during the BAEP recordings, which have been studied in different species of non-human primates (NHP) (Ibánez-Contreras et al., 2012, 2011; Fowler et al., 2010; Torre et al., 2004; Alegre y col 2001; Torre et al., 2000; Torre and Fowler 2000; Laughlin et al., 1999; Lasky et al., 1995; Pineda et al., 1989; Moller y Burgess 1986; Velasco et al., 1984; Doyle et al., 1983; Allen and Starr 1978). However, although these studies are available, in order to obtain valid and reliable recordings it is necessary to define the normal latency values of their major constituents and its relationship toward the various intensities intra-laboratory. (Ibanez-Contreras et al., 2011; Soares et al., 2010; Arai 2008; Delgado et al., 2003). Such approach is a necessity given the great variability of results obtained in relation to the parameters under which the recordings were obtained (Ibanez-Contreras et al., 2011;

Hultcrantz et al., 2006; Henry 2004; Delgado et al., 2003; Yadav et al., 2002; Mc Fadden et al., 1999; Elias 1996). Among the main factors affecting the electrophysiological recordings of latency we can mention the following.

- *Stimulus Intensity:* The intensity of the stimulus is the main parameter that modifies BAEP recordings. A higher intensity is accompanied by lower latency values in all waves making necessary to standardize the values depending on intensity. It has been described that if the intensity of the stimuli is considerably decreased, some waves may not be visible (Ibáñez-Contreras et al., 2012; Alegre and Artieda 2012). According to Ibáñez-Contreras et al., (2012), while studying the effects of intensity under varying stimuli from 70-30 dB in Rhesus macaques (M. mulatta), they observed that in male and female the latency-intensity effect was present, fact that corroborated that under higher intensities latencies were smaller (Figure 7).
- *Temperature:* All absolute latencies and inter-waves tend to increase as the basal body temperature decreases, disappearing completely at temperatures around 20°C. On the other hand, an increase in body temperature of 1°C can elicit a decrease of wave V (Elias 1996). Another factor to be strongly considered is that the higher body temperature of females represents an additional variable that can modify absolute latencies significantly.
- *Sex of the Subject:* It has been previously described that the absolute latency of all waves particularly wave V is higher in men than in women. Similarly, Ibáñez - Contreras et al., (2011) observed this phenomenon in NHP and suggested that were primarily related to differences observed in relation to the size of the skull and the length of the auditory nerve (Ibáñez-Contreras et al., 2011, 2012; Elías 1996; Hecox and Galambo 1974). In addition, the influence of sex hormones on BAEP latencies has been considered as an additional modifying factor (Ibáñez-Contreras et al., 2011; Yadav et al., 2001; Tandon et al., 2001; Poblano et al., 2004).

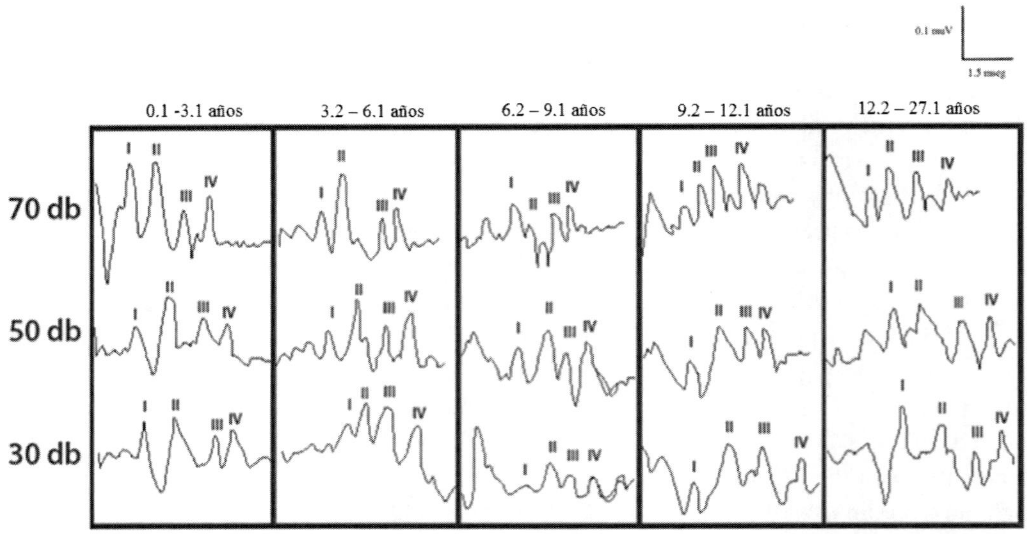

Figure 7. Morphologic features of four *unified* auditory waveforms at different age and stimuli (70-30 dB). Note the reduction in the wave amplitudes and elongation of latencies in response to the level of intensity applied.

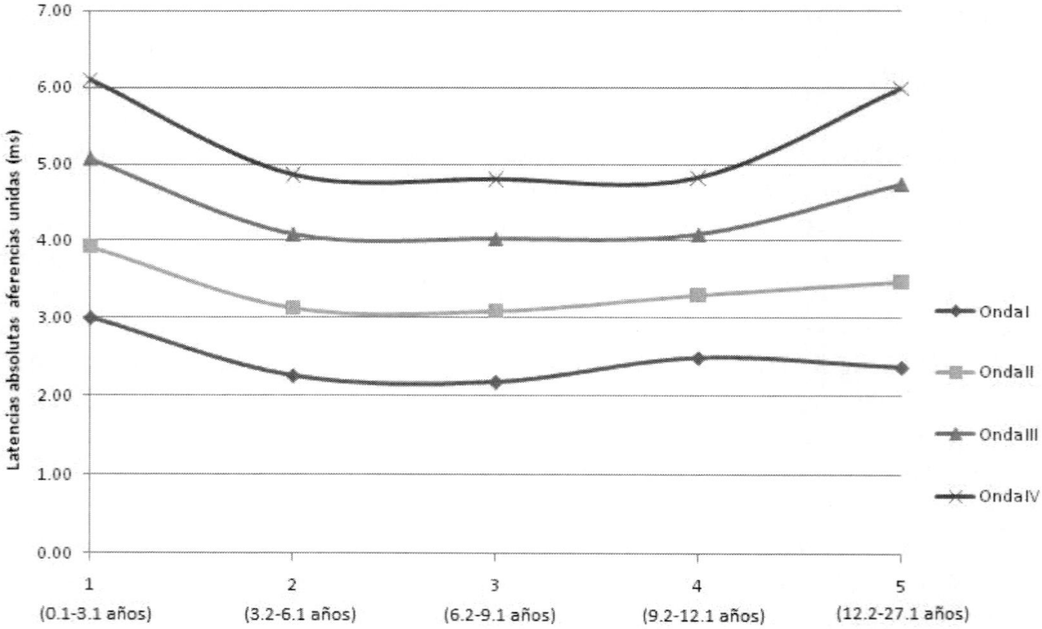

Figure 8. Unified absolute latencies of four auditory waveforms executed in males monkeys at 50 dB (Ibáñez-Contreras et al., 2011).

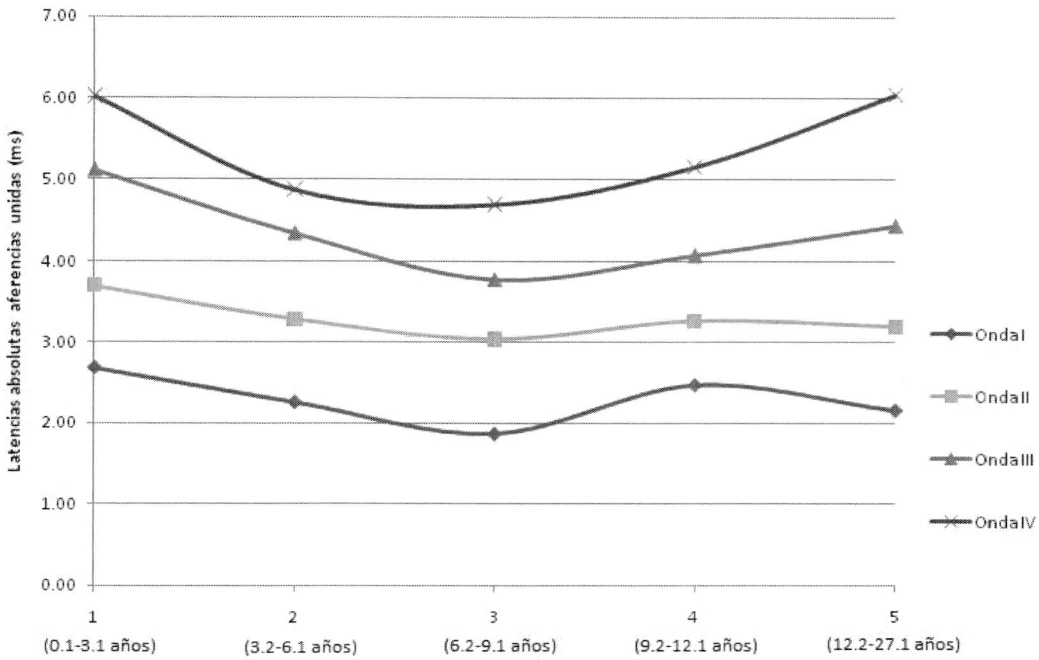

Figure 9. Unified absolute latencies of four auditory waveforms executed in female monkeys at 50 dB. (Ibáñez-Contreras et al., 2011).

- *Age of the Subject:* The characteristic responses from BAEP recordings appear immediately after birth, along with some modifications in the latencies and the morphology of their components due to the sequential myelination process that takes place in infants. (Elias 1996) Therefore, in NHP the maturation of the CNS during the first years of life produces striking changes in BAEP recordings. The waveform morphology of neonates and infant monkeys differs significantly when compared to adults; these changes respond to the incomplete maturation of the cochlear basilar membrane and the lack of myelination that associates hair cells and fibers (Ibáñez-Contreras et al., 2011, 2012; Moore and Linthicum 2007; Oliver et al., 1997; Moore 1985) (Figure 8-9). It has been reported that wave I in rhesus monkeys is more elongated in neonates than adults due to the placement of electrodes closer to the cochlea, consequent with a smaller brain size (Ibáñez-Contreras 2011). Similarly, prolonged latencies have been observed in geriatric NHP finding interpreted as a slow-response in conductivity from the area of the bridge (Ibáñez-Contreras et al., 2012; 2011; Oliver et al., 1997; Elias 1996; Church and Shucard 1986).

Interwave Intervals

Among the assessments that can be performed during BAEP recordings we have found the evaluation of interwave latencies. Generally, these are defined as the time-period between the peaks of one wave and another. Although wave I represents the activity of the VIIIth

cranial nerve, the analysis of the interwave interval provides further information regarding t the synchrony and integrity of the auditory brainstem, from the VIIIth cranial nerve to other nuclei being of great value for neurological diagnosis. Given the large number of factors affecting the absolute latencies during BAEP recordings, it is useful to know that interwave latencies or interwave intervals are relatively stable. There are few factors that can modify interwave latencies such as hypothermia or brain immaturity. (Ibáñez-Contreras et al., 2012; Elias 1996; Hausler et al., 1991).

Table 1. Average Interwave latencies I, II-IV, I-II and I-IV for males (n=7) and females (n=7) at different intensities

Interwave intervals males	70 dB	50 dB	30 dB
I-II	0,93	0,95	0,93
II-IV	2,04	2,03	2,03
I-IV	2,93	2,93	2,92
Interwave intervals females	**70 dB**	**50 dB**	**30 dB**
I-II	0,92	0,91	0,92
II-IV	1,55	1,54	1,54
I-IV	2,53	2,56	2,55

Ibáñez-Contreras et al., 2012.

In human beings the interwave latency I-III represents the conductivity of the proximal end of the VIIIth cranial nerve and the caudal segment of the auditory brainstem, while the interwave latency III-V is an indicator of conductivity for the rostral area of the bridge and the auditory pathway in the mesencephalon. The I-V interval represents the totality of the central conductivity through the brainstem (Elias 1996). This type of assessment differs in the case of NHP given the allocation site of the neural generators existing also scarce papers dealing with this subject in the literature. (Alegre et al., 2000; Velasco et al., 1984; Allen and Starr 1976).

Table 1 shows values obtained by our group (Ibáñez-Contreras et 2012), data demonstrating the lengthening for the intervals in male monkeys, primarily for II-IV and I-IV intervals, in relation to female monkeys. Alegre et al., (2000), Velasco et al., (1984) and Allen and Starr (1976), have reported that the value of the interval I-IV in male rhesus monkeys is 2.90 msec a result in agreement with our report.

It is important to highlight that the majority of BAEP studies carried out in NHP species have been conducted in male subjects; consequently, the recordings and results obtained in this paper using female monkeys provides a new area of research for the standardization for BAEP recordings in the *M. mulatta* species (Ibáñez-Contreras et al., 2012). Noteworthy is the fact that the interwave interval results shown in this Table did not suffered variations regarding different intensities applied, as indicated by Elias (1996) and Soares et al., (2010); these authors pointed out also that the interwave latencies are stable only if subjected to changes in the intensity of stimulus; consequently, this could be a reliable indicator for clinical audiology and neurologic evaluation of BAEPs in humans and NHP.

Conclusion

Throughout numerous studies assessing BAEPs in different animal species and humans, it has long been recognized that BAEP recordings are highly sensitive to differences in parameters being stimulated, the scale of intensity and also the prevalent conditions during registration, including certain characteristics of the subjects such as age and sex. Consequently, in this work and in view of further research, it is estimated that future efforts on this field must be oriented to enhance previous studies on BAEPs in NHP such as the rhesus monkey. The main goal would seek and establish a useful normalization of parameters being applied during the BAEP recordings; such a noninvasive methodology would result in fundamentally recognizing and documenting analogous changes generated by rather complex biologic developmental processes and the maturation and ageing of rhesus monkeys as is seen in humans. It can be concluded also, that enlightening even slight changes related to ongoing biological processes and correlating its impact with development and maturation of NHP, homologies useful to human medicine would enrich previous experimental findings encountered in NHP used as biomodels for the understanding of the physiology of the auditory pathway.

References

Alegre M, Gurtubay IG, Iriarte J, Ciordia E, Manrique M, Artieda J. Brainstem auditory evoked potentials (BAEPs) in the cynomolgus macaque monkey. Equivalence with human BAEPs and proposal of a new nomenclature. *Hear. Res.* 2001; 151: 115-120.

Alegre M, Artieda J. The role of cortical oscillatory activity in information processing mechanisms. *Rev. Neurol.* 2000; 30: 953-8.

Allen AR and Starr A. Auditory brain stem potentials in monkey (M. mulatta) and man. *Electroencephalogr. Clin. Neurophysiol.* 1978; 45: 53-63.

Arai S. Brainstem auditory evoked potentials in cattle sedated with xylazine. *Can. J. Vet. Res.* 2008; 72: 287-290.

Banks W. J. Histología Veterinaria Aplicada. Editorial Manual Moderno. México D.F. 1986; Pp. 680-690.

Butler AB. The evolution of the dorsal pallium in the telencephalon of amniotes: Cladistic analysis and new hypothesis. *Brain Res. Rev.* 1994; 19: 66-101.

Church MW and Shucard DW. Age-related hearing loss in BDF mice as evidenced by brainstem auditory evoked potential. *Audiology.* 1986; 25: 363-372.

Climent S, Sarasa M, Muniesa P, Terrado J. Manual de Anatomía y Embriología de los Animales Domésticos. Conceptos básicos y datos aplicativos. Editorial Acribia, S. A. Zaragoza, España. 1998. Pp 219-246.

Coleman MN, Ross CF. Primate auditory diversity and its influence on hearing performance. *Anat. Rec. A Discov. Mol. Cell Evol. Biol.* 2004; 281: 1123-1137.

Cunningham M. and Acker D. Animal Behavior. In: Animal Science and Industry. (6 ed). New Jersey: Prentice Hall, 2000.

Davis H. Principles of electric response audiometry. *Ann. Otol. Rhinol. Laryngol.* 1976: 85 SUPPL; 1-96.

Delgado HJ, Zenker CF, Barajas de Prat JJ. Normalización de los potenciales evocados auditivos del tronco cerebral 1: Resultados en una muestra de adultos normoyentes. *Auditio* 2003; 2: 13- 17.

Doyle WJ, Webster DB. Neonatal conductive hearing loss does not compromise brainstem auditory function and structure in rhesus monkeys. *Hear. Res.* 1991; 54: 145-151.

Doyle WJ, Rood SR. Anatomy of the auditory tube and related structures in the rhesus monkey (Macaca mulatta). *Acta Anat.* (Basel) 1979; 105: 209-225.

Doyle WJ, Saad MM, Fria TJ. Maturation of the auditory brain stem response in rhesus monkeys (Macaca mulatta). Electroencephalogr Clin Neurophysiol 1983; 56: 210-223

Durand-Rivera A. Bases técnicas y fisiológicas de los potenciales provocados auditivos de tallo cerebral. Para la Salud 1998; 19: 20-30.

Elías Y. Potenciales provocados auditivos del tallo cerebral en el diagnóstico del daño auditivo y neurológico del adulto. En: Hernández OA, Flores R T, Peñaloza LY (eds). Registros electrofisiológicos para el diagnóstico de la patología de la comunicación humana. Instituto de la comunicación humana. México. SSA, 1996. p. 157-170.

Fitzgerald MJH, Gruener G, Mtui E. 2012. Neuroanatomía clínica y neurociencia. Editorial Elsevier Saunder. 6 edició. Barcelona, España.

Fowler CG, Torre P 3rd, Kemnitz JW. Effects of caloric restriction and aging on the auditory function of rhesus monkeys (Macaca mulatta): The University of Wisconsin Study. *Hear. Res.* 2002; 169: 24-35.

Fowler CG, Chiasson KB, Leslie TH, Thomas D, Beasley TM, Kemnitz JW, Weindruch R. Auditory function in rhesus monkeys: effects of aging and caloric restriction in the Wisconsin monkeys five years later. *Hear. Res.* 2010; 261: 75-81.

Gil-Loyzaga P y Poch-Broto J. Fisiología del sistema auditivo periférico. Cap 74. In: Tratado de otorrinolaringología y cirugía de cabeza y cuello. Tomo II. Editors: Suarez C., Gil-Carcedo LM., Marco J, Medina JE., Ortega P, Trinidad J. 2° Edición. Editorial panamericana. 2007.

Gans C and Wever EG. The amphisbaenian ear: Blanus cinereus and Diplometopon zarudyi. *Proc. Nat. Acad. Sci. USA.* 1975; 72: 1487-1490.

Gans C and Wever EG. Ear ad hearing in Sphenodon punctatus. *Proc. Nat. Acad. Sci. USA.* 1976; 73: 4244-4246.

Guerrier Y. and Uziel A. Fisiología Neurosensorial en Otorrinolaringología. Editorial Elsevier Masson, S.A., Barcelona, España. 1985.

Hackett TA, Preuss TM, Kaas JH. Architectonic identification of the core region in auditory cortex of macaques, chimpanzees, and humans. *J. Comp. Neurol.* 2001; 441:197-222.

Hackett TA, Stepniewska I, Kaas JH. Subdivisions of auditory cortex and ipsilateral cortical connections of the parabelt auditory cortex in macaque monkeys. *J. Comp. Neurol.* 1998; 394:475-495.

Harland MM, Stewart AJ, Marshall AE, Belknap EB. Diagnosis of deafness in a horse by brainstem auditory evoked potential. *Can. Vet. J.* 2006; 47: 151-154.

Hartman C. G. The Anatomy of The Rhesus Monkey (Macaca mulatta). Hafner Publishing Co. New York, EUA. 1961; Pp. 339-359.

Hausler R, Cao M, Magnin C, Mulette P. Low frequency hearing threshold determination in newborns, infants and mentally retarded children by middle latency responses. *Acta Otoralyngol. Stockh. Suppl.* 1991; 482: 58-71.

Hecox KE, Galambo R. Brain Stem Auditory Evoked Responses in Human Infants and Adults. *Arrchives of otolaryngology* 1974; 1: 30 – 33.

Henry KR. Males lose hearing earlier in mouse models of late-onset age-related hearing loss; females lose hearing earlier in mouse models of early-onset hearing loss. *Hear. Res.* 2004; 190: 141-148.

Hernández-Godínez B, Ibáñez-Contreras A, Durand-Rivera A, Reyes-Pantoja SA, Ramírez-Hernández R, Rodríguez-Guzmán P, et al. Somatosensory Evoked Potentials SEPs, of Median and Tibial Nerves in Rhesus Monkeys (Macaca mulatta) Under Captivity: Influence of Ontogenic Status in Neonatal, Infant, Young, Adult and Senile Stages. *J. Med. Primatol.* 2011; 40: 79-87.

Hernández-Godínez B, Ibáñez-Contreras A, Durand-Rivera A, Galván-Montaño A, Reyes-Pantoja SA, Cárdenas-Lailson E. Valoración de latencias a través de los potenciales evocados somatosensoriales de nervio medial y tibial en monos rhesus, (Macaca mulatta) sometidos a laminectomía intrauterina para simular mielomeningocele. *Cir. Cir.* 2011, 79: 237-241.

Hildebrand M. and Goslow G.E. Jr. Analysis of Vertebrate Structure. Editorial Wiley. Quinta Edición. U.S.A., 2001.

Hoke M. Appendix: towards a uniform nomenclature of brainstem evoked responses. *Scand. Audiol. Suppl.* 1979; 11:115-7.

Isa T, Yamane I, Hamai M, Inagaki H. Japanese macaques as laboratory animals. *Exp. Anim.* 2009; 58: 451-457.

Ibáñez-Contreras A, Hernández-Godínez B, Durand-Rivera A. Valores de latencia y respuestas electrofisiológicas de los diferentes componentes de los potenciales evocados auditivos del tallo cerebral en mono rhesus a diferentes intensidades (70-30 dB). *Arch. Med. Vet.* 2012; 44: 267-271.

Ibáñez-Contreras A, Durand–Rivera A, Hernández-Godínez B and Reyes-Pantoja S. Potenciales evocados auditivos del tallo cerebral en monos rhesus (Macaca mulatta) en diferentes etapas fisiológicas en condiciones de cautiverio. *Arch. Med. Vet.* 2011; 43: 223-232.

Jewett D.L. y Romano M.N. Neonatal development of auditory system potentials average from the scalp of rat and cat. *Brain Res.* 1972; 36, 101-115.

Jones S, Martin R, and Pilbeam D. The Cambridge Encyclopedia of Human Evolution. Cambridge University Press. United Kindom. 1999.

Kaas JH, Hackett TA, Tramo MJ. Auditory processing in primate cerebral cortex. *Curr. Opin. Neurobiol.* 1999; 9:164-170.

Kaas JH, Hackett TA. Subdivisions of auditory cortex and processing streams in primates. *Proc. Natl. Acad. Sci. U. S. A.* 2000; 97: 11793-11799.

Kandel ER, Schwartz HJ, Jessell MT. Principios de Neurociencia. 4ta ed. Editorial McGraw- Hill Interamericana, Madrid, España. 2001.

Kardong. Vertebrados. Anatomía comparada, Función, Evolución. Editorial Mc Graw Hill Interamericana. Madrid, España. 1999. Pp. 624-663.

Karten HJ. The organization of ascendenting auditory pathway in the pigeon (Columba livia). Diencephalic projections of the inferiorcolliculus (nucleus mesencephali lateralis pars dorsalis). *Brain Res*. 1967; 6: 409-427.

Kemnitz JW, Roecker EB, Haffa AL, Pinheiro J, Kurzman I, Ramsey JJ, Mac Ewen EG. Serum dehydroepiandrosterone sulfate concentrations across the life span laboratory housed rhesus monkeys. *J. Med. Primatol*. 2000; 29: 330-337.

Kirwan JD, Bekaert M, Commins JM, Davies KTJ, Rossiter SJ, Teeling EM. A phylomedicine approach to understanding the evolution of auditory sensory perception and disease in mammals. *Evol. Appl*. 2013;6: 412-422.

Kraus N., Smith D.I., Reed N.L., Stein L. and Cartee C. Auditory brainstem and middle latency responses in nonhuman primates. *Hearing Research*, 1985; 17: 219-226.

Kretzmer EA, Meltzer NE, Haenggeli CA, Ryugo DK. An animal model for cochlear implants. *Arch. Otolaryngol. Head Neck Surg*. 2004; 130: 499-508.

Ladich F. Acoustic communication and the evolution of hearing in fishes. *Philos. Trans. R Soc. Lond. B Biol. Sci*. 2000; 355: 1285-1288.

Lasky RE, Maier MM, Snodgrass EB, Laughlin NK, Hecox KE. Auditory evoked brainstem and middle latency responses in Macaca mulatta and humans. *Hear. Res*. 1995; 89; 212-225.

Laughlin NK, Hartup BK, Lasky RE, Meier MM, and Hecox KE. The development of auditory even related potentials in the Rhesus Monkey (M. mulatta). *Dev. Psychobiol*. 1999, 34: 37-56.

Lawrence M. Structure and Function of the Ear and Auditory Nervous System. *Environ. Health Perspect*. 1982; 44:9-13.

López -Antúnez L. Anatomía Funcional del sistema nervioso. Noriega Editores. Limusa, México. 2003

Lowenstein O and Roberts TDM. The localization and analysis of the responses to vinration from the isolated elasmobranch labyrinth. A contribution to the problem of the evolution of hearing in vertebrates. *J. Physiol*. 1951: 114; 471-489.

Mader, D.R. Reptile medicine and surgery. Saunders. 2nd ed. Canadá. 2006. Pp: 65-66.

Malmierca MS. The structure and physiology of the rat auditory system: an overview. *Int. Rev. Neurobiol*. 2003; 56: 147-211.

Manley GA. Cochlear mechanisms from a phylogenetic viewpoint. *Proc. Natl. Acad. Sci. U. S. A*. 2000; 97: 11736-11743.

Maurer K., Lowitzsch K, Stöhr M. Evoked Potentials. B.C Decker Inc. Toronto Philadelphia, U.S.A. 1989

McFadden SL, Henselman LW, Zheng XY. Sex differences in auditory sensitivity of chinchillas before and after exposure to impulse noise. *Ear Hear* 1999; 20: 164-174.

Meza Ruíz G. Neurobiología de los sistemas sensoriales. Editorial UNAM, México D.F. 1995.

Moiseff A and Konishi M. Binaural characteristics of units in the owl´s brainstem auditory pathway: Precursors of restricted spatial receptive fields. *J. Neurosci*. 1983; 3:2553-2562.

Møller AR, Burgess J. Neural generators of the brain-stem auditory evoked potentials (BAEPs) in the rhesus monkey. *Electroencephalogr. Clin. Neurophysiol*. 1986; 65: 361-372.

Moore JK, FH Linthicum. The human auditory system: a timeline of development. *Int. J. Audio* 2007; 46: 460-478.

Moore DR. Posnatal development of the mammalian central auditory system and the neural consequences of auditory deprivation. *Acta Otolaryngol. Suppl.* 1985; 421: 19-30.

Morales Martínez JJ, Ayala GF, Poblano A. Diferencias auditivas del tallo cerebral en Gallo (Gallus gallus) y el Cuyo (Cavias porcellus). Estudio electrofisiológico comparativo de los potenciales provocados auditivos tempranos. *Acta Zool. Mex.* (n.s.) 2002; 87: 83-97

Northcutt RG, Kaas JH. The emergence and evolution of mammalian neocortex. *Trends Neurosci.* 1995;18: 373-379.

Oliver E, SA Edgley, J Armand, NR Lemon. An electrophysiological study of the posnatal development of the corticospinal system in macaque monkey. *J. Neurosci.*, 1997; 17: 267- 276.

O'Neill P, Mak SS, Fritzsch B, Ladher RK, Baker CVH. The amniote paratympanic organ develops from a previously undiscovered sensory placode. *Nat. Commun.* 2012; 3: 1041.

Picton, T. W., Hillyard, S.A, Krausz, I I. I., Galambos, R.. Human auditory evoked potentials. I. Evaluation of components. *Electroenceph. Clin. Neurophysiol.* 1974; 36: 179- 190.

Pineda JA, Holmes TC, Swick D, Foote SL. Brain-stem auditory evoked potentials in squirrel monkey (Saimiri sciureus). *Electroencephalogr. Clin. Neurophysiol.* 1989; 73:532-543.

Poblano A. Detección y Estimulación tempranas del Niño con Daño Neurológico. Instituto Nacional de Perinatología. Editores de Texto Mexicanos. 2003. p. 113- 118.

Poblano A, Hernández-Godínez B, Arellano A, Arteaga C, Elías Y, Morales JJ, Poblano R, Poblano-Alcalá A. Serum Testosterone and Electroencephalography Spectra in Developmental Male Rhesus Macaca mulatta Monkeys. *Archives of Medical Research* 2004; 35: 406-410.

Poon W.F. Paul and Brugge F. John. Central Auditory Processing and Neural Modeling. Editorial Plenum Press, New York. U.S.A., 1998.

Populin LC. Monkey Sound Localization: Head-Restrained versus Head-Unrestrained Orienting. *J. Neuroscie* 2006; 26: 9820-9832.

Ramsier MA, Dominy NJ. Receiver bias and the acoustic ecology of aye-ayes (Daubentonia madagascarensis). *Commun. Integr. Biol.* 2012: 5: 637-640.

Recanzone GH. Response profiles of auditory cortical neurons to tones and noise in behaving macaque monkeys. *Hear. Res.* 2000; 150: 104-118.

Rowe N. The Pictorial Guide to the Living Primates. Pogonias Press: East Hampton, New York. 1996.

Sánchez MM y Merchant CMA. Estructura y funcionamiento del cerebro auditivo. Cap 75. p. 969. En: Tratado de otorrinolaringología y cirugía de cabeza y cuello. Tomo II. Suarez C., Gil-Carcedo LM., Marco J, Medina JE., Ortega P, Trinidad J. Editores. 2° Edición. Editorial panamericana. 2007.

Sapède D. and Pujades C. Hedgehog Signaling Governs the Development of Otic Sensory Epithelium and Its Associated Innervation in Zebrafish, *J. Neuro*, 2010; 30: 3612-3623.

Satterlie RA. Do jellyfish have central nervous systems?. *Journal of Experimental Biology* 2011; 214: 1215-1223. Schapiro SJ, Bloomsmith MA, Suarez SA, and Porter LM. Effects of social and inanimate enrichment of the behavior of yearling rhesus monkeys. *Am. J. Primatol.* 1996; 40: 247-260.

Scott W. Rogers. Allosaurus,Crocodiles, and Birds: Evolutionary Clues From Spiral Computed Tomography of an Endocast. The Anatomical Record (NEW ANAT.)1999; 257:162–173Segal EE. Housing, Care, and Psychological Well-Being of Captive and Laboratory Primates. Noyes Publications, Park Ridge New 1989.

Shkurovich ZM. Potenciales Evocados Visuales. In: Registro Electrofisiologico para el Diagnostico de la Patologia de la Comunicación Humana. (OA Hernandez, RT Flores, LY Peñaloza eds): Instituto de la comunicación humana. México. SSa: 183-194.

Soares IA, Menezes PL, Carnauba ATL, Pereira LD. Standardization of brainstem auditory evoked potentials using a new device. *Pro Fono* 2010; 22: 421-426.

Sólis-Chávez SA, Ibáñez-Contreras A, Durand-Rivera A, Reyes-Pantoja SA, Rodriguez-Alfaro A, Tena-Betancourt E, Hernández-Godínez B. Morphology of brain waveforms and electrical activity of visual evoked potentials (VEPS) of Iguana iguana species under captivity conditions in Mexico. *Thai J. Vet. Med.* 2013: 43: 113-117.

Starr A. A., Amlie R. M., Martín W. H. and Sanders S. Development of auditory function in newborn infants revealed by auditory brainstem potentials. *Pediatrics*, 1977;60: 831-839.

Steriade, M; Jones, E. G; Mc Cormick, D. A. 1997. Thalamus. Oxford, UK: Elsevier Science.

Striedter, G. F. The telencephalon of vertebrates in evolution. *Brain Behav. Evol.* 1997; 49: 179-213.

Swenson M. J. y Reece W. O. Fisiología de los Animales Domésticos de Dukes. Editorial Uteha Noriega Editores. Segunda Edición. México. 1999. Pp. 816- 835.

Tavolga WN. Mechanisms for directional hearing in the sea catfish (Arius felis). *J. Exp. Biol.* 1977; 67: 97-115.

Torre P. 3rd, Fowler CG. Age- related changes in auditory function of rhesus monkeys (Macaca mulatta). *Hear. Res.* 2000; 142: 131-140.

Torre P. 3rd, Mattison JA, Fowler CG, Lane MA, Roth GS, and Ingram DK. Assessment of auditory function in rhesus monkeys (Macaca mulatta): effects of age and calorie restriction. *Neurobiol. Aging* 2004; 25: 945- 954.

Torre P. 3rd, Lasky RE, Fowler CG. Aging and middle ear function in Rhesus monkeys (Macaca mulatta). *Audiology* 2000; 39: 300-304.

Wilson AC, Sarich VM. A molecular time scale for human evolution. *Proc. Nat. Acad. Sci. USA* 1969; 63: 1088-1093.

Torre P 3rd, Lasky RE, Fowler CG. Aging and middle ear function in Rhesus monkeys (Macaca mulatta). *Audiology* 2000; 39: 300-304.

Velasco M, Velasco F, Romo R, Almanza X. Subcortical correlates of the auditory brain stem potentials in the monkey: bipolar EEG and multiple unit activity responses. *Int. J. Neurosci.* 1984; 22: 235-251.

Yadav A, Tandon OP, Vaney N. Auditory evoked responses during different phases of menstrual cycle. Indian *J. Physiol. Pharmacol.* 2002; 46: 449-456.

Tandon OP, Khaliq F, Goel N. Auditory evoked potential responses in menopausal women: a normative study. *Indian J. Physiol. Pharmacol.* 2001; 45: 361-366.

Toledano A, Alvarez MI, López-Rodríguez AB, Toledano-Díaz A, Fernández-Verdecia CI. Does Alzheimer's disease exist in all primates? Alzheimer pathology in non-human primates and its pathophysiological implications (I). *Neurologia*. 2012;27:354-369.

In: Monkeys
Editor: Hugo A. Barrera Saldaña

ISBN: 978-1-63117-851-1

Chapter 2

The Use of Non-Human Primates As Bio-Models for Aging Brain Studies: Interventional Role of Oxidative Stress in Demyelization Processes

Alejandra Ibáñez-Contreras*[*1,2,3,4,6], Adrian Poblano[5], Braulio Hernández-Godínez[1,2,3,4], Adriana Alarcón-Aguilar[1,6], Marcela Arteaga-Silva[7], Eduardo Tena-Betancourt[8] and Mina Königsberg[6]

[1]Posgrado en Ciencias Biológicas y de la Salud.
Universidad Autónoma Metropolitana (UAM), Unidad Iztapalapa. México D.F.
[2]Applied Research in Experimental Biomedicine S.A. de C.V. (APREXBIO).
Laboratorio de Primatología. Ciudad de México México D.F.
[3] BIOINVERT®, Biología Integral para Vertebrados,
Unidad de Experimentación Animal, Estado de México, México
[4] Centro de Investigación, Proyecto CAMINA, A.C.,
Unidad de Primates no Humanos. Ciudad de México, México D.F.
[5] Laboratorio de Neurofisiología Cognoscitiva.
Instituto Nacional de Rehabilitación. Ciudad de México, México D.F.
[6] Depto de Ciencias de la Salud, Laboratorio de
Bioenergética y envejecimiento celular.
Universidad Autónoma Metropolitana (UAM),
Unidad Iztapalapa. México D.F.
[7] Depto. Biología de la Reproducción,
Universidad Autónoma Metropolitana (UAM), Unidad Iztapalapa. México D.F.
[8] Departamento de Etología Fauna Silvestre y Animales
de Laboratorio, Facultad de Medicina Veterinaria y Zootecnia,
UNAM, México D.F

[*] Corresponding Author: ibanez.alejandra@hotmail.com.

ABSTRACT

Aging has been characterized as a functional decline that impairs an array of physiological functions. It has been stated that many changes that take place at a molecular level during brain aging and neurodegenerative diseases (ND), operate through a common underlying mechanism which has been associated to an oxidative damage to macromolecules. This situation is enhanced by an oxidative homeostatic imbalance which occurs through an increase in reactive oxygen and nitrogen species production, accompanied by a decline in antioxidant systems, thus leading to oxidative damage accumulation. The nervous system (NS) has a high oxidative metabolism being therefore more vulnerable to oxidative injury. Hence, it is important to develop and apply sever diagnostic tools and biomodels, in order to characterize early and progressive stages of ND, and differentiate them from mild-stable alterations of normal aging wear and tear process. As a result, the use of evoked potentials (EP´s) has proved to be a reliable diagnostic indicator, with high sensitivity and specificity to evaluate sensory pathways. The non-human primates (NHP) have been widely used as biomodels for brain aging studies and ND, providing information related to the physiological changes associated to age in non-pathological conditions, hence differentiating them from true pathological processes that occur during ND.

Keywords: Non-human Primate, Oxidative stress, aging, demyelinization

INTRODUCTION

Aging is defined as the accumulation of diverse deleterious changes occurring in cells and tissues in a complex and continual heterogeneous, universal and irreversible process. This process determines a progressive undermining of the organisms adaptive capacity, leading to disease and death (Arcos-Correa and González-Martínez 2009; Harman 2003; OMS 2002). The functional decline that characterizes aging impairs an array of physiological functions; in particular an oxidative homeostatic imbalance that occurs through the elevation in reactive oxygen and nitrogen species (ROS and RNS) production, accompanied by the decline in antioxidant systems, thus leading to oxidative damage accumulation. This imbalance is referred as oxidative stress (OS), and can damage diverse cell biomolecules such as lipids, proteins and DNA (Csiszar et al., 2012; Alarcón-Aguilar et al., 2010; Hartman 1956) (Figure 1). ROS/RNS are known to be associated with the etiology of inflammatory and neurodegenerative diseases, as well as carcinogenesis (Königsberg 2008, 2007; Wei and Lee 2002; González-Fraguela et al., 2000; Cohen and d´Arcy-Doherty 1987). The most important ROS include singlet oxygen $^1O_2$, superoxide anion (O_2•-), hydrogen peroxide (H_2O_2), and the major contributor to oxidative damage the hydroxyl radical (•OH). The RNS are elicited by complex metabolic reactions; these species are nitric oxide (NO•), nitrogen dioxide (NO_2•) and peroxynitrite (ONOO-) (Königsberg 2007; Gonzalez-Fraguelas et al., 2000).

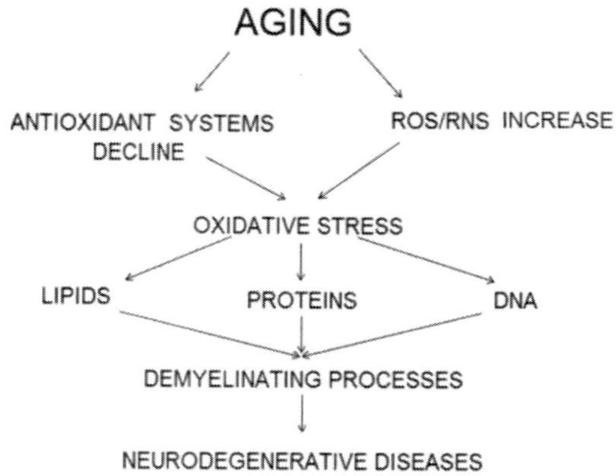

Figure 1. During aging ROS and RNS production increases, generating the loss of oxidative homeostasis and leading to oxidative damage, which has a great impact in the nervous system, and might favor the demyelinating processes, such as the effect observed during neurodegenerative diseases.

Reactive Oxygen and Nitrogen Species (ROS and RNS)

It has been stated that many changes that occur at a molecular level during brain aging and neurodegenerative diseases (ND), operate through a common underlying mechanism which has been identified as oxidative damage to macromolecules; the nervous system (NS) has a high oxidative metabolism which makes it extremely vulnerable to oxidative injury. In addition, the brain is a highly oxygenated structure that receives nearly the fifth part of the total cardiac output and contains elevated concentrations of iron (Fe^{2+}), as well as easily oxidizable substrates, such as carbohydrates and polyunsaturated lipids (Watson et al., 2012; Martínez-Samano 2011; Gonzalez-Fraguela et al., 2000). Moreover, in the central nervous system (CNS), catecholamines are an important ROS/RNS source; an example is the dopamine (DA), serotonin (5HT) and noradrenaline (NA) metabolic catabolism by monoamine oxidase (MAO), which is known to produce H_2O_2 (Santiago-López and Rivas-Aracibia 2008). Another source of ROS, are the ones generated by macrophages, which are known to be demyelination and axonal-damage mediators during Alzheimer (AD) and Parkinson's diseases (PD), as well as amyotrophic lateral sclerosis (ALS) and multiple sclerosis (MS) (Watson et al., 2012; Martínez-Samano 2011; Alarcón-Aguilar et al., 2010; Königsberg 2008, 2007; Carlson and Rose 2006; González- Fraguela et al., 2000; Cohen and d´Arcy-Doherty 1987; Konat and Wiggins 1985).

In order to contend against ROS/RNS, living organisms have a great repertoire of antioxidant mechanisms that constitute important biological defenses and allow their removal (Tabrez and Ahmad 2011). This principal line of protection has been divided into enzymatic and non-enzymatic antioxidants. The first ones are enzymes such as superoxide dismutase (SOD), catalase (CAT), glutathione peroxidase (GPx) and tiorredoxines (Trx) that are produced within the organism. While the non-enzymatic antioxidants, such as vitamins A, C and E, are obtained through dietary means (Königsberg 2008; 2007). ROS/RNS signaling plays an important role in numerous normal physiological processes (Csizar et al., 2012; Konigsberg 2008; Harman 1956). However, an uncontrolled ROS/RNS increase has been reported during aging, which associated to a decrease in the antioxidant systems, an increase in oxidative stress and bimolecular damage (Csiszar et al., 2012; Zorrilla-García 2002; Harman 1956).

In recent years a physiological relationship between ND and aging has been described, establishing an important correlation between gradual neuronal death and physiological events related to oxidative stress. This has been particularly described in various neurodegenerative diseases such as AD, where a decrease in CAT activity leads to an increase in lipid peroxidation (LPX); other alterations reported are iron accumulation in the microglia and changes in other antioxidants, such as α-tocopherol (Toledano et al., 2012; González-Fraguela et al., 2000). Warner (1994) reported a SOD decrease in a wide variety of species, which increased their vulnerability to OS; in particular a genetic alteration of Cu-Zn SOD enzyme, leads to motoneuron premature death during ALS (Muller et al., 2008). Alternatively, it has also been shown that an increase in SOD activity without the compensatory increase in CAT or GPx activities, can lead to an increase in O_2•- production, consequently leading to oxidative damage (González-Fraguela et al., 2000). Moreover, a decrease in GSH levels results in an increase in oxidative damage to lipids, proteins and DNA in old animals' tissues and fluids (González-Fraguela et al., 2000).

It is widely known that the nerve conduction velocity and accuracy depend on the nerve diameter and myelination degree. Myelin allows nerve impulses conduction through the CNS, while oligodendrocytes, a group of highly specialized glial cells, are responsible for the formation of fat-rich myelin sheaths in the nerve fibers. Therefore, it has been proposed that OS associated to age, might cause nerve fibers and certain nuclei demyelination in the cerebral cortex, which in turn could enhance nerve fibers scarring and sclerosis, resulting in nerve conductivity decrease as seen in MS (Watson et al., 2012; Fergestad et al., 2006). Therefore studying brain aging and its relationship with OS will certainly provide vast information about physiological changes caused during non-pathological aging, and might allow its differentiation to the true pathological processes that occur during the course of ND.

Aging, Oxidative Stress and Demyelination

The demyelination processes are known to be systemic changes occurring during aging, which have an important impact on the NS (Von Berhardi 2005; Konat and Wiggins 1985) and have been associated with mitochondrial metabolism defects, ROS generation and excytotoxicity (Escobar-Izquierdo 2001).

During aging, the decrease in the ability to adapt to stress has been particularly associated to an increase in glucocorticoids release (Pichardo-Fuster and Pedrero-Nieto 2009). This fact increases hippocampal neurons vulnerability by enhancing glutamate and other excitatory amino acids deleterious effects (Toussaint 1998) accelerating ATP depletion and Ca^{2+} release, thus stimulating ROS overproduction. It is important to consider that an increase in Ca^{2+} concentrations stimulates the enzyme phospholipase A, which releases araquidonic acid increasing ROS generation, which concomitantly activates nitric oxide synthase (NOS) and thus NO generation. Both, arachidonic acid and ROS, stimulate glutamate release, while inhibiting its recapture, perpetuating the vicious cycle (Tapia-Saavedra 2007; Von Bernhardi 2005; Escobar-Izquierdo 2001).

It is also well known that an increase in cytosolic Ca^{2+} concentration activates several peptidases, such as Calpain I, which catalyzes the enzymatic conversion of Xanthine dehydrogenase into Xanthine oxidase, generating superoxide anion (O_2•-) through purine metabolism. Finally, ROS and Ca^{2+} stimulate the activation of other proteases and phospholipases, and enhance cytokines production, such as tumor necrosis factor α (TNF-α) or lymphotoxin, which are potent mediators of inflammatory and immune response, and might as well induce oligodendrocytes apoptosis favoring the demyelinating process (Von Bernhardi 2005; Díaz - Ruíz et al., 2002; Escobar-Izquierdo 2001; Bruck 1997; Xiao et al., 1996).

Within the main biochemical disorders related to aging, both in human and in animal models, glucose metabolism alteration has a great impact and has been intensely studied (Ibáñez- Contreras et al., 2011a; Wang et al., 2009; Rezzi et al., 2009; Bank et al., 2007; Cantú et al., 2007; Tigno et al., 2004; Escobar-Izquierdo 2001; Toussaint 1998). Some of the metabolic changes caused by hyperglycemia correlate with an antioxidant defense decrease, especially there is a decrease in reduced nicotinamide adenine dinucleotide phosphate (NADPH) and reduced glutathione (GSH) availability. The fall in NADPH concentration affects other antioxidant enzymes activities, such as GR, CAT, and NOX (Díaz-Flores et al., 2004). Based on such information, it has been suggested that a rise in glucose levels generates OS and apoptosis, especially at the spinal cord dorsal root ganglion cells level (Cantú et al., 2007). OS early changes in animal models due to glucose increase, are axonal swelling, and nodal and paranodal regions edema, which decrease nerve conduction velocity, therefore these characteristics are accepted as neuropathy main features. In addition, progressive peripheral myelin sheath detachment in Ranvier node Schwann cells has also been identified (Cantú et al., 2007; Madrigal-Rentería et al., 2007).

From the immunological standpoint, it is well documented that during the aging process an alteration of the immune response is observed. Inflammation and altered cytokines regulation have been linked to various aspects of NS aging (Álvarez-Fernández et al., 2002*)*. It has been described that OS might damage oligodendrocytes, since stimulated macrophages/microglia produce ROS, in particular NO and its derivates, causing an axonal conductivity blockage. In addition, an increase in TNF-α production, as well as the excitotoxicity phenomena has also been implicated in neuronal and oligodendrocyte death (Fuster-Lluch et al., 2004; Bruck 1997; Xiao et al., 1996). Moreover, nowadays it is well known that NO and its derivatives are involved in certain physiopathological conditions, such as hypotension, which accompanies septic shock, and the inflammatory response induced by tisular damage (Saura and Lamas 1995). In animal models, it has been described that ONOO- clearly exerts toxic effects on a variety of CNS cells, including astrocytes and oligodendrocytes (Scott et al., 2003; Touil et al., 2001; Endres et al., 1998). ONOO- is capable of causing cellular damage, and even cell death, through a variety of mechanisms, including tyrosine nitration, lipid peroxidation, as well as mitochondrial respiration impairment. Such scenario can induce primary axonal damage by producing myelin sheath harmful alterations. In addition, *in vitro* studies have demonstrated that ONOO- also reduces oligodendrocytes viability and increases neuronal death (Scott et al., 2004; 2003).

Another important point to be discussed is iron (Fe^{2+}) participation, given that this element actively increases ROS generation, notably •OH during the Fenton reaction. However, Fe results essential as co-factor of several enzymatic reactions and is involved in ND, since the oligodendrocytes, responsible for myelin production are Fe-enriched, in contrast to other NS cells. Hence, it is clear that Fe plays a dual role, and might be related to the demyelination process by generating ROS (Tapia-Saavedra 2007).

NEUROPHYSIOLOGIC EVALUATION: EVOKED POTENTIALS

During normal aging, and even in the absence of ND, the number of synapses decline, and this event is observed along with a permeability and neuronal excitability decrease (Watson et al., 2012; Fergestad et al., 2006). In addition, neuronal membrane oxidation has been related to neurophysiological changes, specifically altering intrinsic electrophysiological parameters (Watson et al., 2012). Hence, it is important to use several diagnostic tools in order to characterize the early and the late stages during ND, and differentiate them from mild-stable alterations of aging wear and tear process. As a result, the use of evoked potentials (EP´s) has proved to be a reliable diagnostic indicator, with high sensitivity and specificity to evaluate sensory pathways. These diagnostic elements are electrical signals recordings generated by the NS during auditory, visual, somatosensory and motor specific pathways, allowing the assessment of cortical and subcortical responses elicited by peripheral sensory fibers stimuli (Sloan 2003; Poblano 2003; Starr et al., 1977; Jewett and Romano 1972). In addition, besides evaluating sensory pathways integrity, the EP´s can characterize and describe the magnitude of a specific impairment and locate the injury to a particular body segment (Hernández-Godínez et al., 2012; 2011).

The parameters recorded by the different sensory pathways (auditory, visual, and somatosensory) are dissimilar and can be observed as changes in the recorded waves number and morphology, including differences in the latencies values and amplitudes. There are several EP´s types depending on the information that is required:

1. Brainstem auditory evoked potentials (BAEPs), are short latency, small electrical voltage potentials, produced by brief sound stimuli, usually as a response to series of 'clicks' of initial rarefaction. They are characterized by having at least four positive permanent waves in animals and five in humans; these waves reflect brainstem auditory specific nuclei electrical activity (Ibáñez-Contreras et al., 2012; 2011b). The BAEPs are a useful diagnostic tool for different clinical situations including coma, brain death, degenerative disorders, demyelinating and infectious diseases, metabolic disorders, tumors, etc. During neurological disorders evaluation, BAEPs help during the diagnosis to differentiate between two events, since they provide information about the topographic location of the damage; the first are those involving the vestibulocochlear nerve (CN VIII) and the caudal brainstem position, while the latter relate to injuries affecting the brainstem rostral portion (Elías 1997). It´s use has also been described in the diagnosis of demyelinating diseases, showing variable anomalies in latency and amplitude (Elías 1997; Chiappa 1980).

2. Visual evoked potentials (VEP´s) allow the electrical activity evaluation of the visual pathway in specific brain areas. This stimulation is performed using short light flashes (stroboscopic lamp) or alternating images (chessboard pattern) and is characterized by at least two negative waves and one positive in animals and in humans (Chiappa 1997). VEP´s are very useful to evaluate eye diseases at the retina, optical nerve and optic chiasm. They are also used in diagnosis and pathophysiologic follow-up in ND such as in MS and PD, where a significant increase in the latency, particularly the P_1 wave of MS and parkinsonian patients, significantly increases as the disease evolves (Nightingale et al., 1986; Delwaide et al., 1980).

3. *Somatosensory evoked potentials (SEPs)* represent a useful and objective tool to assess the somatosensory system functionality and integrity, and can serve to determine and locate neuronal conductivity interruptions or delays; SEPs are non-invasive diagnostic indicators, that allow the study of CNS and peripheral somatosensory pathways (Hernández-Godínez et al., 2012, 2011); helping also in the evaluation of sensory peripheral nerves damage during certain pathologies and metabolic disorders, including diabetes, uremia, MS, nutritional deficiencies, or ND diseases such as PD leading to axonopathies, that include spinocerebellar degeneration, nutritional disorders and polyneuropathies among others (Arezzo et al., 1982; Pedersen and Trojaborg 1981).

As mentioned before, age is a determining factor that affects brain electrical activity. Various reports indicate that in both, humans and animals, the delayed electrical activity obtained from the PE´s, correlate with increasing aging (Hernández-Godínez et al., 2012; 2011; Kohama et al., 2012; Ibánez-Contreras et al., 2012, 2011b; Poblano 2003; Elías1997). This phenomenon is mainly attributed to a defective myelin regeneration or neuroaxonal dystrophy, as well as to cortical neurons and some subcortical nuclei loss, hypertrophic changes in the astrocytary glia, and oligodendrocytes degeneration. Neurotransmitters levels might also contribute to the process, since it has been described that during aging there is a significant decrease in catecholamines, acetylcholine and GABA concentrations. Likewise, microscopic studies have shown a decrease in neurocortical and archicortex neurons during aging without observable pathology association (Kohama et al., 2012; Escobar-Izquierdo 2001; Shkurovich 1997).

Non-Human Primates (NHP) As a Model to Study Cerebral Aging Associated to Oxidative Stress (OS)

To date, numerous papers have dealt with the study of brain aging and its relationship with OS, providing information toward physiological changes associated to age in non-pathological conditions. Hence differentiating this phenomenon from true pathological processes that occur during ND. The aim of these studies was to elucidate factors and changes observed during the aging process, in order to interfere and decrease the negative effects associated to aging (Von Bernhardi 2005). Based on such statement, non-human primates (NHP) have been widely used as biomodels for brain aging studies and ND (Toledano et al., 2012; Kemnitz 2011; Wang et al., 2011; Duncan et al., 2011; Fisher and Austad 2011; Nagahara et al., 2010; Mckiernan et al., 2009; Roth et al., 2004; Gilardi et al., 1997; Cefalu and Wagner 1997; Taylor et al., 1992; Morrison et al., 1990).

The NHP most common species used in neuropathological ailments studies are dwarf and mouse lemurs *(Cherirogaleidae)* and the macaques *(Cercophitecoideae)* families (Toledano et al., 2012). Whilst among the *Cercopithecidae,* the most frequent are green monkeys (*Chlorocebus aethiops*), Rhesus monkeys (*Macaca mulatta*), and the stump-tailed macaques (*Macaca arctoides*) (Hernández Godínez et al., 2012, 2011, Ibáñez-Contreras et al., 2011 b; Isa et al., 2009; Schapiro 2002; Kemnitz et al., 2000; Segal 1989). It is estimated that the lifetime of these animals is approximately 30 years in captivity. Primates described as senile include those that are more than 20 years old, and extremely elderly monkeys include animals close to 30 years old (Toledano et al., 2012; Ibáñez-Contreras et al., 2011 a,b; Fowler et al., 2010).

It is important to mention that O_2•- and H_2O_2 high production has been reported in old stump-tailed *Macaques* (*Macaca arctoides*), as well as a decrease in SOD and CAT enzymatic activities, as occurring in old human beings (González-Fraguela et al., 2000). It has also been shown in *Saimiri sciureus* and *Macaca mulatta* senile animals that caloric restriction (CR), decreased brain amyloid deposits accompanied by reduced cytokines levels (Willette et al., 2010; Qin et al., 2006; Kim et al., 1997; Kemnitz et al., 1994). Fowler and coworkers (Fowler et al., 2010) reported that CR in *Macaca mulatta* decreased animal mortality, associated to smaller amounts of body fat deposits, which were supposed to improve insulin sensitivity and cardiovascular function.

There have also been investigations aimed to understand how the oxidative stress associated with aging might contribute to liver damage and pathology in Rhesus monkeys. Those studies reported a direct age-related correlation between mitochondrial DNA (mtDNA) damage and age (Castro et al., 2012).

Various authors have described that geriatric NHP develop behavioral and morphological alterations, similar to the ones observed in senile human beings (Yue et al., 2012; Csiszar et al., 2012; Toledano et al., 2012; Bendlin et al., 2011; Collier et al., 2011; Valenzuela et al., 2010; Ohnuki et al., 2010; Duce et al., 2008; Sani et al., 2003; González et al., 2000; Brownell et al., 1998; Tumminia et al., 1996; Gearing et al., 1994; Bachevalier et al., 1991; Price et al., 1991). Based on the aforementioned, several NHP animal models have been used as PD models by inducing dopaminergic neuronal injury, using different neurotoxic agents such as the 1- methyl-4-phenyl-l,2,3,6-tetrahydropyridine (MPTP), that is known to induce selective dopaminergic neurons (NDA) degeneration at the substancia nigra level, leading to a reduction in spontaneous activity, rigidity, bradykinesia and trembling, analogous symptoms to human PD (Alarcón- Aguilar et al., 2010; Blesa et al., 2010).

As indicated before, age and demyelinating processes are important factors affecting neurophysiological response and wave recordings. In Rhesus monkeys a 60% decrease in catecholamines biosynthesis has been described (Goldman-Rakic and Brown 1981; Samorajski and Rolsten 1973). This was primarily related to sensorial and cognitive deficiencies, as well as ND diseases, which might delay electrical conductivity (Collier et al., 2011; Grondin et al., 2003). Several studies performed in NHP indicated that recordings showing elongated latencies during EP`s studies, have been interpreted as a delayed response due to degenerative musculoskeletal disorders, and also as an unexpected response associated to neurodegenerative changes in somatosensory pathways (Hernández-Godínez et al., 2012; 2011). Similarly, a progressive decrease in NHP hearing capacity has been observed during primate aging (Fowler et al., 2010; Torre et al., 2004; Fowler et al., 2002; Torre and Fowler 2000; Torre et al., 2000). We have previously reported (Ibáñez-Contreras et al., 2011b) that when studying geriatric monkeys, it is often impossible to identify certain auditory pathway components; and we have proposed that this might be related to some kind of cochlear or retro cochlear injury resulting from degenerative changes associated to the normal aging process. Similarly, it has been observed that in geriatric Rhesus monkeys the absolute latencies of neural auditory pathways components, tend to increase, particularly linked to hormonal changes, mainly due to a decrease of steroid hormones production, as also seen in human beings and other animal species (Ibáñez-Contreras et al., 2011b; Fowler et al., 2010; Hultcrantz et al., 2006; Henry 2004; Tandon et al., 2001).

Finally, according to Solis-Chávez and coworkers (2013), geriatric Rhesus monkeys show VEP´s latencies increase, mainly related to defective myelin regeneration. This effect can also be associated to the ganglion cells degeneration in the retina and to visual pathways alterations. In accordance to previous arguments and the close phylogenetic proximity of NHP to humans, it has been postulated that NHP may exhibit similar pathophysiological patterns to those that have been described in human beings after being subjected to oxidative aggression. However the relationship between systemic oxidative stress and the damage induced to sensory pathways in the *Macaca mulatta* species it has not been documented yet, therefor that feature remains unknown, and might be important during neurodegenerative diseases understanding.

CONCLUSION

The use of NHP as animal models in biomedical research and in favor of public health has constituted a core element in the last years, significantly enhancing the current medical research. Given ethical considerations and difficulties for a systematic evaluation of the aging processes in humans, numerous studies and research have been devise and executed using NHP. To date, a considerable amount of studies dealing with brain aging and its relationship with OS, have provided critical information in order to understand the physiological changes associated to non-pathological aging, and therefore being able to differentiate them form the observable pathological processes that occur during ND.

AKNOWLEDGMENTS

This work was supported by CONACyT grant CB-2012-1-178349. Ibáñez-Contreras A, Alarcón-Aguilar A and Poblano A are CONACyT scholarship holders.

REFERENCES

Alarcón -Aguilar A, Santamaría del Ángel A, Königsberg Fainstein M. 2010. Modelos neurotóxicos de la enfermedad de parkinson y disfunción mitocondrial. *REB* 2010; 29: 92-100.

Álvarez-Fernández B, García-Ordóñez MA, López-Trigo JA, Marín-Carmona JM, Gómez-Huelgas R, Juárez-Fernández C. Modificación de la respuesta inmune en los ancianos con tratamientos nutricionales. *An Med Interna* (Madrid) 2002; 19: 423-429.

Arcos-Correa J.H., González –Martínez J.F. Determinantes de la longevidad humana. En: Geriatría. González-Martínez J.Fco., Pichardo-Fuster A, García L. (Eds), Mc Graw Hill, México D.F., 2009, pp 1-18.

Arezzo J, Schaumburg H, Vaughan HG Jr, Spencer PS, Barna J. Hind limb somatosensory evoked potentials in the monkey: the effects of distal axonopathy. *Ann Neurol* 1982; 12: 24-32.

Bachevalier J, Landis LS, Walker LC, Brickson M, Mishkin M, Price DL, et al. Aged monkeys exhibit behavioral deficits indicative of widespread cerebral dysfunction. *Neurobiol Aging* 1991;12:99-111

Ballou SP, Kushner I. Chronic inflammation in older people: recognition, consequences, and potential intervention. *Clin Geriatr Med.* 1997; 13: 653-670.

Banks WA, Thomas DR, Willoughby LM, Morley JE. Insuline resistance syndrome in the elderly: Assessment of functional, biochemical, metabolic and inflammatory status. *Diabetes Care* 2007: 30; 2369-2373.

Bendlin BB, Canu E, Willette A, Kastman EK, McLaren DG, Kosmatka KJ, et al. Effects of aging and calorie restriction on white matter in rhesus macaques. *Neurobiol Aging.* 2011; 32: 2319. e1-11.

Blesa J, Juri C, Collantes M, Peñuelas I, Prieto E, Iglesias E, et al. Progression of dopaminergic depletion in a model of MPTP-induced Parkinsonism in non-human primates. An (18) F-DOPA and (11) C-DTBZ PET study. *Neurobiol Dis* 2010; 38: 456:463

Brownell AL, Jenkins BG, Elmaleh DR, Deacon TW, Spealman RD, Isacson O. Combined PET/MRS brain studies show dynamic and long-term physiological changes in a primate model of Parkinson disease. *Nat Med* 1998; 4: 1308-12.

Bruck W. 1997. The role of macrophages in Wallerian degeneration. *Brain Patol* 1997; 7: 741-752.

Cantú BC, Baizábal CF, Aldrete VJ, Chávez GE. Aplicaciones clínicas del acido tióctico (ácido α-lipoico) en pacientes con neuropatía diabética sintomática: revison. *Med Int Mex* 2007; 23: 151-158.

Carlson NG, Rose JW. Antioxidants in multiple sclerosis: do they have a role in therapy?. *CNS Drugs* 2006; 20: 433-41

Castro MR, Suarez E, Kraiselburd E, Isidro A, Paz J, Ferder L, Ayala-Torres S. Aging increases mitochondrial DNA damage and oxidative stress in liver of rhesus monkeys. *Exp Gerontol* 2012; 47: 29-37.

Cefalu W, Wagner JD. Aging and atherosclerosis in human and nonhuman primates. *Age* 1997; 20: 15-28.

Chiappa KH. 1997. Brain Stem Auditory Evoked Potentials: Methology. En: Chiappa KH (Eds). Evoked Potentials in clinical medicine. 3a ed. Raven Publishers, Washington Square, Philadelphia, USA, pp 157-198.

Chiappa KH. Brainstem auditory evoked responses in 200 patients with multiple sclerosis. *Annals of Neurology.* 1980: 7; 135-143.

Cohen G.M. d´Arcy Doherty M. Free radical mediated cell toxicity by redox cycling chemicals. *Br J Cancer* 1987; 55: 46-52.

Collier TJ, Kanaan NM, Kordower JH. 2011. Ageing as primary risk factor for Parkinson´s disease: evidence from studies of non-human primates. *Nat Rev Neurosci* 2011; 12: 359-366.

Csiszar A, Podlutsky A, Podlutskaya N, Sonntag WE, Merlin SZ, Philipp EE, Doyle K, Davila A, Recchia FA, Ballabh P, Pinto JT, Ungvari Z. Testing the oxidative stress hypothesis of aging in primate fibroblasts: Is there a correlation between species longevity and cellular ROS production?. *J Gerontol A Biol Sci Med Sci* 2012; 67: 841-852.

Delwaide PJ, Mesraoua B, De Pasqua V. Visual evoked potentials in Parkinson´s disease. *Rev Electroencephalogr Neurophysiol Clin.* 1980; 10: 338-342.

Díaz-Flores M, Baiza- Gutman LA, Ibáñez-Hernández MA, Pascoe-Lira D, Guzmán-Greenfel AM, Kumate- Rodríguez J. Aspectos moleculares del daño tisualr inducido por hiperglucemia crónica. *Gac Méd Mex* 2004: 140: 437-447

Díaz-Ruíz A, Guízar Sahagún G, Ríos-Castañeda C. Estrategias neuroprotectoras después de una lesión traumatica de la médula espinal. *Rev ed IMSS* 2002; 40: 437-455

Duce JA, Podvin S, Hollander W, Kipling D, Rosene DL, Abraham CR. Gene profile analysis implicates Klotho as an important contributor to aging changes in brain white matter of the rhesus monkey. *Glia* 2008; 56:106-117.

Duncan AE, Coman RJ, Kramer PA. Longitudinal study of radiographic spinal osteoarthritis in a Macaque Model. *J Orthop Res* 2011; 29: 1152-1160.

Elías Y. Potenciales provocados auditivos del tallo cerebral en el diagnóstico del daño auditivo y neurológico del adulto. En: Registros electrofisiológicos para el diagnóstico de la patología de la comunicación humana. Hernández OA, Flores R T, Peñaloza LY (Eds). Instituto de la comunicación humana. México. SSA, 1997, pp. 157-170.

Endres M, Scott GS, Salzman AL, Kun E, Moskowitz M, and Szabó C. Protective effects of 5-iodo-6-amino-1,2-benzopyrone, an inhibitor of poly(ADPribose) synthetase against peroxynitrite-induced glial damage and stroke development. *Eur J Pharmacol* 1998; 351: 377–382.

Escobar-Izquierdo A. Envejecimiento Cerebral Normal. *Rev Mex Neuro* 2001; 2: 197-202.

Fergestad T, Ganetzky BY, Palladino MJ. Neuropathology in Drosophila membrane excitabilitymutants. *Genetics.* 2006; 172: 1031–1042.

Fisher KE, Austad SN. The development of small primate models for aging research. *ILAR journal* 2011; 52: 78- 88.

Fowler CG, Chiasson KB, Leslie TH, Thomas D, Beasley TM, Kemnitz JW, Weindruch R. Auditory function in Rhesus monkeys: effects of aging and caloric restriction in the Wisconsin monkeys five years later. *Hear Res.* 2010; 261: 75-81.

Fowler CG, Torre P 3rd, Kemnitz JW. Effects of caloric restriction and aging on the auditory function of rhesus monkeys (Macaca mulatta): The University of Wisconsin Study. *Hear Res* 2002; 169, 24-35

Fuster-Lluch O, Galindo MF, Ceña V, Jordan J. The serine proteases and their function in neuronal death processes". *Rev Neurol.* 2004; 38: 449-457.

Gearing M, Rebeck GW, Hyman BT, Tigges J, Mirra SS. Neuropathology and apolipoprotein E profile of aging chimpanzees: implications for Alzheimer disease. *Proc Natl Acad Sci USA* 1994; 91: 9382- 9386.

Gilardi KVK, Shideler SE, Valverde CR, Roberts JA, Lasley BL. Characterization of the onset of menopause in the rhesus macaque. *Biology of reproduction* 1997; 57: 335-340.

Goldman-Rakic PS, Brown RM. Regional changes of monoamines in cerebral cortex and subcortical structures of aging rhesus monkeys. *Neuroscience* 1981; 6:177-187

González-Fraguela ME, Robinson MA, Delgado R. Estudio de indicadores de estrés oxidativo en monos viejos. *Rev Neurol* 2000; 31: 1012-1015.

Grondin R, Cass WA, Zhang Z, Stanford JA, Gash DM, Gerhardt GA. Glial cell line- derived neurotrophic factor increases stimulus-evoked dopamine release and motor speed in aged rhesus monkeys. *J Neurosci.* 2003; 23: 1974-1980

Harman D. Aging: a theory based on free radical and radiation chemistry. *J Gerontol* 1956; 17: 249-258.

Harman D. The free radical theory of aging. *Antioxid Redox Signal.* 2003; 5: 557-561.

Henry KR. Males lose hearing earlier in mouse models of late-onset age-related hearing loss; females lose hearing earlier in mouse models of early-onset hearing loss. *Hear Res* 2004; 190, 141-148

Hernández-Godínez B, Ibáñez-Contreras A, Reyes-Pantoja SA, Durand-Rivera A, Galvan-Montaño A, Perdigón-Castañeda G, et al. Preliminary Studies of Neurosensory and Cardiopulmonary Health Compromise in Captive– Bred Rhesus Macaques (Macaca mulatta) Suffering Scoliosis. *J Med Primatol* 2012; 41: 163-171.

Hernández-Godínez B., Ibáñez-Contreras A., Durand-Rivera A., Reyes-Pantoja S.A., Ramírez-Hernández R., Rodríguez-Guzmán P., Tena- Betancourt E., Somatosensory Evoked Potentials SEPs, of Median and Tibial Nerves in Rhesus Monkeys (Macaca mulatta) Under Captivity: Influence of Ontogenic Status in Neonatal, Infant, Young, Adult and Senile Stages. *J Med Primatol* 2011; 40: 79-87.

Hultcrantz M, Simonoska R, Stenberg AE. Estrogen and hearing: a summary of recent investigations. *Acta Otolaryngol* 2006; 126, 10-14

Ibáñez-Contreras A, Durand-Rivera A, Hernández - Godínez B, Reyes-Pantoja SA, Morales-Martínez JJ. Potenciales evocados auditivos de tallo cerebral en monos rhesus (Macaca mulatta) en diferentes etapas fisiológicas en condiciones de cautiverio. *Arch Med Vet* 2011; 43: 223-232.

Ibáñez-Contreras A, Hernández-Godínez B, Durand-Rivera A. Valores de latencia y respuesta electrofisiológicas de los diferentes componentes de los potenciales evocados auditivos del tallo cerebral en mono rhesus a diferentes intensidades (70-30 dB*). Arch Med Vet.* 2012: 44: 267-271.

Ibáñez-Contreras A, Hernández-Godínez B, Reyes-Pantoja SA, Jiménez-García A, Morales-Cruz E, Justo-Juárez TP, et al. Clinical Evaluation in geriatric rhesus monkeys (Macaca mulatta): Through serological studies and physiological constants in captivity. *JAVA* 2011: 10: 3281-3286.

Isa T, Yamane I, Hamai M, Inagaki H. Japanese macaques as laboratory animals. *Exp Anim* 2009; 58: 451-457.

Jewett DL, Romano MN. Neonatal development of auditory system potentials average from the scalp of rat and cat. *Brain Res* 1972; 36: 101-115.

Kemnitz JW, Roecker EB, Haffa AL, Pinheiro J, Kurzman I, Ramsey JJ, Mac Ewen EG. Serum dehydroepiandrosterone sulfate concentrations across the life span laboratory housed rhesus monkeys. *J Med Primatol* 2000; 29: 330-337.

Kemnitz JW, Roecker EB, Weindruch R, Elson DF, Baum ST, Bergman RN. Dietary restriction increases Insulin sensitivity and lowers blood glucose in rhesus monkeys. *Am J Physiol.* 1994; 266: E540-E547.

Kemnitz JW. Calorie restriction and aging in nonhuman primates. ILAR. 2011; 52: 66-77.

Kim MJ, Aiken JM, Havighurst T, Hollander J, Ripple MO, Weindruch R. Adult-Onset Energy Restriction of Rhesus Monkeys Attenuates Oxidative Stress–Induced Cytokine Expression by Peripheral Blood Mononuclear Cells. *J Nutr.* 1997: 127: 2293-2301.

Kohama SG, Rosene DL, Sherman LS. Age-related changes in human and non-human primate white matter: from myelination disturbances to cognitive decline. *Age* (Dordr) 2012: 34: 1093-1110

Konat GW, Wiggins RC. Effects of reactive oxygen species on myelin membrane proteins. *J Neurochem* 1985; 45: 1113-1138.

Königsberg M. 2008. Radicales libres y estrés oxidativo. Aplicaciones médicas, 1a ed., México: El Manual Moderno México.

Königsberg M. Nrf2: La historia de un nuevo factor de transcripción que responde a estrés oxidativo. *REB* 2007; 26: 18-25

Madrigal-Rentaría K, Lara-Solares A, Guevara-López U. Alteración de los Estudios de Conducción Nerviosa y su relación con la Intensidad del Dolor en Pacientes con Neuropatía Diabética. *Rev Soc Esp Dolor.* 2007: 14; 284-289.

Martínez-Sámano J, Torres-Durán PV, Juárez-Oropeza MA. El glutatión y su asociación con las enfermedades neurodegenerativas, la esquizofrenia, el envejecimiento y la isquemia cerebral. *REB* 2011: 30: 56-67.

McKiernan SH, Colman RC, Lopez M, Beasley TM, Weindruch R, Aiken JM. Longitudinal Analysis of Early Stage Sarcopenia in Aging Rhesus Monkeys. *Exp Gerontol.* 2009; 44: 170–176.

Morrison JC, Cork LC, Dunkelberger GR, Brown A, Quigley HA. Aging changes of the rhesus monkey optic nerve. *Invest Ophthalmol Vis Sci* 1990; 31: 1623-1627.

Muller FL, Liu Y, Jernigan A, Borchelt D, Richardson A, Van Remmen A. MnSOD deficiency has a differential effect on disease progression in two different ALD mutant mouse models. *Muscle Nerve.* 2008; 38: 1173-1183.

Nagahara AH, Bernot T, Tuszynski MH. Age-related cognitive deficits in rhesus monkeys mirror human deficits on an automated test battery. *Neurobiol Aging* 2010; 31: 1020–1031

Nightingale S, Mitchell KW, Howe JW. Visual evoked cortical potentials and pattern electroretinograms in Parkinson´s disease and control subjects. *J Neurol Neurosurg Psychiatry.* 1986; 49: 1280-1287

Ohnuki T, Nakamura A, Okuyama S, Nakamura S. Gene expression profiling in progressively MPTP-lesioned macaques reveals molecular pathways associated with sporadic Parkinson's disease. *Brain Res.* 2010; 1346: 26-42.

Organización Mundial de la Salud (OMS). Envejecimiento Activo: un marco político. *Rev Esp Geriatr Gerontol* 2002; 37-74-105

Pedersen L, Trojaborg, W. Visual, auditory and somatosensory pathway involvement in hereditary cerebellar ataxia". *Electroencephalogr Clin Neurophysiol* 1981; 52: 283-297

Pichardo- Fuster A. and Pedrero-Nieto L. 2009. "El anciano frágil, cascada de la fragilidad, falla que medra y pérdida de la vitalidad (deslizamiento)". In: Geriatría. Eds. González-Martínez JFco., Pichardo-Fuster A, García L. Ed. Mc Graw Hill, México D.F. p 69-81.

Poblano A. 2003. Detección y Estimulación tempranas del Niño con Daño Neurológico. Instituto Nacional de Perinatología. Editores de Texto Mexicanos. México D.F.

Price DL, Martin LJ, Sisodia SS, Wagster MV, Koo EH, Walker LC, et al. Aged non- human primates: an animal model of age-associated neurodegenerative disease. *Brain Pathol.* 1991; 1:287- 96.

Qin W, Chachich M, Lane M, Roth G, Bryant M, de Cabo R, et al. Calorie restriction attenuates Alzheimer's disease type brain amyloidosis in Squirrel monkeys (Saimiri sciureus). *J Alzheimers Dis.* 2006; 10: 417-22. 52.

Rezzi S, Martin FPJ, Shanmuganayagam D, Colman RJ, Nicholson JK., Weindruch R. Metabolic shifts due to long-term caloric restriction revealed in nonhuman primates. *Exp Gerontol.* 2009; 44: 356-362.

Roth GS, Mattison JA, Ottinger MA, Chachich ME, Lane MA, Ingram DK. Aging in Rhesus monkeys: relevance to human health interventions. *Science.* 2004; 305: 1423-1426.

Samorajski T, Rolsten C. Age and reginal differences in the chemical composition of brains of mice, monkey and humans. *Prog Brain Res* 1973; 40: 253-265.

Sani S, Traul D, Klink A, Niaraki N, Gonzalo-Ruiz A, Wu CK, et al. Distribution, progression and chemical composition of cortical amyloid-beta deposits in aged Rhesus monkeys: similarities to the human. *Acta Neuropathol* (Berl). 2003; 105: 145-56.

Santiago-López D, Rivas Arancibia S. Estrés oxidativo, metabolitos oxidados de dopamina y enfermedad de Parkinson. *Rev Fac Med UNAM.* 2008: 51: 104:107.

Saura M, Lamas S. Regulación de la óxido nítrico sintetasa inducible glomerular. *Nefrología.* 1995: 15: 411- 419.

Schapiro SJ. Effects of social manipulations and environmental enrichment on behavior and cell-mediated immune responses in rhesus macaques. *Pharmacol Biochem Behav.* 2002; 73: 271-278.

Scott GS, Virag L, Szabó C, and Hooper DC. Peroxynitrite-induced oligodendrocyte toxicity is not dependent on poly (ADP-ribose) polymerase activation. *Glia* 2003; 41:105–116.

Scott SG, Kean BR, Mikheeva T, Fabis JM, Mabley GJ, Szabó C, et al. The Therapeutic Effects of PJ34 [N-(6-Oxo-5,6-dihydrophenanthridin-2-yl)-N,N-dimethyl acetamide. HCl], a Selective Inhibitor of Poly(ADP-Ribose) Polymerase, in Experimental Allergic Encephalomyelitis Are Associated with Immunomodulation. *J Pharmacol Exp Ther.* 2004; 310:10: 5310-5361.

Segal EE. 1989. Housing, Care, and Psychological Well-Being of Captive and Laboratory Primates. Noyes Publications, Park Ridge New.

Shkurovich Z M. 1996. Capítulo XIV: Potenciales Evocados Visuales. Registro Electrofisiológico para el Diagnostico de la Patología de la Comunicación Humana. Eds. Hernández, O. A; Flores, R. T; Peñaloza, L. Y. 1996. Instituto de la comunicación humana. México. SSA. pp. 185- 194

Sloan P. 2003. Potenciales evocados en Anestesia y neurocirugía. *Elsevier.* Madrid. 9: 183-200

Solis-Chavez SA, Durand-Rivera A, Ibáñez-Contreras A, Reyes-Pantoja SA, Valderrama K., Heras-Romero Y, Tena-Betancourt E, Galván-Montaño A, Alfaro-Rodríguez A, Hernández-Godínez B. Visual Evoked Potentials to Light Flashes in Captive Rhesus Monkeys: A Study Reflecting Cerebral Cortical Activity and Brain Maturation. *Thai J Vet Med.* 2013: 43: 113-117.

Starr AA, Amlie RM, Martín WH, Sanders S. Development of auditory function in newborn infants revealed by auditory brainstem potentials. *Pediatrics* 1977; 60, 831-839

Tabbal SD, Tian L, Karimi M, Brown CA, Loftin SK, Perlmutter JS. Low nigrostriatal reserve for motor parkinsonism in nonhuman primates. *Exp Neurol* 2012: 237; 355-362.

Tabrez S. Ahmad M. Some enzymatic/nonenzymatic antioxidants as potential stress biomarkers of trichloroethylene, heavy metal mixture, and ethyl alcohol in rat tissues. *Environ Toxicol.* 2011; 26: 207-216.

Tandon OP, F Khaliq, N Goel. 2001. Auditory evoked potential responses in menopausal women: a normative study. *Indian J Physiol Pharmacol* 45, 361-366.

Tapia-Saavedra A. Proposing biomarkers to evaluate the alterations in the brain iron homeostasis and their relation in the physiopathology of Alzheimer´s disease. *Rev Chil Neuro- Psiquiat.* 2007; 45: 29-41.

Taylor LD, Daniels CK, Schmucker DL. Ageing compromises gastrointestinal mucosal immune response in the rhesus monkey. *Immunology* 1992; 75: 614-618.

Tigno X, Gereanich G, Hansen B. Age-related changes in metabolic parameters of Nonhuman primates. *J Gerontol A Biol Sci Med Sci.* 2004: 59; 1081-1088.

Toledano A, Álvarez MI, López-Rodríguez AB, Toledano-Díaz A, Fernández-Verdecia CI. Does Alzheimer's disease exist in all primates? Alzheimer pathology in non-human primates and its pathophysiological implications (I). *Neurologia.* 2012; 27: 354-369

Torre P 3rd, CG Fowler. 2000. Age- related changes in auditory function of rhesus monkeys (Macaca mulatta). *Hear Res* 142, 131-140.

Torre P 3rd, JA Mattison, CG Fowler, MA Lane, GS Roth, and DK Ingram. 2004. Assessment of auditory function in rhesus monkeys (Macaca mulatta): effects of age and calorie restriction. *Neurobiol Aging* 25, 945- 954.

Torre P 3rd, RE Lasky, CG Fowler. 2000. Aging and middle ear function in Rhesus monkeys (Macaca mulatta). *Audiology* 39, 300-304

Touil T, Deloire-Grassin MSA, Vital C, Petry KG, Brochet B. In vivo damage of CNS myelin and axons induced by peroxynitrite. *Neuroreport* 2001; 12: 3637–3644.

Toussaint O. Brain cell death in stress and in aging: role of oxidative stress and energy metabolism. *Ann Ny Acad Sci* 1998; 851: 430-443.

Tumminia SJ, Chambers C, Qin C, Zigler JM Jr, Russell P. A comparison of antioxidant enzyme activities in organ-cultured rhesus monkey lenses following peroxide challenge. *Curr Eye Res.* 1996; 15: 845-851.

Valenzuela R, Barroso-Chinea P, Villar-Cheda B, Joglar B, Muñoz A, Lanciego JL, et al. Location of prorenin receptors in primate substantia nigra: effects on dopaminergic cell death. *J Neuropathol Exp Neurol.* 2010; 11:1130-1142.

Von Bernhardi RM. Envejecimiento: cambios bioquímicos y funcionales del sistema nerviosa central. *Rev Chil Neuro-Psiquiat* 2005; 43: 297-304

Wang M, Gamo NJ, Yang Y, Jin LE, Wang XJ, Laubach M, et al. Neuronal basis of age-related working memory decline. N*ature.* 2011; 27: 476: 210-213.

Wang ZQ, Floyd EZ, Qin J, Liu X, Yu Y et al. Modulation of skeletal muscle insulin signaling with chronic caloric restriction in cynomolgus monkeys. *Diabetes* 2009: 58; 1488-1498.

Warner HR. Superoxide dismutase, aging, and degenerative disease. *Free Radic Biol Med.* 1994; 17: 249-258.

Watson SN, Nelson MA, Wildering WC. Redox agents modulate neuronal activity and reproduce physiological aspects of neuronal aging. *Neurobiol Aging.*2012; 33: 149-161.

Wei YH and Lee HC. Oxidative Stress, Mitochondrial DNA Mutation, and Impairment of Antioxidant Enzymes in Aging. *Exp Biol Med* (Maywood). 2002; 227: 671-682.

Willette AA, Gallagher C, Bendlin BB, McLaren DG, Kastman EK, Canu E, Kosmatka KJ, Fields AS, Alexander AL, Colman RJ, Voytko ML, Weindruch RH, Coe CL, Johnson SC. Homocysteine, neural atrophy, and the effect of caloric restriction in Rhesus monkeys. *Neurobiol Aging*. 2010; 33: 670-680.

Xiao B, Zhang GX, Ma CG, Link H. The cerebrospinal fluid from patients with multiple sclerosis promotes neuronal and oligidendrocyte damage by delayed production of nitric oxide in vitro. *J Neurol Sci.* 1996; 142: 114-120.

Yue F, Zeng S, Wu D, Yi D, Alex Zhang Y, Chan P. Age-related decline in motor behavior and striatal dopamine transporter in cynomolgus monkeys. *J Neural Transm.* 2012; 119: 943-952.

Zorrilla-García A. El envejecimiento y el estrés oxidativo. *Rev Cubana Invest Biomed* 2002: 21; 178-185.

In: Monkeys
Editor: Hugo A. Barrera Saldaña
ISBN: 978-1-63117-851-1

Chapter 3

ONTOGENY OF SOMATOSENSORY EVOKED POTENTIALS OF MEDIAN AND TIBIAL NERVES IN RHESUS MONKEYS (*MACACA MULATTA*): INFLUENCE OF DISSOCIATIVE ANESTHETIC MIXTURES UNDER CAPTIVITY CONDITIONS

***Braulio Hernández-Godínez*[1,2,3,4], *Herlinda Bonilla-Jaime*[6], *Adrian Poblano*[5], *Marcela Arteaga-Silva*[6], *Ma. Alejandra Bautista-Rodríguez*[4], *Salvador Solís-Chávez*[2,3,4], *Yessica Heras-Romero*[8], *Eduardo Tena-Betancourt*[7] and *Alejandra Ibáñez-Contreras*[1,2,3,4]**

[1]Programa de Doctorado en Ciencias Biológicas y de la Salud, Universidad Autónoma Metropolitana-Iztapalapa, Ciudad de México, México
[2]Laboratorio de Primatología, Applied Research in Experimental Biomedicine (APREXBIO), Ciudad de México, México
[3]Unidad de Experimentación Animal, Biología Integral para Vertebrados (BIOINVERT), Estado de México, México
[4]Unidad de Primates no-Humanos, Proyecto CAMINA, Ciudad de México, México
[5]Laboratorio de Neurofisiología Cognoscitiva, Instituto Nacional de Rehabilitación, Ciudad de México, México
[6]Departamento de Biología de la Reproducción, Universidad Autónoma Metropolitana- Iztapalapa, Ciudad de México, México
[7]Departamento de Etología, Fauna Silvestre y Animales de Laboratorio, Facultad de Medicina Veterinaria y Zootecnia-UNAM, Ciudad de México, México
[8]Departamento de Bioterio, Instituto Nacional de Neurología y Neurocirugía, Ciudad de México, México

ABSTRACT

Somatosensory evoked potentials (SEPs) constitute a useful neurophysiological tool commonly used to assess the normal function and developmental degree of the somatosensory system. The aim of this study was to analyze the ontogeny of the somatosensory pathway of the *Macaca mulatta* species throughout Somatosensory evoked potentials (SEPs) of the median and tibial nerves, under the influence of dissociative anesthetics. This study was performed using sixty non-human primates (NHP) of the *Macaca mulatta* species divided into five age-dependent groups, from infantile (0-3 years of age) to senile periods (15 years of age or older) comprising 12 subjects per group, according to the anesthetic agent. Stimulation was performed at specific electrode location sites of the forelegs for the median nerve and in the hindlegs for tibial nerve responses. Two constant, steady and well defined waveforms were observed in all groups at pre-determined ages for upper and lower limbs. We observed that latency values of the SSEP were modified in relation to the anesthetic agent, mainly in the Ketamine-Acepromazine combination in all age groups. Greater variability was observed between groups of age 1 and 5 (younger and older monkeys), which may be due to the processes of myelination and demyelination, respectively.

INTRODUCTION

Non-human primates (NHPs) are the laboratory animals of greatest interest in the field of biomedical investigation and experimentation, due to their phylogenetic similarity to humans (Poblano et al. 2004; Ibáñez-Contreras et al. 2011; Hernández-Godínez et al. 2011a; Hernández-Godínez et al. 2011b). The *Macaque* genus is the most used NHP in biomedical research, in this context the Green Monkey (*Chlorocebus aethiops*), Rhesus Monkey (*Macaca mulatta*), and stump-tailed macaque (*Macaca arctoides*) are the most commonly used species (Toledano et al. 2011). The frequent use of NHPs in biomedical experiments is due to the fact that a significantly higher level of neuronal sophistication is found in their brains. Our best clues to how the human brain works are taken from neurobiological studies in NPH brains, especially in *Macaque*. These NHPs taught us that their sensory and motor systems are extremely complex, in particular the processing that they carried out at the cortical level. In these species, the cortex is divided into several levels with distinct processing *specialization*, where the cortical areas are divided into two or more groups of modules or columns of functionally related neurons. In the processing that takes place at a cortical level in macaque monkeys, 30-40 neuronal groups control the visual area, 15-20 control the auditory areas, 15-20 control the somatosensory areas, and 10 or more control the motor areas (Felleman and van Essen 1991; Picard and Strick 1996; Kaas and Hackett 2000; Luppino and Rizzolatti 2000; Kaas 2007).

Primates' structural and functional organization depends on complex systems which provide permanent information to every one of the cells of the primary sensory cortex, regulating activities at a molecular level for the expression of complex behavior. Registration, codification, and analysis of information in the nervous system are carried-out by structures which extend from the periphery to the cerebral cortex and these together constitute the sensory organization relating living beings with their environment (López-Antúnez 1995). Touch is one of the more developed senses in primates, than in the rest of mammals, because they have a motor and somatosensory cortex, which permits interaction between a variety of sensory systems and component parts of spinal motor systems for handling activities (Ghosh et al. 1987; Drucker-Colín and Anías 2005). On this basis, NPHs have been used in diverse anatomic-physiological studies of the somatosensory pathway and of the motor cortex, such as those that have evaluated the function of distinct anatomical parts of the somatosensory system, and of the neurophysiological basis of the time estimation of interaction among cortico-basal ganglia-thalamic circuit of primates (Merchant et al. 2008; Merchant et al. 2011).

Evolution of the Somatosensory Pathway

The first mammals preserved the majority of the afferent pathways that transmit sensory information to the Central Nervous System (CNS) inherited from their reptilian ancestors. From these evolutionary processes, the first mammals had new sensory opportunities with the appearance of body hair. In this way, afferent receptors, sensitive to touch and to the movement of fine hairs and vibration was developed. On the other hand, mammals had sharpened sensitivity of receptors in muscles and joints that were sensitive to temperature and painful stimulus. It has been documented that the length of body hair evolved in order to detect objects from short distances from the skin through receptors at the base of each hair (Kaas 2010).

The cerebral cortex of the primitive mammals consists, in essence, in sensitive and motor areas, and lacks the pronounced convolutions that augment the surface of the higher mammals. In other higher mammals, and the human in particular, large areas of the cerebral cortex do not have sensitive or motor functions, instead, they are already related to functions of intersensory association, memory and language.

The first mammals had brains that were very small in proportion to their body size and had simple sensorimotor systems. Moreover, in what is actually now a distant relationship, these simple sensorimotor systems have been conserved in today's mammals, with few modifications having been added in the intervening period. Using the methods of modern neuroscience to study the sensorimotor systems of these mammals, along with a cladistic character analysis, the first primates can be distinguished by a more elaborate cortical motor system as opposed to a considerably more complete sensorimotor system. However, primitive primates had an extension to the rear part of the sensorimotor parietal cortex that included various fields involved in visual, auditory and sensorimotor activity.

Somatosensory Pathway

The somatosensory pathway is a complex system that consists of reception and processing centers responsible for receiving the fibers that arrive from the thalamus for the production of sensitive perception (Climent et al. 1998). The areas that make-up the Primary somatosensory cortex are those responsible for the integration of modalities of stimulation such as touch, temperature, proprioceptive sensationsrelating to the position of the body), and nociception (which relates to pain sensory receptors act in the skin, epithelia, skeletal striated muscle, bones and joints, internal organs and the cardiovascular system. The somatosensory pathway reacts to stimuli using different receptors–thermoreceptors, mechanoreceptors and chemoreceptors. The transmission of information from the receptors passes via the spinal nerves through the tracts of the spinal cord, where the axons of the sensory neurons follow two paths, the posterior column and medial spinothalamic tract pathways, both of which send the axons to the thalamus.

During their journey to the thalamus, the axons travelling through the posterior column-medial lemniscus and spinothalamic tract pathways, which have already crossed from the other side of the body into the medial lemniscus, now ascend to the ventral posterior nucleus of the thalamus, where they synapse with the majority of the fibers of the sensory cells. The axons of the thalamus are projected to the primary somatic sensory cortex of the parietal lobe, which at the same time sends axons to the secondary somatic sensory cortex and other areas of association (Tórtora 1981; Anthony 1987; Guyton 1987; Ganong 1990; Ruckebush et al. 1994).

Two concrete regions are relevant: the primary somatic area, found in the posterior portion of the gyrus postcentral and the gyrus ectomarginalis rostralis, which receives impulses from the contralateral side of the body; the secondary somatic area is found in the gyrus ectosylvius rostralis and receives impulses from both sides of the body (Soriano-Mas et al. 2007).

Relay pathways are specific sensory systems that retain through the different thalamic nuclei the separation of the projection in the neocortex. The projections in the sensory cortex come from the opposite side of the organism, except for goats and sheep. Each type of afferent fiber travels to a different region of the cortex. The precise localization and extension of the sensory area depend on the level of development of the cerebral cortex in each species, which is greater in primates. The sensory areas, known as secondary, present a smaller topographical organization and receive a bilateral input of impulses.

Sensations of two types arrive to these areas:

- Tactile and proprioceptive sensations are always contralateral, coming from dorsal roots in the spinal cord. They also reach the ventral nucleus of the thalamus via the medial lemniscus, to which are incorporated the trigeminal nerve fibers, which collect sensations of the same type that arrive in the brain.
- Painful and thermal sensations are bilateral, pertinent to the spinothalamic system. Those which terminate in the Primary somatosensory cortex arising from the ventral posteromedial nucleus of the thalamus, do not constitute the totality of the fibers of this nature that reach the thalamus, remaining in synapse in the neurons of the posterior nuclei (which project onto the parietal cortex, which is not part of the Primary somatosensory cortex) or the intralaminar nuclei, which project themselves diffusely toward regions of the brainstem and other areas of the cerebral cortex (Climent et al.1998).

Somatosensory information is transmitted to the encephalus by various ascending pathways that travel in parallel by means of the spinal cord, the brainstem and the thalamus up to the somatosensory cortex.

The somatosensory system presents a somatotopic organization, where the fibers are spatially organized and finish in a geometric arrangement that maintains the relationships on the body surface. The process of stimulus identification begins with the activation of a diverse population of somatoreceptors, which generate action potentials as a response. The receptors are classified according to the origin of the stimulus to which they respond. If they are found in the skin or in the subcutaneous connective tissue, for example, and respond to external stimulus such as light, sound, or cutaneous sensation, they are referred to as exteroceptors, while those that are found in the visceral organs are known as interoceptors or visceroceptors. These receptors respond to stimuli related to internal conditions, such as blood pressure or blood sugar level. The nociceptors and thermoreceptors are found in the free nerve endings (FNE) that are unencapsulated (those which are not enclosed in a myelin sheath). FNE also serve as receptors, both superficial and profound touch (Soriano-Mas et al. 2007).

Transmission of Sensory Input

The input of sensory information generates neural impulses that are conducted by nerve fibers of diverse diameters. The diameter of the fiber regulates the speed at which the impulses are propagated in the CNS. The spinal nerves conduct information from the somatosensory receptors of the body, up through the spinal cord, by means of the afferent axons that travel via the dorsal ramus. The cell bodies of the neurons from where these axons originate are found in the dorsal root ganglion (Guyton 1987; Climent et al.1998).

Ascending Pathways

The axons that transmit the somatosensory information to the spinal cord and brainstem travel along different pathways. The primary afferent axons that arrive at the dorsal root enter the spine in order of size. The largest fibers enter medially, while the smallest, myelinated or not, do so laterally.

The largest fibers bring the information relating to touch and pressure, while the smallest bring information relating to pain, temperature and protopathic touch. Therefore, the distinct types of axons possess distinct termination patterns in the spinal cord and make up distinct somatosensory pathways within the CNS. The posterior column-medial lemniscus pathway (dorsal column-medial lemniscus pathway or the dorsal white column-medial lemniscus system) transmits information from the mechanoreceptors that are responsible for fine touch and proprioception. The spinothalamic tract pathway is responsible for the sensitivity to pain and temperature. Both pathway systems share a common pattern of organization that is comprised of three synaptic relays between the periphery and the cerebral cortex:

1) The axon of the primary neuron, which has its cell body in the dorsal root ganglion of the spinal nerve, carries out synapse in the ipsilateral grey matter or nuclei in the medulla oblongata.
2) The axon of the second neuron carries out a decussation and terminates in the thalamus.
3) The axon of the third neuron (the sensory neuron of the third order) has its soma in the thalamus and arrives at the contralateral cortex to the place of stimulation.

A) Spinothalamic System

The spinothalmic pathway consists of second order sensory neurons that are projected from laminae I-II, IV-V of the posterior horn towards the contralateral thalamus. The cells of original cells produce excitation or inhibition in the synapse, by neurons of the gelatinous substance in the cell, which possess important modulating effects in process of sensory transmission (Fitzgerald et al. 2012).

The spinal nerves manage the cutaneous sensitivity of the posterior horn, into the spinal cord. The nerves penetrate the spinal cord through the lateral division of the dorsal roots. The central branches of the afferent neurons travel through segments in a cephalic and posterior direction through the posterolateral tract before establishing a synapse with the neurons of the laminae II y IV of the posterior horn, where the level of information at which it is functioning is increasing. They are next submitted to complex integrating processes, which interact in the interneurons by the same laminae. The laminae IV appear to produce a convergence of different modalities of cutaneous sensitivity where those of the laminae V converge impulses with the various cells of the laminae IV. In the integration that is taking place in the posterior horn, the sensory information that is arriving stays under the modulating influence –either facilitating or inhibiting– of the impulses that come from the cerebral cortex and other super-medullary structures (Guyton 1987).

Once it has been processed in the spinal cord, information is projected at higher levels by means of the fibers that, for the most part, pass to the opposite side by the gray commissure, in order to incorporate themselves into the spinothalmic system. Spinothalmic bundles are comprised of both lateral and ventral, which conduct it to the ventral posterolateral nucleus and other thalamic nuclei. However, there are a certain proportion of the fibers that ascend to the thalamus on the same side on which they originated. The lateral spinothalmic bundle ascends through the homonymous spinal cord column, while the ventral is situated in the ventral column of the medulla (Figure 1) (Guyton 1987).

B) Posterior Column-Medial Lemniscus Pathway

The primary afferent axons arrive to the medulla and climb ipsilaterally up the dorsal columns of the medullary white matter. Axons attach themselves, always in a lateral position, as the tracts rises. Therefore, these last axons are organized according to their somatic origin: the axons corresponding to the legs are located in the most medial zone and the axons corresponding to the arms are found more laterally (Soriano-Mas et al. 2007).

Their peripheral prolongations collectively receive information from the largest sensory receptors: Meissner and lamellar corpuscles, the Ruffini endings, Merkel cells, muscle spindles, and Golgi tendon organs. The centripetal prolongations of the cells, which come from the lower limbs and the lower part of the trunk, send branches in the direction of the gray matter before ascending in the fasciculus gracilis in order to arrive at the gracile nucleus in the medulla oblongata.

The corresponding collateral cells run from the upper limb and the upper part of the trunk through the fasciculus cuneatus in order to reach the cuneate nucleus. The second order afferent neurons, which are found in the nuclei of the lower spine, are known as the gracile nucleus and the cuneate nucleus.

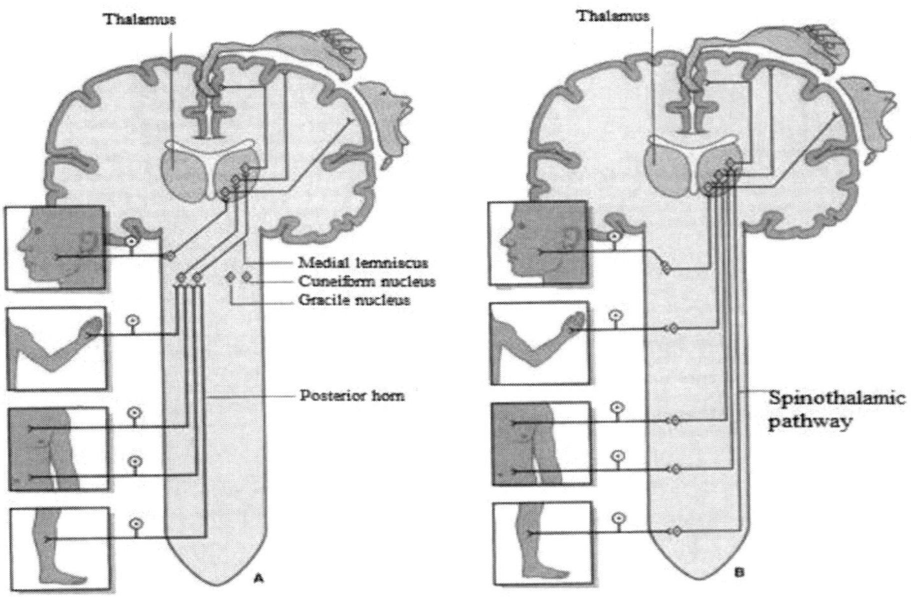

(Fitzgerald 2011).

Figure 1. Basic diagram of the (A) posterior column-medial lemniscus pathway, B) and the spinothalmic pathway.

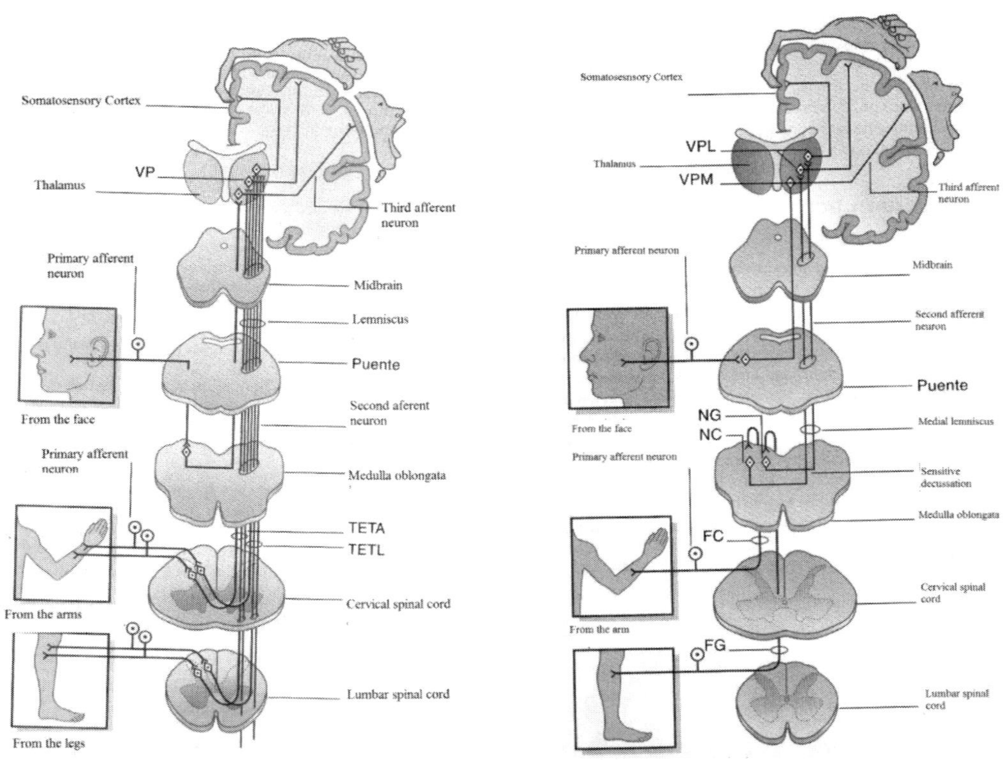

(Fitzgerald 2011).

Figure 2. A) Posterior column-medial lemniscus pathway. B) The spinothalmic pathway.

Their prolongations pass through the segment of the medulla oblongata before crossing with their direct counterparts in the great sensory decussation before which, the fibers, having crossed medial longitudinal fissure, turn face forward into the posterior column-medial lemniscus. The third order afferents are projected from the thalamus to the somatosensory cortex (Figure 2) (Fitzgerald et al. 2012).

SOMATOSENSORY COMPONENTS OF THE THALAMUS

All the pathways that carry somatosensory information (originate in the medulla, or the brainstem) converge in the ventral posterior nucleus in the thalamus, which comprises one medial and one lateral nucleus:

- The nucleus that is more laterally situated, the ventral posterolateral nucleus that receives projections from the posterior column-medial lemniscus and from the spinothalmic division, transmits the body's somatosensory information.
- The more medially situated nucleus, the ventral posteromedial nucleus, receives projections from the trigeminal lemniscus and transmits somatosensory information from the face (Soriano-Mas et al. 2007).

Primary Somatosensory Cortex

There is a proportional relationship between the cortical area dedicated to the somatic processing and the fine touch capacity in the contralateral part of the body corresponding to this cortical area. The cortical direct stimulation of an individual produces a referred sensation to the part of the body whose information produces such a cortex. The primary somatosensory cortex can be divided into four parallel groups, two of which process subtle tactile information, and another the information relating to pressure, while the last processes a combination of different types of somatic information (Fernández 2005). The axons that are born in the ventral posterior nucleus of the thalamus project towards the cortical neurons that are fundamentally localized in the IV layer of the somatosensory Cortex.

The Secondary Somatosensory Cortex

The secondary somatosensory cortex is found in a position lateral and posterior to the primary somatosensory cortex, in the interior of the lateral sulcus. The secondary somatosensory cortex receives information from the primary somatosensory cortex and in the same way directly from the thalamus, specifically, from the ventral posterior nuclei. For this reason, there is a process parallel to the somatosensory stimuli in the cortex. The secondary somatosensory cortex sends projections to the limbic system, which has been related to the learning and memory involved in the somatosensory experiences (Soriano-Mas et al. 2007).

The areas in the brain responsible for motor activity also receive information from the somatosensory cortex. The areas of the motor cortex include the primary, premotor and supplementary cortices. The motor cortex in mammals is one of the most developed at birth (Fox 1965). The motor cortex is a fine strip of tissue that extends through the cerebral surface, directly in front of the somatosensory cortex. The neurons of the motor cortex form an inverted topographical map of the body. Here exists a complex map of the body: the nerve cells that cause movement (by means of connections with the motor neurons of the spinal cord (Morris and Fillenz 2003).

The Primary Motor Cortex

The primary motor cortex is a site named this way because it presents the lowest threshold for the production of movements through the direct stimulation of the cortex. Additionally, it is an area of interaction of motor and somatosensory systems circuits. It receives organized somatotopical impulses from the premotor and supplementary cortices of the primary somatosensory cortex (Brodmann areas 1, 2, and 3), from the auditory and visual cortex and from the premotor and supplementary motor areas adjacent to the frontal cortex, and from the posterior parietal areas 5 and 7 (Ghosh et al. 1987; Drucker and Anías 2005). This is the principal origin of the corticospinal tract. The primary motor cortex is the place of origin in which the motor system commands, generated by voluntary movements, are produced. Its neurons project directly towards the limb and trunk muscles, represented by a surface proportional to the precision or accuracy of the movements that each can produce. The musculature of the leg and foot are represented dorsally, the hand and arm laterally, and the face and tongue centrally in the gyrus.

Pre-Motor Cortex

This is found in front of the primary motor area, occupying the inferior thirds in the 6th Brodmann area. It is a small extension situated in the lateral face of the hemisphere, between the motor area and the gyrus proteus, and is found to be organized somatotopically. Its function is far from being understood, although its afferent nerve fibers from the posterior parietal area and those that move toward the motor area to the sub-cortical centers, in which the descending ventromedial pathways originate, help us to understand more of its function; that is, related to the orientation of the body and the contraction of the axial and proximal muscles necessary to prepare for the realization of a particular movement towards an object that has attracted the attention of the subject (Climent et al. 1998). To provoke movement, more powerful electric stimuli are required than in the primary motor area, with movements being complex ones, which already involve various joints. They also have a topographical organization of the body, where the more complex movement patterns are found, as would be the postural position for the execution of specific movements. Part of the signal that activates the circuits is found in the sub-cortical structures such as the basal ganglia and the thalamus. It comes from this region and then terminates in the primary motor area in order to control movements, which participate in various planes of the body (Ellaway 2000; Hildenbrand and Goslow 2001; Drucker and Anías 2005).

Supplementary Motor Cortex

In primates, this is found on the medial face of the hemisphere and occupies the rest of the 6th Brodmann area, above the pre-motor area and fundamentally, its afferents come from the striatum and are activated in the planning of movements, more than just the production of simple signals destined for the motor-neurons. It is found to be somatotopically organized and projects toward the motor area (to control the distal muscles of the limbs), or even towards the subcortical centers of the brainstem or the spinal cord. It is bilaterally located and as a consequence, causes a unilateral injury in primates; the incapacity to realize movements that demand the coordination of both hands (Climent et al. 1998; Hildenbrand and Goslow 2001; Drucker and Anías 2005).

Myelination of the Somatosensory Pathway

There is a notable sensory and motor immaturity present at birth, as a result of ontogeny development. Brain development follows a characteristic pattern that permits organisms to have a greater chance of survival. In this way, some structures of the CNS, which from the moment of birth onwards are necessary for the control of its functions, mature more quickly, while in other structures the processes are prolonged (Hernández et al. 2008). The post-natal development of the somatosensory nervous system is highly complex due to the changes generated through the course of the development of the organism, such as the length of the nerve tract, the degree of myelination of the neurosensory pathways, and finally the increase in the number of synapses in the neurosensory pathway (Hernández-Godínez et al. 2011a).

While the formation of myelin is present in only a small amount before the fourth month in the intrauterine life of humans, this process continues up until after birth. However, after birth only a few areas of the brain are found myelinated, a process which only begins once the neural proliferation and migration has finished. Once the axons have been myelinated, the neurons can reach a level of complete functionality as the myelin permits fast and efficient conduction (Roselli 1997). There is evidence indicating that the myelinic sheath augments in thickness from infancy up until maturity (López-Antúnez 1995; Poch-Olivé 2001).

Throughout the development that takes place in the last stage of gestation and in the postnatal period, there is an additive phase of the super production of synapse and fibers, which follows a regressive phase of elimination. The synaptic connections specialize as they develop. This phase of neuronal expansion, known as hodogenesis (the formation of pathways), is made by means of the growth of the neural prolongations in search of the target cells in order to establish their connections (Poch-Olivé 2001).

The period in which the different systems of nerve fibers are myelinated is highly variable, although in general, the myelin is deposited first in the pathways of the greatest phylogenetic maturity (López-Antúnez 1995), beginning with the sensorimotor system, followed by the secondary cortical areas, and finally by the cortices of association of the frontal, parietal and temporal lobes. The myelination of the association fibers continues throughout adult life (Hernández et al. 2008).

The motor pathways are myelinated before the sensory pathways in the peripheral nervous system, while the opposite takes place in the CNS. The areas of association are the last to be myelinated and continue on this process throughout the second decade of life (Hernández-Godínez et al. 2011a).

Myelination begins at distinct moments and possesses a variable rhythm and duration for each region of the nervous system. For example, the spinothalmic and spinocerebellar bundles already present myelination from the beginning of the fourth month of intrauterine life, while the corticospinal bundles begin to acquire myelin towards the end of gestation; the process takes many years to complete. Myelination reaches the brain at birth with the myelination of cortex later, presenting an anatomical sequence. In the brainstem, the medial longitudinal fasciculus appears to be one of the areas that myelinates earliest (López-Antúnez 1995; Poch-Olivé 2001; Hernández et al. 2008).

It has been described that the thalamus is a critical element in the conduction of signals to the brain cortex. The development of the evoked potentials had been found to be in close relationship between maturity and thalamic myelination (Ysunza 1996). In the case of NPHs (Oliver 1997), the myelination process in *Macaca mulatta*, shows the existence of a developmental change from low to high speed of cortico-spinal conduction in the spinal cord of neonatal to adult monkeys associated to the degree of myelination, suggesting that the myelination process in Rhesus monkeys culminates at the age of 36 months. Moreover, an alteration and absence of important corticospinal endings, has been observed in the motor nuclei of neonatal monkeys (Kuypers 1962; Debecker et al. 1973; Oliver et al. 1997), due to the immaturity of the cortico motoneuronal projections; as the subjects mature these projections show an increase between 6-8 months of age in rhesus monkeys (Hinde et al. 1964; Lawrence and Hopkins 1974).

SOMATOSENSORY EVOKED POTENTIAL (SSEP)

Evoked potential is a bioelectrical signal produced from the activity presented in the neural pathway, peripheral and central, as a response to external stimuli. It presents as oscillations masked by the electroencephalography (EEG) signal and is described, generally, in terms of maximum and minimum magnitude (peak-valley amplitude) and its presentation after stimulus onset (latency). Evoked potential is a tool that allows the collection of functional information related to specific neural structures and constitutes an objective measurement of nervous activity (Shkurovich 1997; Durand-Rivera 1998; Sloan 2003; Poblano 2003). To estimate the result obtained from the SSEPs of the median and tibial nerves, the latencies and amplitudes are evaluated off-line. Latency is the interval between the beginning of the stimulation and each positive or negative deflections of the potentially evoked muscular action measured in milliseconds (ms) (Samorajski 1977; Ordy and Brizzee 1979; Chiappa 1983; Allison et al. 1983; Shkurovich 1997). The amplitude or voltage of the potential for muscular action is the sum of the amplitude of all of the muscular fibers and indicates axonal damage, given that low amplitude is directly proportional to the number of lost axons (Chiappa 1983).

The Rhesus monkey *(Macaca mulatta)* is an ideal model animal for the study of the fundamental neural mechanisms of the somatosensory pathway, as well as for preclinical studies (Courtine et al. 2007). The use of evoked potential studies as a diagnostic tool offers an integral evaluation of the function of the somatosensory pathway, as well as the advantage of having a low level of invasiveness, which can be used as a technique in longitudinal studies (Hernandez-Godínez et al. 2011b).

NEURONAL GENERATORS OF SSEPS

The median nerve, the first positive potential known as P9 (for the latency with which it normally appears, 9 ms), is generated in the distal portion of the brachial plexus in humans. The second positive potential, P11, is generated pre-synaptically in the dorsal root entry zone. The third positive component, P13, is formed probably in lemniscal pathways (medial lemniscus) at the level of the brainstem (Ysunza: The neuronal generators of SSEP of the median nerve and the SSEP of the tibial nerve have been studied by diverse authors for decades in humans (Allison et al. 1983; Desmedt 1985; Moller et al. 1986; Ysunza 1996). It was therefore known in the evaluation of 1996). While, it is not clear if the generator of the first negative component of N20 has a thalamic, thalamo-cortical or even cortical generator, they have been suggested. The next negative component, P23, has its origin in the somatosensory cortex (Ysunza 1996). In the same way, Shimazu et al. (2000) investigated the origin of the high frequency components through the SSEP of the median nerve of the cortex of six monkeys and analyzed the laminar field potentials in the areas 3b (N10; which corresponds to N20 in humans) and 1 (P12; which corresponds to P25 in humans).

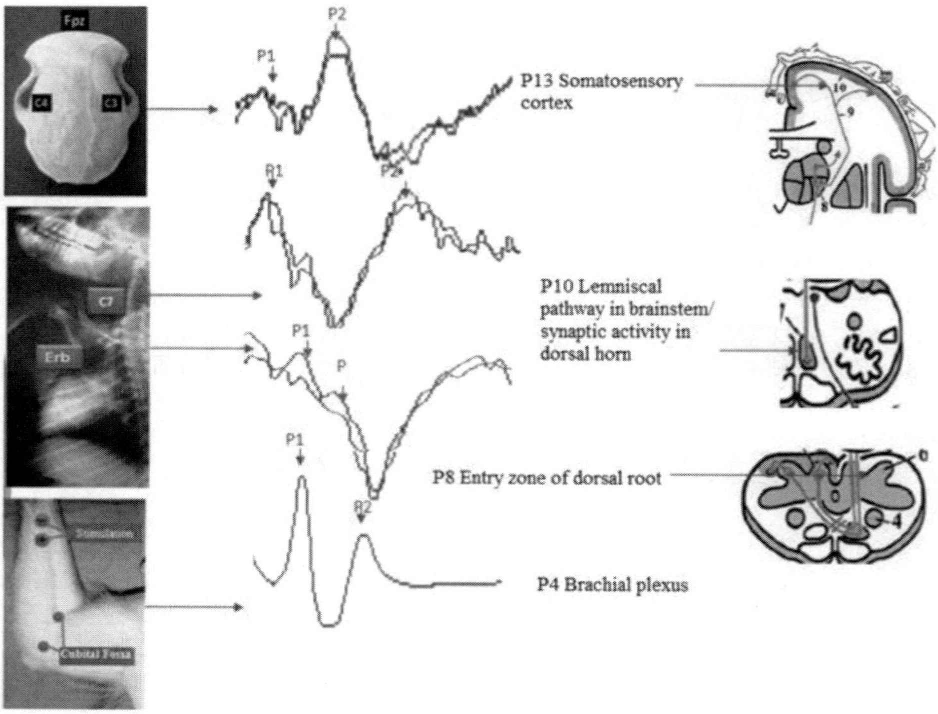

Figure 3. Neural generators of the somatosensory evoked potentials (SSEP) of the median nerve in Rhesus monkey.

Identifying between 4-6 components, where the 1st to the 4th peaks were identified in area 3b (latency of 7-11 ms) and the 3rd to the 5th in area 1 (9-13 ms), a reverse polarity was identified between the surface and the depth of the cortex (Figure 3).

The generators that have been designated by the SSEP of the tibial nerve are not as uniform as the results obtained for the median nerve (Ysunza 1996). In the evaluation of the posterior tibial nerve, the derivation of the popliteal fossa records the electronegative peripheral nerve action potential from the tibial nerve N8. The electrode derivation of the lumbar column registers 2 electronegative potentials: the first, N19, represents the conduction of the afferent stimulus in the cauda equina, while the second, N22, represents a stationary potential that reflects the postsynaptic activity of the interneurons of the gray matter of the lumbar spinal cord (Ysunza 1996). The derivation of the cervical column registers a later stationary potential, named N27. The cranial derivation gives the first major component registered in the surface of the cranium, P37, which reflects the electropositivity of the cortical surface, oriented ipsilaterally. The next potential is electronegative, being generated in each cortex contralateral to the stimulus and is known as P45 (Ysunza 1996) (Figure 4). While these neural generators are described in humans (Moller et al. 1986; Ysunza 1996), in the case of NHPs, and in particular the rhesus monkey, there are only two papers in the literature, one of which describes the neural generators of only 6 young male Rhesus monkeys (Moller et al. 1986), and the second studied 5 Japanese macaques with an age range between 4-6 years (Hayashi 1995). It is remarkable that in the last two studies described above, craniotomies were carried out in order to affix the electrodes.

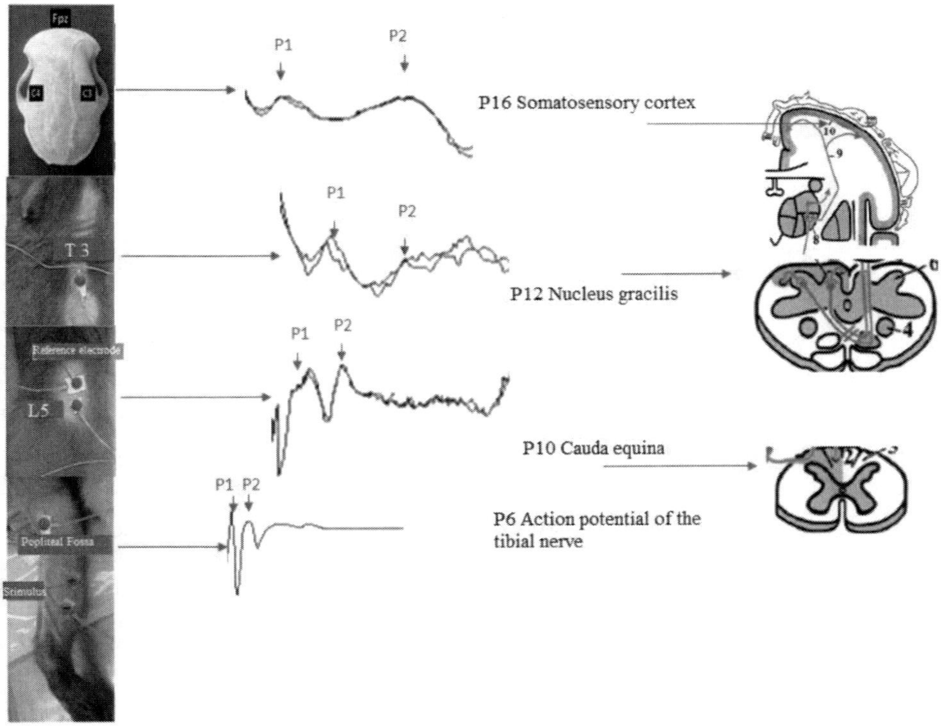

Figure 4. Neural generators of the somatosensory evoked potentials (SSEP) of the tibial nerve in Rhesus monkey.

It is widely documented that among the diverse factors that affect SSEPs waves' identification, one is using only male animals of a limited age range.

In our opinion, it is not possible to identify and measure the changes produced in the neural generators using short age period and only one gender.

Factors that Affect the SSEP Register

Indispensable within the laboratory is gathering data that permits an accurate interpretation of the recorded results. The clinical management and monitoring of monkeys are basic in these experiment protocols (Ibáñez- Contreras et al. 2011). It has been described that factors such as age, height, weight and total weight of subjects are significant variables in the peripheral conduction of nerve impulses in SSEPs conducted on the median and tibial nerves (Hernandez-Godínez et al. 2011a; Hernández-Godínez et al., 2011b). Some authors have examined both, the hematological values and the blood chemistry of the Rhesus monkey, including some factors, such as the type of confinement, food, and state of health, that intervene in the viability of the results, such as in several pharmacological studies of chemical exposure, for example (Buchl and Howard 1997; Woodward and Weld 1997; Hom et al. 1999; Ibáñez-Contreras et al. 2011; Ibáñez-Contreras et al. 2012). Further to the diverse factors already mentioned in the evaluation of the somatosensory pathway using SSEP, it is important to consider the effect of the anesthetics required by the procedure. For this reason the question that emerges is: whether electrophysiological changes could be observed as a result of using different dissociative agents.

Age: As has been described in both humans and animals, the process of myelination during the development of the CNS occurs in the first few years of life; on the other hand, regarding demyelination in older animals, conditions result in a decrease in the former, and an increase the last age groups, of the conduction speed in the somatosensory pathway, lengthening the absolute latencies of the synaptic relays of the neural components (Oliver et al. 1997; Hernández-Godínez et al. 2011; Ibáñez-Contreras et al. 2011). In the same way, Hernández-Godínez et al. (2011) evaluated a population of 20 rhesus monkeys of different ages, using SSEP-the somatosensory pathway and concluded that, as in humans, age and the myenlinic processes are the main factors in the determination of latency in SSEPs.

Size: There is a close correlation between fiber size and the functional properties of the peripheral nerve in development. The length of the nerve also increases in parallel with the increase in length of the limbs; those results in latency increase the wave components of SSEPs (Ysunza 1996; Hernández-Godínez et al. 2011a; Hernández-Godínez et al. 2011b).

Gender: According to what has been written about humans, significant statistical changes between men and women have not been observed. However, an increase in absolute latencies in the case of men is noted, which is related to the length of the limbs, principally the arms, on a central level, owing to the size of the cranium (Ysunza 1996). As a general rule, the cranial proportions correspond to body size in primates, where the males have greater proportions than the females (Woodward and Weld 1997).

Temperature: It has been described that, in those neural generators of the sensory pathways in humans, the absolute latencies have a tendency to increase when the temperature falls (Hernández- Godínez et al. 2011a, Hernández-Godínez et al. 2011b).

Use of Anesthetics: As non-collaborative patients, the use of dissociative anesthetics on NHPs are essential. In veterinary practice, three drugs are commonly used to induce dissociative anesthesia: phencyclidine, ketamine and tiletamina (Booth 1988; Pérez-Guille et al. 2007).

According to Barajas (1985) and Rosete (1996), SSEPs do not seem to be affected by the states of awareness, sleep, or sedation; however, with the use of a dissociative anesthetic there exists electroencephalographic evidence indicating a dissociation of functions between the thalamus and the limbic system.

These mental absences have been linked to the inhibition of the neuronal transport of the cerebral monamines (serotonin, noradrenaline, and dopamine). Furthermore, it is known that the inhibition of the synaptic recapture carried out by γ-Aminobutyric acid (GABA) is associated with the occurrence of muscular rigidity. In the same way, it has been written that all anesthetic agents suppress the synaptic function in the brain and in the gray matter of the spinal cord, decreasing the amplitude of the electric signal and increasing latency. For this reason, any signal decrease caused by an anesthetic agent can compromise the correct interpretation of the electric signal of evoked potentials (Figure 5).

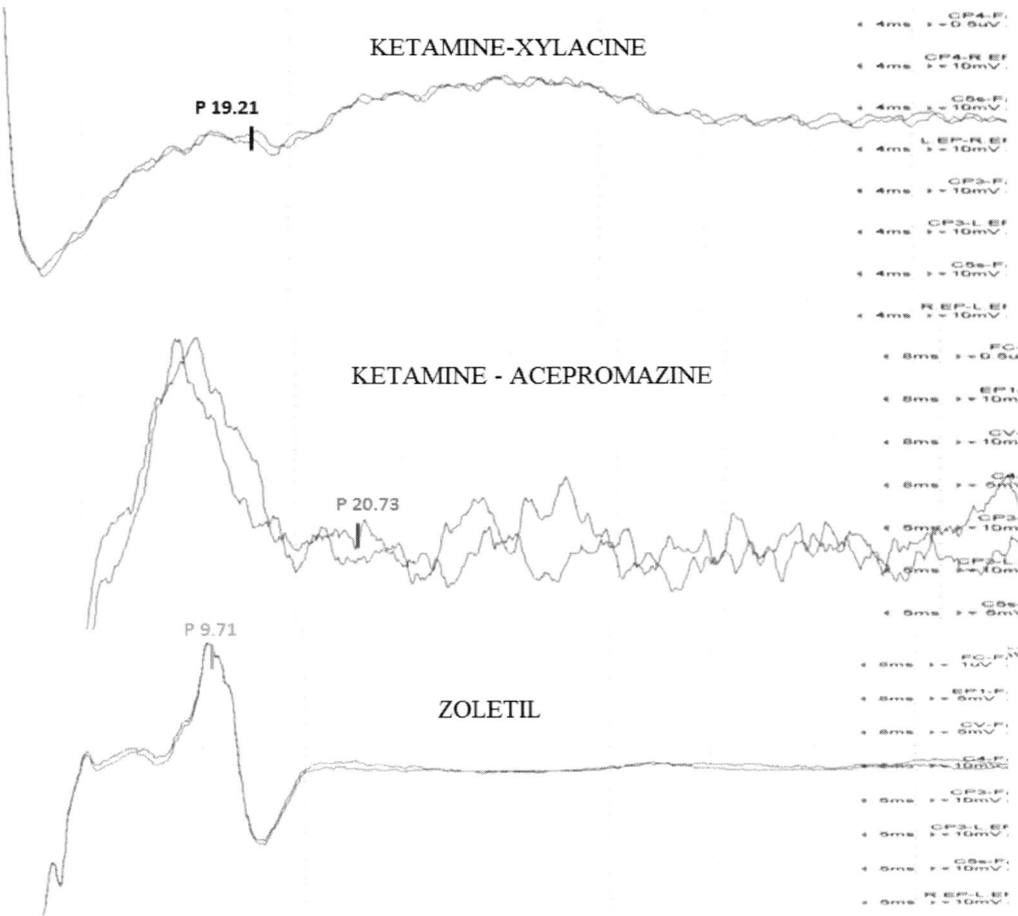

Figure 5. Changes in the morphology of the waves of the SSEP with different anesthetic agents in Rhesus monkeys.

DISSOCIATIVE ANESTHETICS IN THE APPLICATION OF SSEPS

It is well known that anesthetic agents alter neuronal excitability, impeding synaptic function or slowing down axonal conduction (Sloan 1990). The mechanism of neuronal depression varies depending on the anesthetic agent used and of the administration method. Many anesthetic agents are administered intravenously, such as the barbiturates (i.e. etomidate, propofol) and benzodiazepines, and augment the inhibiting effects of the γ-Aminobutyric acid. The result is a hyperpolarization of the axonal membrane, which, therefore, inhibits synaptic transmission (Table 1) (Sloan 1990).

- *Tiletamine-Zolazepam*

  Commercially available as Zolteil®, it is derived from zolazepam and sold under proportion 1:1. Is a combination of anesthetic and tranquilizer (Tiletamine + Zolazepam). The tiletamine is a cyclohexylamine with dissociative effect, neither opiate nor barbiturate, and is more potent that ketamine, which does have a more prolonged action. Zolazepam is a benzodiazepinic tranquilizer with notable effects as anti-convulsive and a muscle relaxant (Sumano and Ocampo 1992). The combination of the drug is reconstituted in a quantity of 50mg/ml (1:1) of sterile distilled water for each combination and each one is dosed in mg/kg (Lee et al. 2003).

  The anesthetic results in dissociation between the thalamus and the limbic system (tiletamine), combined with an anticonvulsive effect and a benzodiazepine tranquilizer which diminishes the action of the dopamine in the limbic system and the striatal areas of the brain (Pérez-Guille et al. 2007).

  Tiletamine-zolazepam does not produce significant changes in heart frequency, and does not result in respiratory depression. The combination of drugs offers a wide margin of security, being well tolerated intramuscularly, and proving a good anesthetic effect with rapid anesthetic induction, and a gentle recuperation (Sumano and Ocampo 1992). Following prior stabilization, a moderate two-phase decrease in arterial pressure is induced. The pressure decrease is due to the fact the tiletamine acts on the myocardial contractility, while a certain increase is caused by an augmentation in the adrenergic tone via the CNS (Sumano 1999). On entering the circulatory system, tiletamine-zolazepam is rapidly metabolized in the liver, with a half-life of 3.2 hours (Lee et al. 2003).

- *Ketamine*

  Ketamine is metabolized by hepatic enzymes. The major biotransformation route is N-demethylation by cytochrome P_{450}, which hydroxylated the drug. The maximum drug levels are achieved within one minute of being intravenously administered, and 5 minutes of being intramuscularly administered. Initially, the drug is distributed in the most highly perfused tissues, and is then redistributed into the more or less perfuse organs. After initial phase of the distribution, the drug has a half-life of 7-11 minutes, with the elimination phase occurring within 2-3 hours.

  Ketamine use is recommended in combination with other drugs, and not as a single agent, as it has been observed following a dose of 4 mg/kg in NHPs. The animal stays with its eyes open, pupils dilated, with excessive salivation, rigidity or extension of the lower limbs, opisthotonus and possibly tonic-clonic seizures (Sumano and Ocampo1992). Combinations of ketamine and different chemical substances (anesthetics, tranquilizers, or preanesthesics) have been described in the literature as having been used with the objective of decreasing the undesirable effects, or to increase its anesthetic effects. The main combination in NHPs used is xylazine in a dose of 0.5 mg/kg via intramuscular injection (Buchl and Howard 1997; Hom et al. 1999; Lee et al. 2003), in order to prevent muscular hypertonicity. This combination results in animal sedation, prolongs the anesthesia, reduces the necessary quantity of ketamine by 50%, and shortens the recuperation time. The combination used with acepromazine maleate in a dose of 0.5-1 mg/kg via intramuscular injection, produces sedation with a minimal degree of muscle relaxation, although the analgesic level is not enough for major surgeries (Sumano and Ocampo 1992).

TRANQUILIZERS AND SEDATIVES

- *Acepromazine maleate*

  Acepromazine maleate is a derivative of the phenothiazines, which are used on animals for minor surgery, and clinical and diagnostic procedures due to its neuroleptic effects. Acepromazine suppresses brainstem and cerebral cortex connections. Blocking dopamine, which is widely distributed cortical neurotransmitter, exerts a sedative action. It has a moderate blocking effect on the alpha adrenergic receptors (Sumano and Ocampo 1992).

  The effects of acepromazine administered via intravenous injection are observed within 1-3 minutes, and within 10-25 minutes when administered via intramuscular injection. It produces a significant increase of central venous pressure 90 minutes after administration, and provokes a decrease in hemoglobin concentration 45 minutes after administration. It also causes muscle relaxation, has an antiemetic effect, is hypotensive, and has hypothermic properties (Sumano and Ocampo 1992).

  Acepromazine is slowly metabolized in the liver, has a half-life of 6 hours, and its metabolites can be detected in urine and other tissues for an extended period (Gross and Booth, 1995). The use of acepromine in combination with ketamine is recommended in order to obtain the best anesthetic and analgesic effect (Sumano and Ocampo 1992).
- *Xylazine*

  Xylazine is a sedative, a muscle relaxant and a non-narcotic analgesic of short duration. Xylacine is an alpha-2 ($\alpha 2$) adrenergic receptor and acts as a genuine neuroleptanalgesic agent producing analgesia and sedation. However, it has been related to CNS depression (Pérez et al. 1999) and its effects as a muscle relaxant are based on impulse transmission inhibition to muscles at CNS level (Sumano and Ocampo 1992; Rey et al. 1998). The immobilization occurs between 3-5 minutes after intravenous injection, or between 10-15 minutes after intramuscular administration, while the analgesic effect lasts for 15-30 minutes. A somnambulant state can last for up to 2 hours. It has been described that xylazine can produce hyperglycemia and has a markedly diuretic effect. Occasionally, it also produced muscle tremors, bradycardia and arteriovenous blocking within therapeutic doses (Sumano and Ocampo 1992).

Xylazine is metabolized rapidly in the liver, producing 20 metabolites. The metabolites peak excretion occurs between 2-4 hours after administration (Gross & Booth 1995). It has been reported that in NHPs, the combination of xylacine (0.6 mg/kg IM) and ketamine (0.6 a 2 mg/kg) results in a good anesthetic effect; a certain level of relaxation and a fast return to consciousness level. However, respiratory frequency and body temperature decrease has also been observed (Clifford 1984; Lee et al. 2003).

Material and Methods

The subjects of this study were NHPs of the species *Macaca mulatta*, kept in captivity in the Research Center, for the CAMINA Project to cure Paralysis in Mexico City. The NHP facility at the CAMINA Research Center is registered with the number DGVS-PIMVS-CR-IN-1014-D.F./08 in the national Environment and Natural Resources Ministry. The research was carried out in accordance with NOM-062-ZOO-1999 of the World Organization for Animal Health, International Animal Health Code (for mammals, birds and bees) and was approved by the Internal Committee for the Care and Use of Laboratory Animals and by the Ethics and Research Committee of the Research Center, CAMINA. As described by Ibáñez-Contreras et al. (2011) and Hernández-Godínez et al. (2011).

This study used 60 NPHs of the *Macaca mulatta* species, divided in groups of 4 males, his female and toddlers.

Table 1. Dissociative anesthesia, tranquilizers and sedatives

Drug	Derived	Receptor	Effect time (min)	Administration	CNS effect
Ketamine	Fencilidine	NMDA	30 - 40	IV, IM	Cortical Thalamic System/ stimulates limbic system
Tiletamine	Fenciclidine	NMDA	30 - 40	IV, IM	Cortical Thalamic System
Zolacepam	Benzodiazepine	Benzodiazepine	20 – 30	IV, IM	Cortical thalamic system depression/ neuronal inhibition at the spinal level
Acepromazine	Fenotiacine	Dopaminergic	20 - 30	IV, IM	Motor area
Xylazine	Tiacine	alpha-2 ($\alpha2$) adrenergic Cholinergic Serotonergic Histamine Opiates			Motor area

For the registration of the Somatosensory Evoked Potentials SSEP, a specialized computer, Viasis Healthcare Niccolet® was used in this study, and a filter settings band-pass was set up in the range of 5-2,500 Hz with a sensitivity of 10 μV. Recordings of lower limb SEPs were done at the tibial nerve by placing disk electrodes at the popliteal fossa, lumbar (L5S) and thoracic (T3S) locations, and in the opposite somatosensory cortical hemisphere to the stimulated limb (C3 or C4); Fpz (frontopolar) electrode was used as a reference, and the mastoid location was used as a ground. The upper limb was recorded at the median nerve by placing electrodes at the antecubital fossa, Erb's point, cervical (C7S), and the somatosensory cortical hemisphere opposite to the stimulated limb (C3 or C4), and an Fz electrode was used as a reference and similarly grounded to the mastoid. The selected nerves were stimulated through monophasic square pulses of 100 ms of duration using a frequency of 0.5 to 1.1 stimuli/second, applied behind the internal malleolus at an average of 250 stimuli. Although minimal intensity was applied it was consistent enough to induce an observable digital contracture by the use of 1.5 milliamps. Data were evaluated using common measurements of central tendency: arithmetic mean and standard deviation for all groups of the same age considering the two main waves.

Three experimental groups were formed depending on the anesthetic used:

- Group A: Tiletamine and Zolazepam (4mg/kg IM) (Hernández-Godínez et al.
- 2011a, 2011b, and Ibáñez-Contreras et al. 2011).
- Group B: Ketamine and Xylazine (4mg/kg and 0.5-1.0mg/Kg IM).
- Group C: Ketamine and Acepromazine (4mg/kg and 0.5-1.0mg/Kg by IM).

Each experimental group was formed by 20 males. Five subgroups were formed according to age, comprising 4 males as follows: Infant group 1 (0-1 year); Infant group 2 (1-3 years); Young group (3-5 years); Adult group (5-15 years); Senile group (15 years and older) (Ibáñez-Contreras et al. 2011; Hernandez-Godínez et al. 2011).

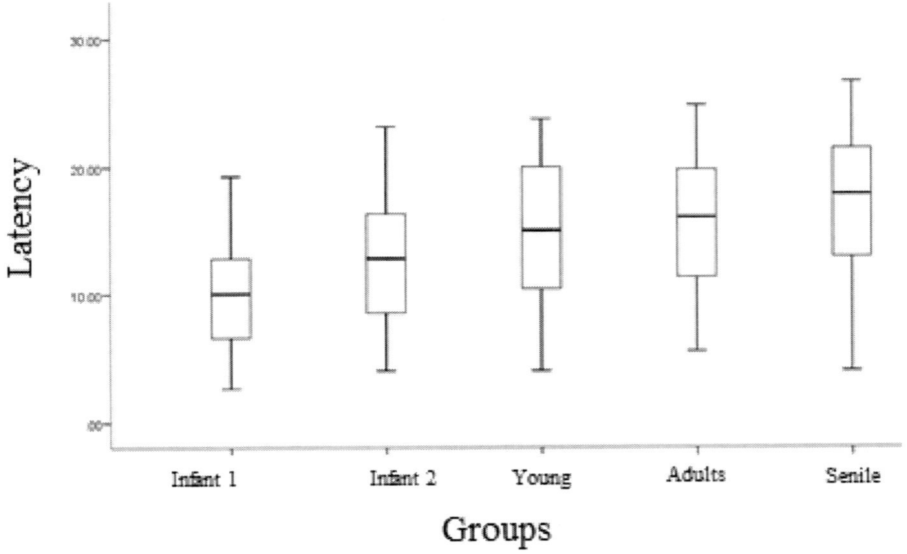

Figure 6. Comparison between age groups with Zoletil Median Nerve.

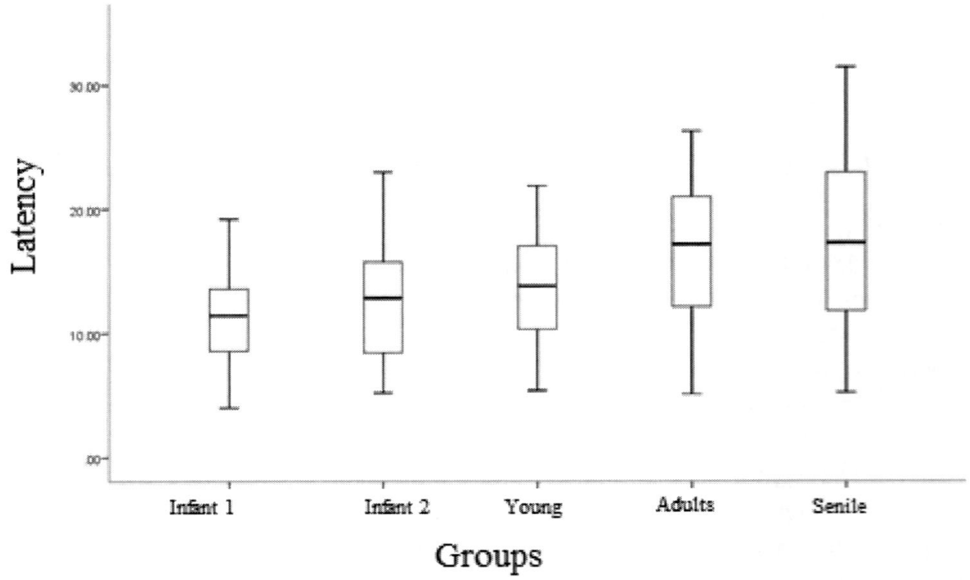

Figure 7. Comparison between age groups with Zoletil Tibial Nerve.

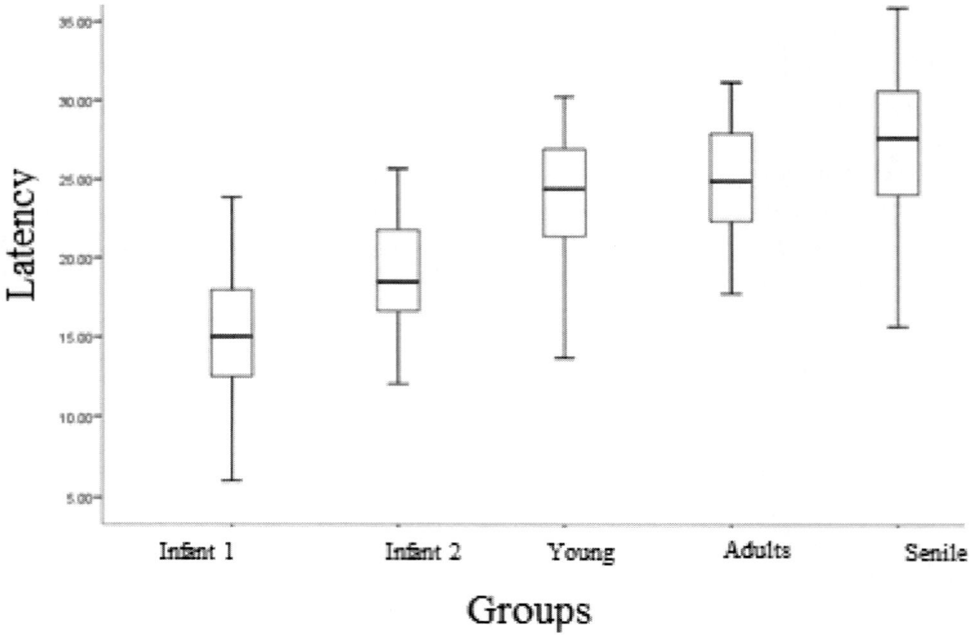

Figure 8. Comparison between age groups with Ketamine-xylacine Median Nerve.

Preliminary Results

Two constant, steady and well-defined waveforms were observed in all groups at pre-determined ages for upper and lower limbs. We observed that latency values of the SSEP were modified in relation to the anesthetic agent, mainly in the Ketamine-Acepromazine combination in all age groups. Greater variability was observed between groups of age 1 and 5 (younger and older monkeys), which may be due to the processes of myelination and demyelination respectively.

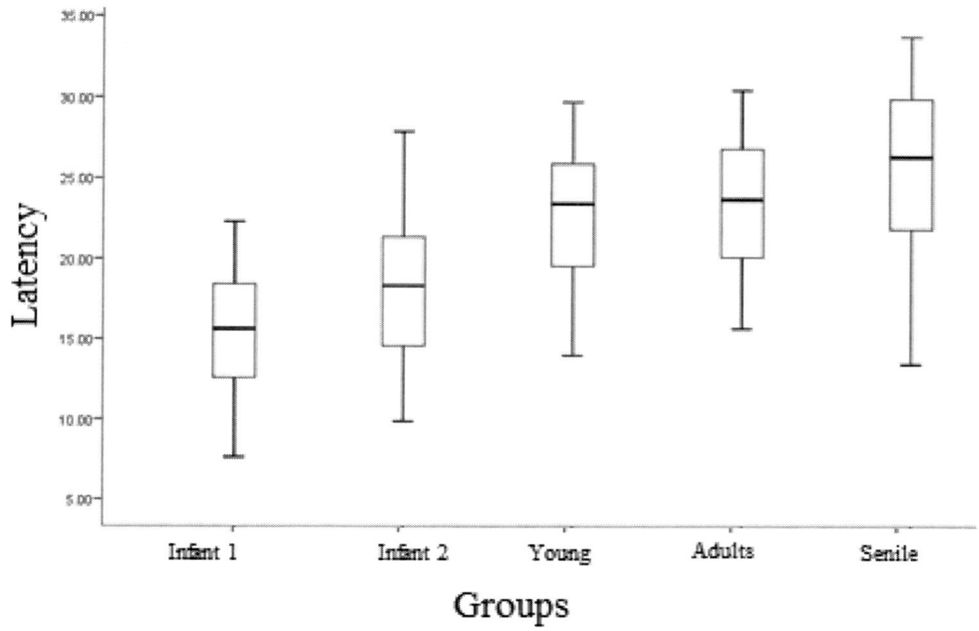

Figure 9. Comparison between age groups with Ketamine-xylacine Tibial Nerve.

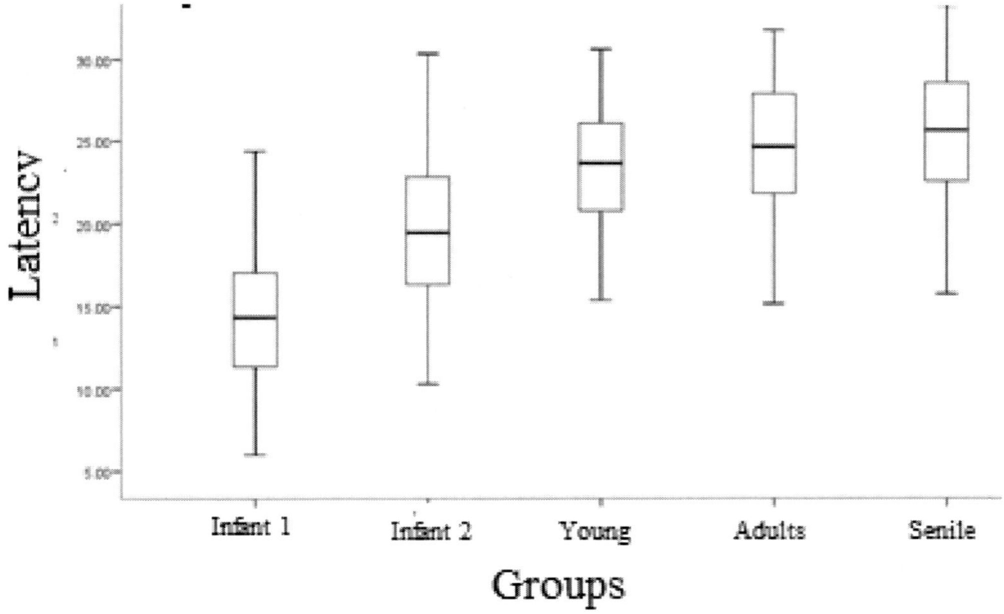

Figure 10. Comparison between age groups with Ketamine-Acepromazine Median Nerve.

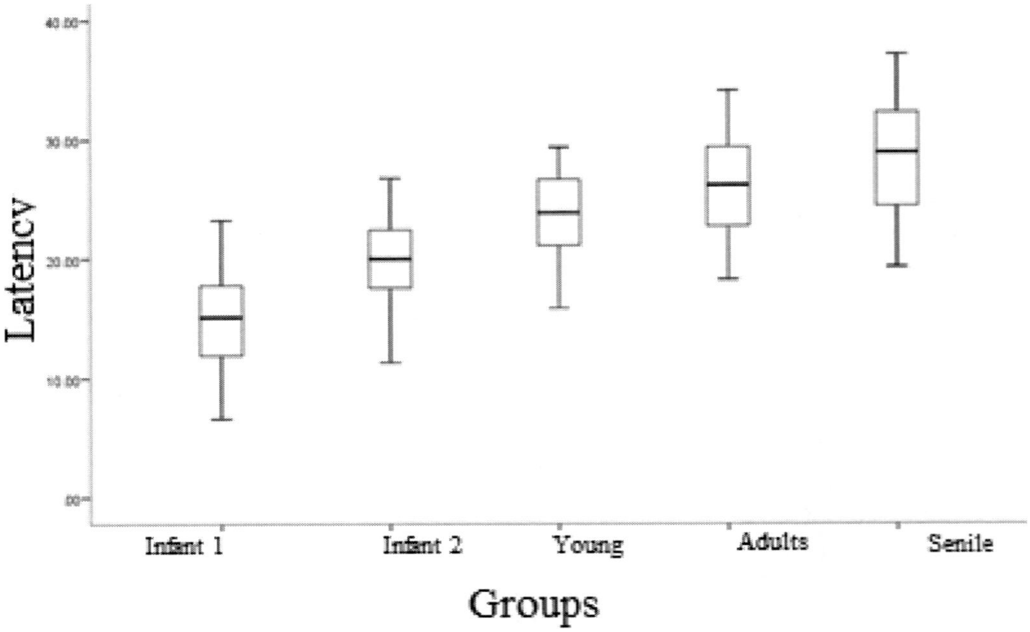

Figure 11. Comparison between age groups with Ketamine-Acepromazine Tibial Nerve.

CONCLUSION

Based on the preliminary results obtained in this investigation we concluded that SEPs components were deeply influenced by the physiologic status of the subjects and the dissociative anesthetic drug; affecting typical peak latencies and spinal conduction pathway responses during the recordings, emphasizing age as critically variable influencing the sensory responses of P1 and P2; also associated with the myelination degree along the cortex and spinal cord in immature and in aged encephalon, when analyzed in direct comparison of the fiber tracts of adult monkeys with fully myelinized systems.

Based on the above data, in the case of non-human primates, Oliver et al. (1997) described the myelination process of rhesus macaque monkeys (Macaca mulatta), indicating a slow neural corticospinal speed conductivity at the spinal cord level for neonates, related to their myelination degree, suggesting that the myelination process in rhesus monkeys culminates at 36 months of age.

A lack of important corticospinal fibers in motor nuclei has been observed in neonates (Kuypers 1962; Debecker et al. 1973; Oliver et al. 1997) due to incomplete maturation of the cortico-motoneuronal projections as indicated by Hinde et al. (1964), and Lawrence and Hopkins (1976), who describe the full maturity of these projections as occurring at 6-8 months of age in monkeys.

Finally it is known that all anesthetic agents suppress the synaptic function in the brain and the gray matter of the spinal cord, decreasing amplitude of the electric signal and increasing latency, as observed in this preliminary study, in which there exists a trend to an increase of latencies with the Ketamine-Acepromazine mixture. Perhaps this is due to the increase of temperature with the use of this anesthetic mixture.

REFERENCES

Allen D. G., Dyson D. H., Pascoe P. J., O'Grady M. R. (1986). Evaluation of a xylazine-ketamine hydrochloride combination in the cat. *Canadian Journal of Veterinary Research,* 50, p. 23-26.

Allison T., Wood C. C., Goff W. R. (1983). Brainstem auditory, pattern reversal, and short latency somatosensory evoked potentials: latencies in relation to age, sex, and brain and body size. *Electroencephalography and Clinical Neurophysiology*, 55, p. 619-636.

Anthony C. P. (1987). Anatomía y fisiología [in Spanish]. 2nd edition. Ciudad de México, Editorial Interamericana.

Barajas J. J. (1985). Uso de los potenciales evocados auditivos en el diagnóstico audiológico [in Spanish]. In: Bogacz J. (ed.) Los potenciales evocados en el hombre. Buenos Aires, El Ateneo. p. 34- 57.

Booth, N. H., McDonald L. E. (1988). Farmacologia y terapeutica veterinaria [in Spanish]. 5th edition. Zaragoza, Editorial Acribia.

Buchl S. J., Howard B. (1997). Hematologic and serum biochemical and electrolyte values in clinically normal domestically bred rhesus monkeys (Macaca mulatta) according to age, sex, and gravidity. *Laboratory Animal Sciences,* 47, p. 528-533.

Chiappa K. H. (1983). Evoked potentials in clinical medicine. 2nd edition. New York, Raven Press.

Clifford D. H. Pre-anaesthesia, anaesthesia, analgesia and euthanasia. (1984). In: Fox J., Cohen B., Loew F. (eds). Laboratory Animal Medicine. New York, Academic Press. p. 530-551.

Climent S., Sarasa M., Muniesa P., Terrado J. (1998). Manual de anatomía y embriología de los animales domésticos. Conceptos básicos y datos aplicativos [in Spanish]. Zaragoza, Editorial Acribia, p. 219-246.

Courtine G., Bunge M. B., Fawcett J. W., Grossman R. G., Kaas J. H., Lemon R., Maier I., Martin J., Nudo R. J., Ramon-Cueto A., Rouiller E. M., Schnell L., Wannier T., Schwab M. E., Edgerton V. R.(2007). Can experiments in nonhuman primates expedite the translation of treatments for spinal cord injury in humans? *Nature Medicine*, 13, p. 561-566.

Debecker J., Desmedt J. E., Chorazyna H. (1973). Somatosensory evoked potential newborn and premature baboon. *Archives of International Physiology and Biochemistry*, 81, p. 554-556.

Desmedt E. J. (1985) Generadores neuronales de los componentes precoces de los potenciales evocados somatosensitivos en el hombre [in Spanish]. In: Bogacz J (ed.). Los potenciales evocados en el hombre. Buenos Aires, El Ateneo. p. 100-119.

Durand-Rivera A. (1998). Bases técnicas y fisiológicas de los potenciales provocados auditivos de tallo cerebral [in Spanish]. Revista para la Salud, 19. p. 20-30.

Drucker-Colín R., Anías C. J. (2005). Fisiología médica [in Spanish]. Ciudad de México. Manual Moderno. p. 821-879.

Dolman C. L., McCormick A. O., Drance S. M. (1980). Aging of the optic nerve. *Archives of Ophthalmology*, 98, p. 2053-2058.

Fernández R. J. (2005). Fisiología médica [in Spanish]. Ciudad de México, Manual moderno. p. 893-896.

Felleman D. J., van Essen D. C. (1991). Distributed hierarchical processing in the primate cerebral cortex. Cerebral Cortex. 1, p. 1–47.

Fitzgerald M. J. H., Gruener G., Mtui E. (2012). Neuroanatomía clínica y neurociencia [in Spanish]. 6th edition. Barcelona, Elsevier-Saunders.

Fox M. W. (1965). Canine behavior. Springfield, Charles C. Thomas Publisher.

Ganong W. F. (1990). Fisiología médica [in Spanish]. 1st edition. Ciudad de México, Manual Moderno.

Gross M. E., Booth N. E. (1995). Tranquilizers, alfa 2-adrenergic agonists and related agents. In: Richard A. H. (ed). Veterinary pharmacy and therapeautics. Iowa State University Press. p. 325-333.

Gosh S., Brinkman C., Porter R. A. (1987). A quantitative study of the distribution of neurons projecting to the pre-central motor cortex in the monkey (M. fascicularis). *Journal of Comparative Neurology*, 259, p. 424-44.

Guyton A. C. (1987). Fisiología humana [in Spanish]. Ciudad de México, Interamericana.

Haskins S. C., Farver T. B., Patz J. D. (1985), Ketamine in dogs. *American Journal of Veterinary Research,* 46, p. 1855-1860.

Hayashi N. H., Nishijo T., Ono S., Endo, E., Tabuchi M. (1995). Generators of somatosensory evoked potentials investigated by dipole tracing in the monkey. *Neuroscience,* 68, p. 323-338.

Hernández C. A. M., Galván G. J. A., López L. R. (2008). Maduración neurológica en lactantes productos de madres con embarazo de alto riesgo [in Spanish]. *Revista Mexicana de Medicina Física y Rehabilitación,* 20, p. 37-42.

Hernández-Godínez B., Ibáñez-Contreras A., Durand-Rivera A., Reyes-Pantoja S. A., Ramírez-Hernández R., Rodríguez-Guzmán P., Tena-Betancourt E. (2011a) Somatosensory evoked potentials SEPs, of median and tibial nerves in Rhesus monkeys (Macaca mulatta) under captivity: influence of ontogenic status in neonatal, infant, young, adult and senile stages. *Journal of Medical Journal of Primatology*, 40, 79-87.

Hernández-Godínez B., Ibáñez-Contreras A., Durand-Rivera A., Galván-Montaño A., Reyes-Pantoja S. A., Cárdenas-Lailson E. (2011b). Latencies using somatosensory evoked potentials of the median and tibial nerve in rhesus monkey undergoing intrauterine laminectomy to simulate myelomeningocele. *Cirugía y Cirujanos*, 79, p. 237-241.

Hernández-Godínez B. (2011). La Neurofisiología como herramienta antropofísica en la ontogenia de las vías visuales, auditivas y somatosensoriales con un biomodelo evolutivo: El caso de los monos rhesus (Macaca mulatta) [in Spanish]. Tesis de Licenciatura, Escuela Nacional de Antropología e Historia (ENAH), Ciudad de México.

Hinde R. A., Rowell T. E., Spencer-Booth Y. (1964). Behavior of socially living rhesus monkeys in their first six months. *Proceedings of the Zoological Society of London*, 143, p. 609-649.

Hilibrand A., Schwartz D., Sethuraman V. (2004). Compilation of electric motor somatosensory evoked potential monitoring during cervical spine surgery. *Journal of Bone and Joint Surgery*, 86, p. 1248-1253.

Hildebrand M., Goslow G. E. Jr. (2001) Analysis of Vertebrate Structure. 5th edition. New Jersey, John Wiley and Sons.

Hom G. J., Bach T. J., Carroll D., Forrest M. J., Mariano M. A., Trainor C. E., Wang P. R., MacIntyre D. E. (1999). Comparison of cardiovascular parameters and/or serum (Macaca mulatta). *Contemporary Topics of Laboratory Animal Sciences*, 38, p. 60-64.

Ibáñez-Contreras A., Durand-Rivera A., Hernández-Godínez, B., Reyes-Pantoja S. A. (2011). Potenciales evocados auditivos del tallo cerebral en monos rhesus (Macaca mulatta) en diferentes etapas fisiológicas en condiciones de cautiverio [in Spanish]. *Archivos de Medicina Veterinaria,* 43, p. 223-232.

Ibáñez-Contreras A., Hernández-Godínez B., Reyes- Pantoja S. A., Jiménez-García A., Galván-Montaño A. (2012). Changes in blood parameters in rhesus monkeys (Macaca mulatta) during the first trimester of the gestation. *Journal of the American Veterinary Association,* 10, p. 3281-3286.

Inde R. A., Rowell T. E., Spencer-Booth Y. (1964). Behavior of socially living rhesus monkeys in their first six months. *Proceedings of the Zoological Society of London*, 143, p. 609-649.

Kaas J. H. (2010). Sensory and Motor Systems in Primates. In: Platt M. L., Ghazanfar A. A. (eds.). Primate Neuroethology. New York, Oxford University Press. p. 177-200.

Kaas J. H. (2007). The evolution of sensory and motor systems in primates. In: Evolution of the Nervous System. Kaas J. H. (ed.). London, Elsevier. p. 35-57.

Kaas J. H., Hackett T. A. (2000). Subdivisions of auditory cortex and processing streams in primates. *Proceedings of the National Academy of Science USA*, 97, p. 11793–11799.

Kuypers H. G. J. M. (1962). Corticospinal connections: postnatal development in the rhesus monkey. *Science*, 138, p. 678-680.

Lawrence D. G., Hopkins D. A. (1976). The development of motor control in rhesus monkey: evidence concerning the role of corticomotoneuronal connections. *Brain*, 99, p. 235-254.

Lee J., Hong S., Lee S., Kim Y., Kim M. (2003). Immobilization with ketamine HCI and tiletamine-zolazepam in cynomolgus monkeys. *Journal of Veterinary Sciences*, 4, p.187-191.

López-Antúnez L. (1995). Anatomía funcional del sistema nervioso [in Spanish]. Ciudad de México, Limusa.

Luppino G., Rizzolatti G. (2000). The organization of the frontal motor cortex. *News in Physiological Sciences,* 15, p. 219–224.

Merchant H., Naselaris T., Georgopoulos A. P. (2008). Dynamic sculpting of directional tuning in the primate*ournal of Neurosciences,* 28, p. 9164-917.

Merchant H., Zarco W., Pérez O., Prado L., Bartolo R. (2011). Measuring time with different neural chronometers during a synchronization*roceedings of the National Academy of Sciences U S A,* 108, p. 19784-19789.

Morris R., Fillenz M. (2003). Neurociencias: La ciencia del cerebro [in Spanish]. London, British Neuroscience Association.

Møller A. R., Jannetta P. J., Burgess J. E. (1986). Neural generators of the somatosensory evoked potentials: recording from the cuneate nucleus in man and monkeys. *Electroencephalography and Clinical Neurophysiology*, 65, 241-248.

Oliver E., Edgley S. A., Armand J., Lemon N. R. (1997). An electrophysiological study of the postnatal development of the corticospinal system in macaque monkey. *Journal of Neuroscience,* 17, p. 267- 276.

Ordy J. M., Brizzee K. R. (1979). Functional and structural age differences in the visual system of man and non-human primate models. In: Ordy J. M., Brizzee K. R. (eds.). Sensory systems and communication in the elderly. New York, Raven Press. p. 13-50.

Pérez R. A., Parra M. J., Merino D. J. (1999). Tranquilización [in Spanish]. In: Pérez R. A., Parra M. J., Merino D. J. (eds). Manual de analgesia y anestesia en el perro. Madrid, McGraw-Hill. p. 8, 25, 26, 40.

Pérez-Guillé B., Soriano-Rosales E., Villanueva-Sánchez O., Villegas-Álvarez F., Hernández-González R., Carmona-Mancilla A. (2007). Guía de los anestésicos más utilizados en animales de laboratorio [in Spanish]. Ciudad de México. Instituto Nacional de Ciencias Médicas y Nutrición Salvador Zubirán.

Picard N., Strick P. O. (1996). Motor areas of the medial wall: a review of their location and functional activation. *Cerebral Cortex*, 6, p.342–353.

Pomeranz, B., Wall P. B., Weber W. V. (1968). Cord cell responding to fine myelinated afferents from viscera, muscle and skin. *Journal of Physiology London*, 199, p. 511-532.

Poblano A. (2003). Exámenes neurofisiológicos auxiliares en el diagnóstico del daño neurológico temprano. En Poblano A. (ed.). Detección y estimulación tempranas del n iño con daño neurológico [in Spanish]. Ciudad de México, Editores de Textos Mexicanos. p. 113- 118.

Poblano A., Hernández-Godínez B., Arellano A., Arteaga C., Elías Y., Morales J., Poblano R., Poblano-Alcalá A. (2004). Serum testosterone and electroencephalography spectra in developmental male rhesus Macaca mulata monkey. *Arch. Med. Res.*, 35, p. 406-410.

Poch- Olivé M. L. (2001). Neurobiología del desarrollo temprano [in Spanish]. *Contextos Educativos,* 4, p. 79 -94.

Rey R. A., Olmos Z. R., Sostres V. A. (1998). Cuidado y uso de animales de laboratorio [in Spanish]. Ciudad de México. p. 37.

Rosete M. (1996). Potenciales de tallo cerebral, fundamento y aplicaciones en la hipoacusia infantil [in Spanish]. In: Hernández-Orozco F., Flores-Rodríguez T., Peñaloza-López Y. Registros electrofisiológico para el diagnostico de la patología de la comunicación humana. Ciudad de México, Instituto Nacional de la Comunicación Humana, p. 143-151.

Rosselli M., Ardila A., Lopera F., Pineda D. (1997). Neuropsicología Infantil [in Spanish]. Medellín, Prensa Creativa.

Ruckebusch, Y., Phaneuf L. P., Dunlop R. (1994). Fisiología de pequeñas y grandes especies [in Spanish]. Ciudad de México, Manual Moderno.

Samorajski T. (1977). Central neurotransmitter substances and aging: a review. *Journal of the American Geriatric Society,* 25, p.337-347.

Shimazu H., Kagi R., Tsujimoto T., Kohara N., Ikeda A., Kimura J., Shibasaki H. (2000). High-frequency SEP components generated in the somatosensory cortex of the monkey. *Neuroreport,* 21, p. 2821-2826.

Sloan T. B., Fugina M. L., Toleikis J. R.(1990) Effects of midazolam on median nerve somatensory evoked potentials. *British Journal of Anesthesia*, 64, p. 590-593.

Soriano-Mas C., Guillazo–Blanch G., Redolar-Ripoll D. A., Torras-García M., Vale-Martínez A. (2007). Fundamentos de neurociencia [in Spanish]. Madrid, Editorial UOC.

Sumano L. H. (1999). Experiencias clínico farmacológicas de la mezcla tiletamina-zolacepam en pequeñas especies [in Spanish]. Ciudad de México, Documento Técnico Virvac de México. p. 1-5.

Sumano L. H., Ocampo C. L. (1992).Tranquilizantes [in Spanish]. In: Sumano L. H., Ocampo C. L. (eds.). Farmacología veterinaria. Ciudad de México, *McGraw-Hill*, p. 417-438.

Swenson M. J., Reece W. O. (1999). Fisiología de los animales domésticos de Dukes [in Spanish]. 2ª Edición. Ciudad de México, Uteha-Noriega. Ciudad de México. p. 816- 835.

Shkurovich M. (1997). Potenciales evocados visuales [in Spanish]. In: Hernández-Orozco F., Flores-Rodríguez T., Peñaloza-López Y. Registros electrofisiológicos para el diagnostico de la patología de la comunicación humana. Ciudad de México, Instituto Nacional de la Comunicación Humana, p. 185- 194.

Tórtora N. P. (1981). Principios de anatomía y fisiología [in Spanish]. Ciudad de México, Harla.

Wall P. B. (1962). The origin of spinal cord slow potential. *Journal of Physiology London,* 164, p. 508-526.

Woodward R. A., Weld K. P. (1997). A comparison of ketamine, ketamine-acepromazine, and tiletamine-zolazepam on various hematologic parameters in rhesus monkeys (Macaca mulatta) *Contemporary Topics of Laboratory Animal Sciences*, 36, p. 55-57.

Ysunza A. (1996). Potenciales somatosensoriales [in Spanish]. In: Hernández-Orozco F., Flores-Rodríguez T., Peñaloza-López Y. Registros electrofisiológicos para el diagnostico de la patología de la comunicación humana. Ciudad de México, Instituto Nacional de la Comunicación Humana, p. 197-201.

SOCIAL ASPECTS

In: Monkeys
Editor: Hugo A. Barrera Saldaña
ISBN: 978-1-63117-851-1

Chapter 4

How New World Monkeys See the Social World

Amici Federica*
Department of Developmental and Comparative Psychology, Max Planck Institute for Evolutionary Anthropology, Leipzig, Germany

Abstract

In this chapter, I review some of the most recent findings on social cognition in New World monkeys. Referring to the theoretical background that has been briefly outlined in our previous chapter on physical cognition, I discuss the different paradigms that have been used to test social cognitive skills in captivity and, to a minor extent, in the wild. In particular, I will describe how New World monkeys recognize other individuals, how they understand their behaviour and what they understand of direct and third-party relationships. Moreover, we will see whether New World monkeys reciprocate and cooperate, and we will briefly analyze other complex interactions that monkeys have with other group members. I will provide information on their gestural and vocal communication, on their gaze following skills and on their ability to use tools. Finally, I will provide some hints to culture and theory of mind in New World monkeys. The literature cited allows me to depict a general picture of how monkeys perceive the social world and discuss whether specific evolutionary factors might have shaped their social cognitive skills. Finally, I conclude by suggesting some possible directions for future research.

Introduction

Cognition has been defined as the ability to acquire, retain and process information (Shettleworth, 2001). When this information mainly consists in knowledge about the social world, scientists talk about social cognition (see Tomasello & Call, 1997). The distinction between social and physical cognition is of course artificial, as animals usually need to use physical skills also when solving problems in the social arena, or vice-versa, and the intertwinement of different cognitive skills required to solve a task is not so easy to disentangle (e.g., Tebbich et al., 2007). However, it is generally accepted that animal cognition is divided in the physical and the social domain, containing different cognitive skills partially connected to each other (see Deaner et al., 2006; Amici et al., 2012).

The social environment in which all monkeys live is complex. Monkeys have to relate to their conspecifics, making complex decisions and using flexible cognitive strategies to deal with the socio-ecological environment surrounding them (Tomasello & Call, 1997). Monkeys need to recognize others, mate, communicate and form a variety of long-term social relationships with group-mates. Moreover, they need to track complex webs of relationships through time, maintain cohesion, coordinate behaviour, cooperate and allow the formation of long-term coalitions and alliances (Dunbar, 1992, 1998; Kudo & Dunbar, 2001).

Moreover, group living also imposes important costs in term of competition costs (van Schaik, 1983; Dunbar, 1988; van Schaik & Janson, 1988; van Schaik & Kappeler, 1997). Primates, for example, also compete for food resources and need to track, predict and manipulate others' behaviour (Humphrey, 1976; Byrne & Whiten, 1988). Living in long-term social groups, New World monkeys might have therefore evolved strategies and tactics to outwit other group members, with whom competition for scarce resources seems to be extremely high, triggering a sequence of strategies and counter-strategies of increasing complexity (Byrne & Whiten, 1988). More than the static complexity of the physical world, the fluidity and reactivity of other group members would therefore constitute a special cognitive challenge for animals to deal with (Chance & Mead, 1953; Byrne & Whiten, 1988). Indeed, contrary to food resources, conspecifics are able to make decisions of their own in response to others' behaviour, so that keeping track of whatever happens in the social arena is much more complex than any other kind of contingent monitoring that primates do (Milton, 1988).

In brief, social complexity creates challenges that can only be solved through the use of flexible cognitive strategies (Dunbar, 1998). Therefore, it makes sense to predict that all monkey species share some basic social cognitive skills that allow them to understand, predict and possibly manipulate their partners' behavior. The basic question is to what extent these cognitive skills are distributed across monkey species and whether this distribution can be better explained in terms of specific socio-ecological demands that the species are facing. Indeed, primates have a large brain relative to their body size and an expanded neocortex (Dunbar, 1992), which is assumed to be one of the executive substrates for higher cognitive skills (see Barton, 1999; Barton & Harvey, 2000).

As we will see, New World monkeys can behave flexibly, have mental representations and make complex decisions. Below, I will report the most important studies that have been recently done on New World monkey social cognition. This review does not aim to be exhaustive, as priority has been given to more recent, significant studies and to controlled experimental paradigms rather than behavioural observations. Furthermore, invasive paradigms have been excluded for ethical reasons.

RECOGNIZING OTHER INDIVIDUALS

Primates live in complex social groups and therefore need to interact with other group members. In particular, New World monkeys might benefit from individually recognizing other group members, understanding and predicting their behaviour, and remembering past interactions with them and among other individuals (i.e. third-party relationships; Tomasello & Call, 1997).

The most basic of these cognitive skills is naturally the ability to recognize others as specific individuals. Humans rely on visual, auditory and vocal cues to recognize other individuals and obtain information on their sex, age, intentions and emotional state (see Blank et al., 2011). Similarly, New World monkeys recognize others when looking at them or when listening to their vocalizations. The first studies that have shown some evidence of individual recognition in New World monkeys were done using infants' vocalizations as stimuli and monitoring their mothers' behaviour in response. In a classical study on squirrel monkeys (*Saimiri sciureus*), mothers could discriminate their infants on the basis of auditory cues, and they increased their vocalizations and spent more time near the stimulus when they heard their own infant vocalizing (Kaplan et al., 1978). However, it is not just mothers who can recognize their own offspring. Pigmy marmosets (*Cebuella pygmaea*), spider monkeys (*Ateles geoffroyi*) and squirrel monkeys, for example, can identify the contact call of individual group members (Biben & Symmes, 1991; Masataka, 1986; Snowdon & Cleveland, 1980). Playback experiments showed that pygmy marmosets, for example, responded differently depending on the type of contact call and on the individual vocalizing (Snowdon & Cleveland, 1980).

Other studies have tried to experimentally investigate whether New World monkeys can also recognize other individuals visually. A recent study, for example, used a computerized oddity task to investigate whether capuchin monkeys (*Cebus apella*) can discriminate the faces of group mates based on their identity (Pokorny & de Waal, 2009). In this task, subjects were first trained to solve oddity paradigm tasks, in which subjects were presented four two-dimensional images, three of which were identical and the fourth one was not. If the subject chose the odd image, the experimenter provided her with a food reward. After the training, subjects went through the oddity paradigm task by choosing among images of conspecifics. In some conditions, subjects were presented with three identical images of a conspecific and had to select the image with a different conspecific (either a group member or not). In other conditions, subjects were presented with three images of the same individual from different views and had to select the image with a different individual (either a group member or not). Capuchin monkeys were clearly able to individually recognize faces of both familiar and unfamiliar conspecifics, confirming the results of another study in which only unfamiliar conspecifics were used as stimuli (Dufour et al., 2006). Similarly, cotton-top tamarins (*Saguinus oedipus*) could recognize conspecifics from humans, attending more to conspecific than human faces (Neiworth et al., 2007).

Understanding Others' Behaviour, Direct and Third-Party Relationships

There are several occasion in which primates might need to predict and anticipate the behaviour of their group mates. Primates, for example, would benefit from predicting when others' aggressive behaviour might escalate into an attack, or when other group members might compete over a concealed food resource. In this case, for example, subjects would need to monitor the others' position (see below) and attentional state toward the food resource, only accessing it when the competitor cannot subtract it. In a recent study, spider monkeys and capuchin monkeys were exactly tested in a withholding of information task, in which subjects were informed on the location of hidden food, while dominant partners were not (Amici et al., 2009). Subjects were then allowed access to the box hiding the food and could retrieve it. However, naïve dominants were housed in the same room as the subjects, so that the latter should have better waited to retrieve the food until the dominant was far enough from the box and could not eventually steal the food. Both species were able to withhold information, with subjects refraining from interacting with the box when the dominant was nearby. However, spider monkeys were more efficient at getting the food, retrieving it mostly when the dominant was far from the box. Capuchin monkeys were also quite efficient when being alone at the box, but they lost much of the food when manipulating the box with the dominant nearby. Similarly, capuchin monkeys basically failed at deceiving partners in a similar set-up by Fujita and colleagues (2002) and by Hare and colleagues (2003).

Slightly different results were instead obtained with another set-up, in which capuchin monkeys showed communicative and deceptive pointing when they benefited by indicating, accurately or falsely, the location of hidden food (Mitchell & Anderson, 1997). In this study, subjects indicated the right location to the cooperative experimenter, withheld pointing to a competitive experimenter or learned to deceptively point to the competitive experimenter. Also squirrel monkeys learned to accurately indicate hidden food to a cooperative experimenter, but only after an extensive training and with several limitations (Anderson et al., 2001). Moreover, several events of deceptive behaviour have been observed in the wild, including acts of active concealment or active misleading by the subject (e.g., Byrne & Whiten, 1988; 1992). These anecdotic reports are important to evidence behavioural patterns that are usually hard to detect in the wild, but they should of course be followed by more consistent observations or experimental approaches in order to be fully confirmed.

However, primates do not simply have to recognize individuals. New World monkeys live in long-term social systems and they interact with their group mates repeatedly through time. After a certain number of interactions, individuals start forming relationships that allow them to better predict each other's behaviour in a variety of contexts, including agonistic and affiliative ones (see Tomasello & Call, 1997). Therefore, they might behave differently depending on the partner. Individuals, for example, differentially interact with kin or friends, exchanging with them more grooming and support doing aggressions as compared to other group members. Kin and friends, moreover, often reciprocate exchanges. Reciprocal exchanges do not need to be cognitively complex: they can be simply based on symmetries inherent in dyadic relationships ("symmetry-based reciprocity"), or they can be based on the mirroring of social attitudes between partners ("attitudinal reciprocity", Brosnan & de Waal, 2002). Captive capuchin monkeys, for example, share food and they do so in a reciprocal way (de Waal, 2000). These monkeys seem to share food depending on the quality of their own and the partner's food, using memory to mirror social attitudes between partners (de Waal, 2000).

These results have also been confirmed in the wild. Non kin, for instance, distribute altruistic behavior in a reciprocal way, with subjects choosing the partner based on the benefits received in the past (Tiddi et al., 2011). In particular, wild capuchin monkeys exchange grooming and tolerance over food resources in a reciprocal way. This implies that subjects can not only recognize individuals with which they have interacted in the past, but can monitor their behaviour and keep track of their previous interactions with multiple group members. Whether this bookkeeping is emotional rather than cognitive, however, is still a matter of debate (Brosnan & de Waal, 2002; Schino & Aureli, 2010). Similarly, captive capuchin monkeys allow more food thefts from those partners with whom they have stronger long-term social bonds (Sabbatini et al., 2012).

Even more cognitively demanding is the ability to understand relationships between other individuals ("third-party relationships"). These relationships are cognitively more demanding because they do not directly involve the subject. Moreover, by monitoring the relationship among group members, subjects need to increase the amount of information processed in a geometrical way (Whiten & Byrne, 1988; Tomasello & Call, 1997). Humans, for example, change their behaviour toward conspecifics not only depending on their direct past experience with them, but also based on their interactions with third parties. Recently, a study on captive capuchin monkeys showed that these New World monkey species also use information on third- party relationships to make decisions (Anderson et al., 2013). In particular, capuchin monkeys were allowed to observe two humans interacting with each other, one reciprocating and the other one not. In this study, subjects accepted food less frequently from humans who refused to help their conspecifics, suggesting that the ability to observe third-party interactions and adjust behaviour based on this information is also present in primates (Anderson et al., 2013).

RECIPROCAL BEHAVIOUR, COOPERATION AND OTHER COMPLEX INTER- INDIVIDUAL INTERACIONS

By living in complex socialities, New World monkeys continuously relate to each other, exchanging grooming, support in coalitions, social tolerance around food resources and sex, for instance. All these services are exchanged in a complex way, which we are still far from completely understanding. Some authors, for example, suggest that when more partners are available, individuals exchange commodities depending on the demand and supply, which vary through time (Noë & Hammerstein, 1994, 1995). Consequently, commodities can also be exchanged for other services. For example, grooming can be provide in exchange for grooming, but also for food, social tolerance or sex (Schino et al., 2003; Ventura et al., 2006; Gumert et al., 2007; Fruteau et al., 2009). A recent study, for example, showed that spider monkeys exchange embraces for infant handling, probably as a result of increased interest from the other group members (Slater et al., 2007). Also female capuchin monkeys exchange grooming for handling infants, although these exchanges are not affected by the number of infants in the group (i.e., they do not provide support for the existence of a biological market in capuchin monkeys; Tiddi et al., 2010).

A highly debated topic at the moment is whether primates are prosocial, i.e., whether they also spontaneously donate services to their partners, without necessarily expecting their service to be paid back. Prosociality can be defined as any behaviour performed by one actor to alleviate others' need or improve their welfare, at no extra cost (Schino & Aureli, 2010; Cronin, 2012). Experimental studies on captive monkeys show contrasting results, with species behaving prosocially in some studies, but not in others. When tested with sliding tables, marmosets (*Callithrix jacchus*; Burkart et al., 2007) and capuchin monkeys

(Lakshminarayanan & Santos 2008; Takimoto et al., 2010) are prosocial, providing food to non-reciprocating and genetically unrelated individuals. Some authors have therefore proposed that other-regarding preferences need not be related to complex cognitive skills, but depend on prosocial predispositions emanating from cooperative breeding (Burkart et al., 2007). However, we should note that cotton- top tamarins, despite being cooperative breeders, are not prosocial in an experimental set-up, although they donate to each other in reciprocative contexts (Cronin et al., 2009, 2010; Stevens, 2010).

Other authors have therefore proposed that some set-ups might simply not be natural enough to elicit prosocial behaviour. Few years ago, capuchin monkeys were tested by providing them with two different tokens, one always providing food to the subject, and the other one also providing food to the partner (de Waal et al., 2008). If subjects were prosocial, they would have to significantly prefer the token also providing food to the partner. With this set up, capuchin monkeys behaved prosocially (de Waal et al., 2008). However, in this study no test was done to ensure subjects' understanding of the set-up, and control conditions in which partners had no access to food were not provided, or only provided at the end. In a recent comparative investigation, Amici, Visalberghi and Call (in preparation) therefore aimed to compare several primate species, including two New World monkeys (capuchin and spider monkeys) using the two different set-ups (i.e. tokens and sliding tables). Their aims were (i) to compare the two procedures and see whether they elicited comparable responses across species, and (ii) to see whether species differed in terms of prosocial behaviour. In this study, they found that species performed similarly across both tasks, showing no evidence of prosocial behaviour. Even more interestingly, however, is probably the fact that the token procedure, which seemed to evidence prosociality in capuchin monkeys, was not understood by the 43 dyads we tested, with the exception of 3 dyads, who did not show prosocial behaviour.

These differences across studies require an explanation. Some authors have proposed that some species are self-regarding because they don't have the motivational attitude (Burkart et al., 2007; Rekers et al., 2011), the social tolerance levels (Hare et al., 2007) or the communicative skills for that (Yamamoto et al., 2012). Other authors have argued that primates are prosocial, with empathy promoting prosocial behavior and sense of fairness stabilizing its emergency (Yamamoto & Takamoto, 2012). The reason why prosocial behaviour is only elicited in few studies would rather be that some set-ups are unnatural and elicit competitive attitudes, thus failing to evidence prosociality (Horner et al., 2011). While the debate is still on, more studies are needed to really understand to what extent, if ever, primates are prosocial.

In contrast to prosociality, altruistic behaviour implies that the actor incurs in some extra costs to provide benefits to non-kin partners (Schino & Aureli, 2010). Recently, capuchin monkeys have been tested on their propensity to help partners obtain an out-of-reach object (Barnes et al., 2008). In this study, subjects altruistically helped human experimenters even in absence of a food reward, but with important limitations. In particular, capuchin monkeys gave more importance to the potential costs required to help their partner, or to their own benefits, rather than to the benefits they could provide. When helping was costly, moreover, very few subjects spontaneously helped the experimenter to obtain the object, and when they helped in the absence of a reward, they probably did so expecting to be rewarded (Barnes et al., 2008). Similarly, capuchin monkeys were not altruistic when provided with the chance to share tokens with conspecifics (Skerry et al., 2011). Instead of providing tokens to their partner, who could have therefore accessed food, subjects were not sensitive to their partners' welfare, suggesting that variation in contextual demands can importantly affect the cognitive processes involved in altruistic behaviour (Barnes et al., 2008; Skerry et al., 2011). Moreover, New World monkey species might even differ from Old World monkey species and great apes by lacking important cognitive skills required for altruistic behaviour, like the ability to understand beyond their own interest in order to altruistically help others (Barnes et al., 2008).

However, it is important to note that more basic forms of altruism and reciprocation do not necessarily require complex cognitive skills. Calculated reciprocity, for instance, requires subjects to keep track of previous exchanges and is probably only found in humans, but emotionally based reciprocity is simply based on an emotionally based bookkeeping of previous exchanges, with altruism received triggering subsequent altruism both in the short and in the long term (Schino & Aureli, 2010). In this respect, there is no reason to predict that New World monkeys should differ from great apes in their reciprocation levels. Indeed, capuchin monkeys seem to reciprocate food transfers, preferring partners with whom they have stronger long-term social bonds (Sabbatini et al., 2012).

New World monkeys also interact with each other to form coalitions and alliances, for example during competitive interactions over food resources or sex mates (Tomasello & Call, 1997). In some cases, New World monkeys also cooperate, i.e., they incur in common costs to obtain a common benefit. However, it is not yet clear to what extent individuals understand each other's role during cooperative events, and to what extent they use communicative strategies to better coordinate their behaviour (Tomasello & Call, 1997). Moreover, evidence of cooperative behaviour in New World monkeys is still controversial. Capuchin monkeys, for example, are known to cooperatively hunt in the wild, with several individuals pursuing a prey but only one capturing it (Perry & Rose, 1994; Rose, 1997). This behaviour is not common among non-human primates, and has so far been extensively documented only in chimpanzees (Boesch & Boesch, 1989; Boesch, 1994).

In captivity, however, capuchin monkeys perform very differently, depending on the set-up used. In one task, dyads were required to simultaneously pull two handles (which were far from each other and could not be pulled simultaneously by only one subject) in order to access a platform with food (Chalmeau et al., 1997). Although subjects understood how the apparatus worked and some of them were successful at pulling together, subjects did not take into account the partner's behaviour in order to pull (Chalmeau et al., 1997). Similarly, their behaviour was only partially affected by the partner's position and experience did not improve performance (Visalberghi et al., 2000). In other words, capuchin monkeys did not seem to understand the partner's role, but they were probably successful thanks to their manual dexterity and high levels of social tolerance (Chalmeau et al., 1997). This form of cooperation, therefore, needs not be cognitively grounded (Visalberghi et al., 2000). When using the same paradigm, however, Mendres and de Waal (2000) found that capuchin monkeys monitor their partner and adjust their pulling behaviour depending on the partner's presence (pulling more when the partner is present, and looking more at the partner in cooperative conditions). In another task, moreover, capuchin monkeys showed some form of cognitively more complex cooperation. After being trained, capuchin monkeys could divide the sequence of actions required to obtain a reward (Hattori et al., 2005). However, subject did not show spontaneous division of labour, and they did not communicate to improve their performance. Other studies show that depending on the context capuchin monkeys might be more successful at cooperating, for example when food rewards are not clumped and can be more easily shared (de Waal & Devis, 2003). In a certain sense, capuchin monkeys might have some predictions on the outcome of cooperative interactions, and might therefore decide to cooperate only when the chance to obtain food rewards is higher (de Waal & Devis, 2003). Consistently, capuchin monkeys also share food rewards more readily when a cooperative effort is required to obtain them (Brosnan et al., 2006).

Since they are the only New World monkey species showing some evidence of cooperative hunting in the wild, capuchin monkeys are clearly the best candidate to study cooperative skills in captivity. However, cooperative breeding species also seem good candidates for the study of cooperation. Cooperative breeders are primates in which individuals other than parents help to care for and provision offspring (Wilson, 1975). Marmosets and tamarins, for example, usually live in social groups composed of a single breeding pair and its offspring (Garber et al., 1995). Within the groups, all individuals systematically engage in cooperative behaviors, carrying and feeding infants, as well as contributing to communal group defense (Burkart et al., 2000). For this reason, they might also show cooperative behaviour when tested in captivity with experimental paradigms. Cotton-top tamarins, for instance, were tested with a set- up similar to that used by Chalmeau and colleagues (1997). In this study, tamarins managed to simultaneously pull the handles to retrieve the food, performing better than other species previously tested (Cronin et al., 2005). Moreover, tamarins adjusted their behaviour to the partner's presence, developing sustained pulling techniques to solve the task. The authors therefore suggested that the high levels of cooperation shown by tamarins might depend on their cooperative social system (Cronin et al., 2005). Indeed, another New World monkeys characterized by high levels of cooperative breeding, common marmosets, also performed cooperatively in a similar task (Wardenich & Huber, 2002). In particular, half of the tested dyads were successful, with one subject pulling the food for the partner and both sharing the food afterwards. The authors argued that cooperative behaviour in marmosets therefore depends on a specific distribution of roles during the task and on the social tolerance levels naturally displayed by the species (Wardenich & Huber, 2002).

Although some studies seem to suggest that at least some New World monkeys are able to substantially cooperate in an experimental set-up, more studies are required to understand the evolutionary reasons for inter-specific differences and, especially, to what extent these forms of cooperation are cognitively complex.

Gestural Communication

Primates use gestures and vocalizations to communicate with each other. In intentional communication, the sender communicates something to a potential receiver, in order to reach a specific goal, and does so with some degree of flexibility (Tomasello & Call, 1997). In humans, pointing becomes intentional around 12 months and has a key role in the development of cognitive skills to understand others' intentions and cooperate (Bates et al., 1975; Bullinger et al., 2011). The complexity of gestural communication greatly varies across non-human primate species, with great apes showing the most flexible and wide gestural repertoire (Pika et al., 2005). This high variability and flexibility in terms of gestural communication has led some authors to hypothesize that gestural communication in great apes is a precursor of human language, and the modality within which symbolic communication possibly evolved (Arbib et al., 2008; Pika et al., 2005).

In contrast to great apes, however, New World monkeys show a rather limited range of gestures, which they use in a less flexible way. Consequently, it is hard to say whether gestures in New World monkeys are simply controlled conditioned responses, or whether they can also be used in a referential way (i.e. to refer to external events or objects), like it partially happens in Old World monkeys and great apes (see Liebal et al., 2007). Finally, gestures in New World monkeys clearly lack the symbolic value typical of human communication.

New World monkeys are clearly able to understand pointing as a referential gesture. Capuchin monkeys and cotton-top tamarins, for instance, can easily locate hidden food when provided with a pointing cue by a human experimenter (Anderson et al., 1995; Neiworth et al., 2002). More contrasting evidence has been collected on New World monkeys' ability to produce pointing as a referential gesture. Mitchell and Anderson (1997), for example, have reported that a human-reared capuchin monkey could spontaneously and communicatively point to a hidden food reward in presence of a cooperative trainer. In this study, two other capuchin monkeys could also learn to point after operant conditioning. In another study, capuchin monkeys' pointing was sensitive to the experimenter's eye direction, but not to the attentional state of the experimenter (Hattori et al., 2007). However, the same monkeys showed the ability to understand human attentional states to a certain extent (Hattori et al., 2010). Finally, trained squirrel monkeys learned to use pointing in a flexible way, monitoring the recipient in order to take into account its attentional state (Anderson et al., 2001, 2007, 2010).

VOCAL COMMUNICATION

In contrast to what one might expect, vocal communication in primates seems to be less complex and flexible than gestural one. Similarly, vocal communication in non-human primates is also much less complex than vocal communication in birds. One reason for that might be that non- human primates do not travel much and are therefore less likely to encounter individuals from other populations, making vocal learning less adaptive than in other species (see Snowdon, 2009). Another reason might be that non-human primate vocalizations have no role in sexual selection, in contrast to what happens in birds (see Snowdon, 2009). Whatever the reason, primates show a limited repertoire of vocalizations, although they can partially adjust vocalizations to different social and environmental circumstances, and they partially modify them through development.

The majority of studies on vocal communication in New World monkeys has been done on marmosets and tamarins, aiming to investigate the plasticity of their vocal production, usage and response to calls (i.e. whether vocalizations change as a possible consequence of social interactions). Studies on the plasticity of vocal production, for example, show that vocalizations in New World monkeys can change across populations, depending on environmental factors and social influences. De la Torre and Snowdon (2009), for instance, studied different populations of pygmy marmosets (*Callithrix pygmaea*) and found both inter-individual and inter-population variation in vocal production. Although some acoustic differences in call structure were better explained in terms of adaptations to local acoustic environment, other differences were explained in terms of genetic drift or possibly even vocal learning. Environmental differences thus also partially affect vocal production in New World monkeys, suggesting that monkeys can control and flexibly adjust vocal production to environmental noise. Common marmosets, for example, change the amplitude and duration of twitter calls with increased ambient noise (Brumm et al., 2004). Similarly, cotton-top tamarins disturbed by specific noise interrupt the production of long calls, produce shorter notes and change the amplitude and intervals between calls (Miller et al., 2003; Egnor et al., 2006). Social influences also affect vocal production in New World monkeys. Depending on social status, for example, cotton-top tamarins seem to change the structure of their calls (Roush & Snowdon, 1999). Similarly, changes in the structure of vocalizations were detected in pygmy marmosets and black-tufted marmosets (*Callithrix kuhlii*) when altering the groups' social structure (Elowson & Snowdon, 1994; Ruckstalis et al., 2003). These results suggest that also New World monkey vocalizations are not completely stereotyped (Snowdon, 2009). Vocal production is also affected by deve lopmental factors. Some of the chirps used by cotton-top tamarins in feeding contexts, for example, seem to be better formed in adults than in immature individuals (Roush & Snowdon, 1994). Similarly, infant pygmy and common armosets use the same vocalizations as adults, but with a different frequency and in different contexts, possibly to better draw adults' attention (Snowdon & Elowson, 2001). Even though there is not yet evidence that vocal learning is involved in vocal production, vocal practice and physical maturation are surely needed for infants to acquire the same vocal repertoire of adults (Snowdon, 2009).

New World monkeys also show some plasticity in the way they use vocalizations. The use of vocalizations is affected by developmental influences. Young cotton-top tamarins and pygmy marmosets, for example, cannot differentiate between contexts in the use of most calls, suggesting that the connection between call type and context needs to be acquired through some developmental process (Castro & Snowdon, 2000; Snowdon & Elowson, 2001). The use of vocalizations seems to be also affected by the presence of other individuals. Capuchin monkeys, for example, are more likely to give food vocalizations in presence of other group members (Pollick et al., 2005). In contrast, cotton-top tamarins do not give more food calls depending on the presence of other group members (Roush & Snowdon, 2000). Cotton-top tamarins raised in captivity also react differently than their wild counterparts to predators, showing different alarm responses and vocalizations (see Snowdon, 2009). Marmosets can also systematically vary call structure depending on the distance from other group members, producing more easily detected calls when other individuals are further (Snowdon & Hodun, 1981). Finally, the use of food- related calls is not automatic, but it seems to be affected by the individual preference for that food, by the quality of the food and by food being animate or not (spider monkeys: Chapman & Lefebvre, 1990; cotton-top tamarins: Elowson et al., 1991; white-faced capuchin monkeys, *Cebus capucinus:* Gros-Luis, 2004; pygmy marmosets: Snowdon, 2009).

New World monkeys can also comprehend and respond to vocalizations in a plastic way, as it is suggested by those studies in which captive cotton-top tamarins do not respond to playback of predator cues (Hayes & Snowdon, 1990; Friant et al., 2008; Campbell & Snowdon, 2009). The clearest evidence of primate flexibility in comprehension of vocalizations comes from studies of cross-species communication, which however have been mainly conducted on species other than New World monkeys (Snowdon, 2009).

Finally, vocalizations can affect imitation and other forms of social learning, although the role of vocalizations seems to vary across contexts and tasks. When presented with adulterated food, for example, cotton-top tamarins, but not capuchin monkeys, learn from each other to avoid the food (Visalberghi & Addessi, 2000; Snowdon & Boe, 2003). In contrast to capuchin monkeys, tamarins trying adulterated food also produce less food vocalizations and more alarm calls, allowing the other group members to learn to avoid the food without necessarily having to try it. Moreover, tamarins use special food vocalizations to transfer food to inexperienced infants, and this behaviour seems to be related to the infants' acquisition of foraging skills (Rapaport, 2006). Lion tamarins (*Leontopitechus* spp.), for example, but not common marmosets, are more likely to transfer unfamiliar food to infants, suggesting that they understand something about the infants' knowledge of food (Rapaport, 1999; Brown et al., 2005). Similarly, lion tamarins also vocalize more and transfer food to juveniles, when the food is an animate prey and thus harder for the juveniles to obtain it (Rapaport, 2006; Rapaport & Ruiz-Miranda, 2006). In an experimental version of this study, cotton-top tamarin parents adjusted their vocalizations and food transfers when the juvenile successfully learned to forage alone, with greater parental response to juvenile vocal begging leading to a better performance in food acquisition (Humle & Snowdon, 2008). Finally, vocalizations seem to play a major role also in food exchanges among conspecifics. In contrast to common marmosets, for example, cotton-top tamarins do not learn to cooperate to obtain food rewards (Cronin et al., 2005; Burkart et al., 2007), and donations from one individual to another decrease when partners show interest for the food by for example giving food vocalizations (Cronin et al., 2009). This suggests that tamarins are attending to the partner's vocalizations, although their reaction is not prosocial. For an excellent review on primate vocalizations, see Snowdon (2001) and Snowdon(2009).

Perspective Taking, Gaze Following

Another important cognitive skill in the social domain is the ability to follow the direction in which other individuals are looking, or gaze following. Gaze following can indeed allow individuals to gain information on the social and physical environment by monitoring others' behaviour (Tomasello et al., 1998).

The ability to follow others' gaze does not necessarily imply high cognitive skills. Individuals, for example, can simply shift their gaze as an innate response triggered by something shifting their attention, like another individual's head turn, with this behaviour being reinforced through time by the presence of external visible targets (Povinelli & Eddy, 1996; Corkum & Moore, 1998). However, gaze following might also imply that individuals understand what others see, like children do. In this case, primates would understand others' perceptions and voluntarily shift their attention as a result of their own expectancies and intentions, and they would learn to understand the intentional and referential nature of looking, the mentalistic experience of seeing and the role of the eyes (Butterworth & Jarrett, 1991; Corkum & Moore, 1998; Brooks & Meltzoff, 2002). Although some studies conducted on great apes seem to suggest that their gaze following behaviour is partially of the more complex type (e.g. Hare et al., 2000), evidence on New World monkeys is more limited and controversial.

In the first study on monkey gaze following, Itakura (1996) failed to find evidence of gaze following in capuchin monkeys and squirrel monkeys. Further studies, however, sowed that capuchin monkeys can follow others' gaze. In particular, both capuchin monkeys and spider monkeys show evidence of spontaneous gaze following, when a human experimenter looks at a target location, and they also engage in gaze following behaviour around barriers when the human experimenter looks behind an opaque barrier placed outside the testing room (Amici et al., 2009). However, neither species performed gaze alternation, i.e., they did not look back at the human's face when they did not find anything interesting in the target location, which is considered evidence that the performer has expectancies and understands that gaze is about something specific in the environment (Butterworth & Cochran, 1980). These results therefore suggest that capuchin monkeys and spider monkeys show some indication of perspective-taking abilities, but potential differences from the abilities shown by the great apes (Call et al., 1998; Amici et al., 2009). Similarly, limited understanding of gaze following comes from a different experiment on capuchin monkeys. When tested with a competitive paradigm, subordinate capuchin monkeys preferred to retrieve the food that was hidden from the dominant (Hare et al., 2003). However, subjects were not selecting the food reward based on their understanding of what the dominant could or not see, but they were simply monitoring the dominant's behaviour, avoiding the piece of food first approached by the dominant (Hare et al., 2003). Squirrel monkeys have also been tested on their ability to follow others' gaze. Interestingly, they performed gaze alternation when reaching toward out-of-reach food in the presence of a human partner, although this behaviour was not influenced by changes in the partner's focus of attention (Anderson et al., 2007). Finally, marmosets can follow humans' gaze past distractors, but they do not follow the gaze around barriers (Burkart & Heschl, 2006, 2007).

Social Learning and Culture

Group living in primates probably represents an evolutionary compromise between the high fitness benefits it provides and the competition costs that it unavoidably imposes (e.g. van Schaik, 1983; Dunbar, 1988; van Schaik & Janson, 1988; van Schaik & Kappeler, 1997; Silk, 2007). Apart from defense against predation and infanticide, one of the most important benefits of living in group is that social life allows individuals to use the expertise already acquired by others to gain information about the surrounding environment. This expertise can be used in a variety of ways, from simply scrounging others' resources to learning something from others, by for example trying to reproduce or match their behaviour (Bandura, 1986; Russon, 1997; Tomasello & Call, 1997).

This kind of social learning can have different degrees of cognitive complexity (Whiten & Ham, 1992; Whiten, 2000). Primates, for example, can simply orient their behaviour and attention toward an object or a location, as a result of stimulus enhancement. They can learn the context in which a specific behaviour should be used, through observational conditioning. They can emulate others' goals, but not the actions required to pursue them. Or they can imitate others, i.e., by seeing an action performed by another individual and reproducing it.

Differences across these forms of social learning are not dichotomous, as they rather change across a continuum (Whiten, 2000). In all these cases, subjects are actively learning from others.

So far, very contrasting evidence has been collected on the ability of New World monkeys to learn from others. The first experiments on New World monkeys generally failed to evidence social learning, except for the easiest form of stimulus enhancement (see Visalberghi & Fragaszy, 1990). Indeed, most species seem to avoid specific foods by learning it individually, not socially (Visalberghi, 1994). Juvenile squirrel monkeys, for example, need to individually learn to rub noxious spines off caterpillars before eating them, although they often see adults doing that. Similarly, capuchin monkeys do not learn to open nuts by watching others (Visalberghi, 1987), nor do they show evidence of imitative learning by copying a skillful conspecific retrieving food (e.g. Fragaszy & Visalberghi, 1989; Adams-Curtis & Fragaszy, 1995). As far as we know, New World monkeys also do not intentionally pass on their knowledge to others. Spider monkeys, for example, fail to prevent infants from eating noxious food (Fairbanks, 1975).

Other studies, however, have shown some evidence of social learning in New World monkeys. Capuchin monkeys, for instance, were tested by presenting them with a human experimenter opening an artificial fruit and then giving them the opportunity to open the fruit themselves (Custance et al., 1999). Capuchin monkeys did not extensively copy the human model when opening the fruit, but the overall approach to the fruit was similar to the one performed by the human model, suggesting some form of emulation or basic imitation. Finally, two marmosets showed evidence of imitating the way in which another marmoset opened a doorway to retrieve the food (Bugnyar & Huber, 1997).

If evidence of social learning is contrasting, even more problematic is the question of whether New World monkeys show evidence of culture. The first problem is that the concept of culture has been defined in several different ways. In a broad sense, culture is intended as behavioral conformity spread or maintained by non-genetic means, through social learning or social influence (see Whiten, 2000). If we define social culture in this way, also New World monkeys have culture, in that they show evidence of social learning, at least in its most basic forms. Stimulus enhancement, for instance, seems to be the way in which most juveniles learn about what and where to eat (Whiten, 1989; Visalberghi, 1994; Fragaszy & Visalberghi, 1996). Other authors, however, have highlighted that human culture is much more complex than this, relying on complex forms of social learning, like imitation and teaching (Galef, 1992). For this reason, it would be misleading to talk about primate culture, if behavioural transmission in primates occurs through mechanisms less complex than imitation and teaching. Lefebvre (1995) has consequently analyzed the time course of several documented cases of diffusions of novel behaviors in primates. Across all cases, chimpanzees were the only ones showing the accelerating curve predicted if some sort of social transmission was taking place.

THEORY OF MIND

Subjects can understand others as animate agents, intentional agents or mental agents (Tomasello & Call, 1997). In the first case, subjects understand that others generate their own behaviour spontaneously, and thus differ from inanimate objects. In the second case, subjects understand that others engage in flexible decision-making processes. Therefore, they choose among different strategies to obtain goals, and pay selective attention to specific objects in their perceptual field. In the third case, subjects understand that others have their own thoughts and beliefs about the external world, which might conflict with their own ones. If primates had a theory of mind, they would be able to understand others as intentional or mental agents, understanding others' psychological states and others' behaviour as intentional (Tomasello & Call, 1997).

Several paradigms have been used to test how primates perceive others. Some studies have investigated how primates understand others' behaviour, by for example investigating gaze following behaviour and perspective taking (see above), self-recognition in a mirror, and understanding others' intention or attentional state. Most of these studies, however, have been done on great apes and macaques, but not on New World monkeys. The few studies that have directed tested New World monkeys' understanding of others suggest that their understanding of others as intentional or mental agents is in the best case very limited (e.g. Hare et al., 2000; 2003), but no conclusions can be drawn unless more studies are conducted across different species.

CONCLUSION: CURRENT PROBLEMS AND FUTURE RESEARCH

New World monkeys have a complex understanding of the social world surrounding them. They live in complex social groups and regularly interact with their group-mates. All species that have been studied so far clearly recognize other individuals using visual and/or auditory cues. However, not all monkey species seem equally good at predicting others' behaviour, with capuchin monkeys for example failing to withhold information from dominant partners and anticipate their behaviour in competitive tasks. However, anecdotic reports of deceptive behaviour in the wild are rather common across New World monkey species.

More complex results have been obtained in studies on reciprocal behaviour. In general, monkeys seem to reciprocate quite systematically in the wild, but not in captivity. Capuchin monkeys are a possible exception, keeping track of others' behaviour and reciprocal exchanges, although this bookkeeping could be emotional rather than cognitive. Moreover, capuchin monkeys also seem to understand third-party relationships and adjust their behaviour accordingly. Similarly, several species of New World monkeys exchange services and commodities in complex ways, but mainly in the wild. Evidence of prosocial behaviour is much scantier, with marmosets but not tamarins spontaneously donating food to others. Capuchin monkeys also donate food prosocially, but only in some studies, and they are altruistic to a very limited extent. In cooperative tasks, capuchin monkeys also seem to have a limited understanding of each other's role, and do not spontaneously divide labour or communicate with each other to better coordinate their efforts. With similar paradigms, tamarins and marmosets seem to perform better. Possible evolutionary explanations for the emergence of prosocial, altruistic, and cooperative behaviour in these species are that tamarins and marmosets are cooperative breeders, and thus have higher degrees of social tolerance and better social predispositions toward others (Burkart et al., 2007). However, given the contrasting results so far evidenced in capuchin monkeys and the lack of tasks across several species, more studies are surely required in order to understand the evolutionary reasons for inter-specific differences and, especially, to what extent these forms of prosociality, altruism and cooperation are cognitively complex.

New World monkeys also use gestures and vocalizations to communicate. Their gestural repertoire is much less complex than the one known for great apes, but New World monkeys are still able to understand and partially use pointing gestures in a referential way. Moreover, several monkey species can be trained to understand human attentional states. Vocal communication in New World monkeys is probably even less complex and flexible than the gestural one, although monkeys can partially adjust vocalizations to different social and environmental circumstances, partially modifying them through development. In marmosets and tamarins, for example, vocalizations partially change across populations and depending on the socio-environmental context, suggesting that New World monkey vocalizations are not completely stereotyped.

New World monkeys also spontaneously follow others' gaze to monitor their behaviour.

Capuchin monkeys and spider monkeys, for example, follow others' gaze even behind opaque barriers, but they don't perform gaze alternation, suggesting some limits to their gaze following skills. In contrast, marmosets don't follow gaze around barriers, but squirrel monkeys do, and they even perform gaze alternation.

To date, contrasting evidence has been collected on the ability of New World monkeys to learn from others. The first experiments on New World monkeys generally failed to evidence social learning, except for the easiest form of stimulus enhancement. Recent studies have however shown that New World monkeys can learn from others to a certain extent. Anyway, individual learning seems to play a major role across all studied monkey species. Similarly, there is so far no convincing evidence of the existence of culture in New World monkeys, at least not in the more complex sense of the term.

Finally, the very few studies that have so far directly tested New World monkeys suggest that their understanding of others as intentional or mental agents is in the best case very limited, but no conclusions can be drawn unless more studies are conducted across different species.

Given these premises, it seems again hard to hint to specific socio-ecological factors that might explain the distribution of social cognitive skills across New World monkeys. In general, all New World monkey species seem to have a basic understanding of the social world, recognizing individuals through visual and auditory cues. However, the ability to predict others' behaviour is not uniformly distributed across studied species, and might be a reflex of the complexity of the social structures they live in. Fission-fusion dynamics have been proposed as a possible explanation for inter-specific differences in this respect, but more comparative studies are required to confirm this hypothesis. Prosocial, altruistic and cooperative tendencies might be affected by several factors. Cooperative breeders, for example, might have a better social predisposition toward others that might explain their partially better performance in these tasks. However, inter-specific differences are not so clear, and important differences are observed when other paradigms are used. Similarly, all New World monkey species communicate with each other through gestures and vocalizations. Their communication seems to be generally less rich and flexible than great apes, but more studies are required to really map inter-specific differences in communication across New World monkeys. New World monkeys also follow others' gaze to monitor their behaviour, but with important limitations, and they learn from each other to a limited extent. Similarly, they don't seem to have culture, at least in the more complex sense of the term, and their understanding of others as intentional or mental agents also seems to be restricted. All these data seem to suggest that New World monkeys have a somehow more limited understanding of others, although more studies are required to draw convincing conclusions on inter-specific differences in primate social cognitive skills. Moreover, it makes sense to think that more evolutionary factors will simultaneously predict the distribution of social cognitive skills, with a complex interaction of social, ecological and phylogenetic factors on the evolution of these skills (e.g., Dunbar, 1992; Amici et al., 2012).

As we have already outlined in our previous chapter, the interaction of all these evolutionary factors can be only disentangled by addressing some crucial questions in future studies on comparative cognition. Firstly, future studies will need to include more species. If these species differ enough in terms of socio-ecological characteristics, we will be able to experimentally contrast the different evolutionary hypotheses and finally understand which socio-ecological factors best predict the distribution of social cognitive skills. Secondly, future studies should include more subjects, in order to control for intra-specific variability in terms of cognitive skills. This is especially important in studies on social cognition, in which subjects are often tested together with partners, so that the different quality of relationships might add further intra-specific variability. Since socio-ecological factors can dramatically vary also across conspecific populations, the inclusion of multiple populations for each species would also be beneficial, or at least the simultaneous collection of socio-ecological and cognitive data from each of the tested populations. Similarly, the inclusion of both wild and captive individuals would ideally allow controlling for the effects of ontogenetic experience on cognition. Thirdly, experimental paradigms should be used consistently across species, to ensure that inter-specific differences in social cognitive skills are not merely a reflex of differences in the experimental procedures that have been used. Addressing these controversial aspects will provide further strength to the future study of comparative cognition and will allow us to better understand the evolutionary factors explaining the distribution of social cognitive skills across New World monkeys.

Acknowledgments

This chapter was written while the author held a Humboldt Research Fellowship for Postdoctoral Researchers (Humboldt ID number 1138999). Thanks to Bino Majolo for revising a previous draft of the manuscript.

References

Adams-Curtis, L. E., & Fragaszy, D. (1995). Influence of a Skilled Model on the Behavior of Conspecific Observers in Capuchin Monkeys. *American Journal of Primatology, 37*, 65-71. Amici, F., Aureli, F., & Call, J. (2008). Fission-Fusion Dynamics, Behavioral Flexibility, and Inhibitory Control an Primates. *Current Biology, 18*, 1415-1419.

Amici, F., Aureli, F., & Call, J. (2010). Monkeys and Apes: Are Their Cognitive Skills Really So Different? *American Journal of Physical Anthropology, 143,* 188–197.

Amici, F., Barney, B., Johnson, V. E., Call, J., & Aureli, F. (2012). A Modular Mind? A Test Using Individual Data from Seven Primate Species. *Plos ONE, 7*(12), e51918.

Amici, F., Call, J., & Aureli, F. (2009). Variation in Withholding of Information in Three Monkey Species. *Proceeding of The Royal Society B: Biological Sciences, 276*, 3311-3318.

Amici F, Aureli F, Visalberghi E, & Call J. (2009). Spider Monkeys (*Ateles geoffroyi*) and Capuchin Monkeys (*Cebus apella*) Follow Gaze around Barriers: Evidence for Perspective Taking? *Journal of Comparative Psychology, 123*, 368-374.

Amici, F., Call, J., & Aureli, F. (2012). Aversion to Violation of Expectations of Food Distribution: The Role of Social Tolerance and Relative Dominance in Seven Primate Species. *Behaviour, 149*, 345–368.

Anderson, J. R., Sallaberry, P., & Barbier, H. (1995). Use of Experimentergiven Cues During Object- Choice Tasks by Capuchin Monkeys. *Animal Behavior, 49,* 201–208.

Anderson, J. R., Kuroshima, H., Kuwahata, H., Fujita, K., & Vick, S. J. (2001). Training Squirrel Monkeys (*Saimiri sciureus*) to Deceive: Acquisition and Analysis of Behavior Toward Cooperative and Competitive Trainers. *Journal of Comparative Psychology, 115,* 282–293.

Anderson, J. R., Kuwahata, H., & Fujita, K. (2007). Gaze Alternation during ''Pointing'' by Squirrel Monkeys (*Saimiri sciureus*)? *Animal Cognition, 10,* 267–271.

Anderson, J. R., Kuroshima, H., Hattori, Y., & Fujita, K. (2010). Flexibility in the Use of Requesting Gestures in Squirrel Monkeys (*Saimiri sciureus*). *American Journal of Primatology, 72*, 707–714.

Anderson, J. R., Kuroshima, H., Takimoto, A., & Fujita, K. (2013). Third-Party Social Evaluation of Humans by Monkeys. *Nature Communications, 4,* doi:10.1038/ Ncomms2495.

Arbib, M. A., Liebal, K., & Pika, S. (2008). Primate Vocalization, Gesture, and the Evolution of Human Language. *Current Anthropology, 49*, 1053-1076.

Bandura, A. (1986). *Social Foundations of Thought and Action.* Princeton NJ: Prentice-Hall

Barnes, J. L., Martinez, M., Langer, M., Hill, T., & Santos, L. R. (2008). Helping Behavior and Regard for Others in Capuchin Monkeys (*Cebus apella*): An Evolutionary Perspective on Altruism. *Biology Letters, 4,* 638–640.

Barton, R. A. (1999). The Evolutionary Ecology of the Primate Brain. In: P.C. Lee (Ed.), *Comparative Primate Socioecology* (pp. 167-203). Cambridge: Cambridge University Press.

Barton, R. A., & Harvey, P. H. (2000). Mosaic Evolution of Brain Structure in Mammals. *Nature, 405*, 1055–1058.

Bates, E., Camaioni, L., & Volterra, V. (1975). The Acquisition of Performatives Prior to Speech. *Merrill-Palmer Quarterly, 21,* 205–226.

Biben, M., & Symmes, D. (1991). Playback Studies of Affiliative Vocalizing in Captive Squirrel Monkeys: Familiarity As a Cue to Response, *Behaviour, 117,* 1-19.

Blank, H., Anwander, A., & von Kriegstein, K. (2011). Direct Structural Connections between Voice- and Face-Recognition Areas. *Journal of Neuroscience, 31,* 12906-12915.

Brooks, R., & Meltzoff, A. (2002). The Importance of Eyes: How Infants Interpret Adult Looking Behavior. *Developmental Psychology, 38,* 958–966.

Brosnan, S. F., & de Waal, F. B. M. (2002). A Proximate Perspective on Reciprocal Altruism. *Human Nature, 13*, 129–152.

Brown, G. R., Almond, R. E. A., & Bates, N. J. (2005). Adult–Infant Food Transfer in Common Marmosets: An Experimental Study. *American Journal of Primatology, 65*, 301–312. Boesch, C. (1994).

Cooperative Hunting in Wild Chimpanzees. *Animal Behaviour, 48*, 653-667. Boesch, C., & Boesch, H. (1989). Hunting Behavior of Wild Chimpanzees in the Taï National Park. *American Journal of Physical Anthropology, 78*, 547-573.

Brosnan, S. F., Freeman, C., & de Waal, F. B. M. (2006). Partner's Behavior, Not Reward Distribution, Determines Success in an Unequal Cooperative Task in Capuchin Monkeys. *American Journal of Primatology, 68*, 713-724.

Brumm, H., Voss, K., Koellmer, I., & Todt, D. (2004). Acoustic Communication in Noise: Regulation of Call Characteristics in a New World Monkey. *Journal of Experimental Biology, 207*, 443–448.

Bugnyar, T., & Huber, L. (1997). Push or Pull: An Experimental Study on Imitation in Marmosets. *Animal Behaviour, 54*, 817–831.

Bullinger, A. F., Zimmermann, F., Kaminski, J., & Tomasello, M. (2011). Different Social Motives in the Gestural Communication of Chimpanzees and Human Children. *Developmental Science, 14,* 58–68.

Burkart, J. M., Fehr, E., Efferson, C., & van Schaik, C. P.(2007). Other-Regarding Preferences in a Non- Human Primate: Common Marmosets Provision Food Altruistically. *Proceedin of The National Academy of Sciences, 104*, 19762–19766.

Burkart, J., & Heschl, A. (2006). Geometrical Gaze Following in Common Marmosets (*Callithrix jacchus*). *Journal of Comparative Psychology, 120*, 120–130.

Burkart, J., & Heschl, A. (2007). Perspective Taking or Behaviour Reading? Understanding Visual Access in Common Marmosets (*Callithrix jacchus*). *Animal Behaviour, 73*, 457–469.

Burkart, J. M, Hrdy, S. B., & van Schaik, C. P. (2009). Cooperative Breeding and Human Cognitive Evolution. *Evolutionary Anthropology, 18*, 175-186.

Butterworth, G. (2003). Pointing Is the Royal Road to Language for Babies. In: Kita, S. (Ed) *Pointing: Where Language, Culture, and Cognition Meet* (pp. 9-33). Erlbaum, Mahwah.

Butterworth, G., & Cochran, E. (1980). Towards a Mechanism of Joint Visual Attention in Human Infancy. *International Journal of Behavioral Development, 3*, 253–272.

Butterworth, G., & Jarrett, N. (1991). What Minds Have in Common Is Space: Spatial Mechanisms Serving Joint Visual Attention in Infancy. *British Journal of Developmental Psychology, 9*, 55–72.

Byrne, R.W., & Whiten, A. (1988). *Machiavellian Intelligence: Social Expertise and The Evolution of Intellect in Monkeys, Apes and Humans.* Oxford: Oxford University Press.

Byrne, R.W., & Whiten, A. (1992). Cognitive Evolution in Primates: Evidence from Tactical Deception. *Man, 27*, 609–627.

Call, J., Hare, B., & Tomasello, M. (1998). Chimpanzee Gaze Following in an Object Choice Task. *Animal Cognition, 1*, 89–100.

Campbell, M. W., & Snowdon, C. T. (2009). Can Auditory Playback Condition Predator Mobbing in Captive-Reared Cotton-Top Tamarins? *International Journal of Primatology, 30*, 93–102. Castro, N. A., & Snowdon, C. T. (2000). Development of Vocal Responses in Infant Cotton-Top Tamarins. *Behaviour, 137*, 629–646.

Chalmeau, R., Visalberghi, E., & Gallo, A. (1997). Capuchin Monkeys, *Cebus apella*, Fail to Understand a Cooperative Task. *Animal Behaviour, 54,* 1215-1225.

Chance, M., & Mead, A. (1953). Social Behaviour and Primate Evolution. Symposia of the Society for Experimental Biology. *Evolution, 7,* 395–439.

Chapman, C. A., & Lefebvre, L. (1990). Manipulating Foraging Group Size: Spider Monkey Food Calls at Fruit Trees. *Animal Behaviour, 39*, 891–896.

Corkum, V., & Moore, C. (1998). The Origins of Joint Visual Attention in Infants. *Developmental Psychology, 34*, 28–38.

Cronin, K. A. (2012). Prosocial Behaviour in Animals: The Influence of Social Relationships, Communication and Rewards. *Animal Behaviour, 84*, 1085-1093.

Cronin, K. A., Kori, K. E. S., & Snowdon, C. T. (2010). Prosocial Behavior Emerges Independent of Reciprocity in Cottontop Tamarins. *Proceedings of the Royal Society, London B, Biological Science, 1701,* 3845–3851.

Cronin, K. A., Kurian, A.V., Showdon, C. T. (2005). Cooperative Problem Solving in a Cooperatively Breeding Primate (*Saguinus oedipus*). *Animal Behaviour, 69*, 133-142.

Cronin, K. A., Schroeder, K. K. E., Rothwell, E. S., Silk, J. B., & Snowdon, C. T. (2009). Cooperatively Breeding Cottontop Tamarins (*Saguinus oedipus*) Do Not Donate Rewards to Their Long-Term Mates. *Journal of Comparative Psychology, 123*, 231–241.

Custance, D. M., Whiten, A., & Fredman, T. (1999). Social Learning of an Artificial Fruit Task in Capuchin Monkeys (*Cebus apella*). *Journal of Comparative Psychology, 113*, 1–11.

De La Torre, S., & Snowdon, C.T. (2009). Dialects in Pygmy Marmosets? Population Variation in Call Structure. *American Journal of Primatology, 71*, 333–342.

Deaner, R. O., van Schaik, C. P., & Johnson, V. (2006). Do Some Taxa Have Better Domain-General Cognition Than Others? A Meta-Analysis of Nonhuman Primate Studies. *Evolutionary Psychology, 4*, 149-196.

Dufour, V., Pascalis, O., & Petit, O. (2006). Face Processing Limitation to Own Species in Primates: A Comparative Study in Brown Capuchins, Tonkean Macaques and Humans. *Behavioural Processes, 73*, 107–113.

Dunbar, R. I. M. (1988). *Primate Social Systems*. London: Croom Helm.

Dunbar, R. I. M. (1992). Neocortex Size as a Constraint on Group Size in Primates. *Journal of Human Evolution, 22*, 469-493.

Dunbar, R. I. M. (1998). The Social Brain Hypothesis. *Evolutionary Anthropology, 6,* 178–190. Egnor, S. E. R., Iguina, C., & Hauser, M.D. (2006). Perturbation of Auditory Feedback Causes Systematic Perturbation in Vocal Structure in Adult Cotton-Top Tamarins. *Journal of Experimental Biology, 209*, 3652–3663.

Elowson, A. M., & Snowdon, C.T. (1994). Pygmy Marmosets, *Cebuella pygmaea*, Modify Vocal Structure in Response to Changed Social Environment. *Animal Behaviour, 47*, 1267–1277.

Elowson, A. M., Tannenbaum, P. L., & Snowdon, C. T. (1991). Food-Associated Calls Correlate with Food Preferences in Cotton-Top Tamarins. *Animal Behaviour, 42,* 931–937.

Fairbanks, L. (1975). Communication of Food Quality in Captive *Macaca nemestrina* and Free-Ranging *Ateles geoffroyi*. *Primates, 16*, 181-190.

Fragaszy, D. M., & Visalberghi, E. (1989). Social Influences on the Aquisition and Use of Tools inTufted Capuchin Monkeys (*Cebus apella*). *Journal of Comparative Psychology, 103*, 159–170.

Fragaszy, D. M., & Visalberghi, E. (1996). Social Learning in Monkeys: Primate 'Primacy' Reconsidered. In C. M Heyes, & B. G Galef, Jr. (Eds.), *Social Learning in Animals: The Roots of Culture* (pp. 65–84). London: Academic Press.

Friant, S. C., Campbell, M. A., & Snowdon, C. T. (2008). Captive-Born Cotton-Top Tamarins (*Saguinus oedipus*) Respond Similarly to Vocalizations of Predators and Non-Predators. *American Journal of Primatology, 70*, 707–710.

Fruteau, C., Voelkl, B., van Damme, E., & Noë, R. (2009). Supply and Demand Determine the Market Value of Food Providers in Wild Vervet Monkeys. *Proceedings of The National Academy of Sciences, 106,* 12007–12012.

Fujita, K., Kuroshima, H., & Masuda, T. (2002). Do Tufted Capuchin Monkeys (*Cebus apella*) Spontaneously Deceive Opponents? A Preliminary Analysis of an Experimental Food- Competition Contest between Monkeys. *Animal Cognition, 5*, 19-25.

Galef, B. G., Jr. (1992). The Question of Animal Culture. *Human Nature, 3*, 157–178.

Garber, P. A., Rosenberger, A. L., & Norconk, M. A. (1995). Marmoset Misconceptions. In: Norconk, M.A, Rosenberger, A. L., & Garber, P. A. Editors. *Adaptive Radiations of Neotropical Primates.* New York: Plenum Press.

Gros-Louis, J. (2004). The Function of Food Associated Calls in White Faced Capuchins, *Cebus capucinus*, from the Perspective of the Signaler. *Animal Behaviour, 67*, 431–440.

Gumert, M. D. (2007). Payment for Sex in a Macaque Mating Market. *Animal Behaviour, 74*, 1655–1667. Hare, B., Addessi, E., Call, J., Tomasello, M., & Visalberghi, E. (2003). Do Capuchin Monkeys (*Cebus qpella*) Know What Conspecifics Do and Do Not See? *Animal Behaviour, 65*, 131-142.

Hare, B., Call, J., Agnetta, B., & Tomasello, M. (2000). Chimpanzees Know What Conspecifics Do and Do Not See. *Animal Behaviour, 59*, 771–785.

Hare, B., Melis, A., Woods, V., Hastings, S., & Wrangham, R. (2007). Tolerance Allows Bonobos to Outperform Chimpanzees in a Cooperative Task. *Current Biology, 17*, 619–623.

Hattori, Y., Kuroshima, H., & Fujita, K. (2005). Cooperative Problem Solving by Tufted Capuchin Monkeys (*Cebus apella*): Spontaneous Division of Labor, Communication, and Reciprocal Altruism. *Journal of Comparative Psychology, 119*, 335-342.

Hattori, Y., Kuroshima, H., & Fujita, K. (2007). I Know You Are Not Looking at Me: Capuchin

Monkeys' (*Cebus apella*) Sensitivity to Human Attentional States. *Animal Cognition, 10*, 141–148.

Hattori, Y., Kuroshima, H., & Fujita, K. (2010). Tufted Capuchin Monkeys (*Cebus apella*) Show Understanding of Human Attentional States When Requesting Food Held by a Human. *Animal Cognition, 13*, 87–92.

Hayes, S. L., & Snowdon, C.T. (1990). Predator Recognition in Cotton-Top Tamarins (*Saguinus oedipus*). *American Journal of Primatology, 20*, 283–291.

Horner, V., Cartera, J. D., Suchak, M., & De Waal, F. B. M. (2011). Spontaneous Prosocial Choice by Chimpanzees. *Proceedings of the National Academy of Sciences, 108*, 13847–13851.

Humle, T., & Snowdon, C. T. (2008). Socially Biased Learning in the Acquisition of a Complex Foraging Task in Juvenile Cottontop Tamarins (*Saguinus oedipus*). *Animal Behaviour, 75*, 267–277.

Humphrey, N. K. (1976). The Social Function of Intellect. In: P. Bateson, & R. A. Hinde (Eds.), *Growing Points in Ethology* (pp. 307-317). Cambridge: Cambridge University Press.

Itakura, S. (1996). An Exploratory Study of Gaze-Monitoring in Non-Human Primates. *Japanese Psychological Research, 38*, 174–180.

Kaplan, J. N., Winship-Ball, A., & Sim, L. (1978). Maternal Discrimination of Infant Vocalizations in Squirrel Monkeys. *Primates, 19*, 187-193.

Kudo, H., & Dunbar, R. I. M. (2001). Neocortex and Social Network Size in Primates. *Animal Behaviour, 62*, 711–722.

Lakshminarayanan, V., & Santos, L. R. (2008). Capuchin Monkeys Are Sensitive to Others' Welfare. *Current Biology, 18,* R999-R1000.

Lefebvre, L. (1995). Culturally Transmitted Feeding Behaviour in Primates: Evidence for Accelerating Learning Rates. *Primates, 36*, 227–239.

Liebal, K., Mueller, C., & Pika, S. (2007). *Gestural Communication in Nonhuman and Human Primates.* John Benjamins Publishing Co, Amsterdam.

Masataka, N. (1986). Rudimentary Representational Vocal Signalling of Fellow Group Members in Spider Monkeys. *Behaviour, 96,* 49-61.

Mendres, K. A., & de Waal, F. B. M. (2000). Capuchins Do Cooperate: The Advantage of an Intuitive Task. *Animal Behaviour, 60*, 523-529.

Miller, C. T., Flusberg, S., & Hauser, M. D. (2003). Interruptibility of Long Call Production in Tamarins: Implications for Vocal Control. *Journal of Experimental Biology, 206*, 2629–2639.

Milton, K. (1988). Foraging Behaviour and the Evolution of Primate Intelligence. In: R. Byrne, & A. Whiten (Eds.), *Machiavellian Intelligence* (pp. 285-305). Oxford: Oxford University Press.

Mitchell, R. W., & Anderson, J. R. (1997). Pointing, Withholding Information, and Deception in Capuchin Monkeys (*Cebus apella*). *Journal of Comparative Psychology, 111*, 351–361.

Neiworth, J. J., Burman, M. A., Basile, B. M., & Lickteig, M. T. (2002). Use of Experimenter-Given Cues in Visual Co-Orienting and in an Object-Choice Task by a New World Monkey Species, Cotton Top Tamarins (*Saguinus oedipus*). *Journal of Comparativa Psychology, 116,* 3–11.

Neiworth, J. J., Hassett, J. M., & Sylvester, C. J. (2007). Face Processing in Humans and New World Monkeys: The Influence of Experiential and Ecological Factors. *Animal Cognition, 10*, 125–134.

Noë, R., & Hammerstein, P. (1994). Biological Markets: Supply and Demand Determine the Effect of Partner Choice in Cooperation, Mutualism and Mating. *Behavioral Ecology and Sociobiology,35*, 1–11.

Noë, R., & Hammerstein, P. (1995). Biological Markets. *Trends in Ecology & Evolution, 10,* 336–339. Perry, S, & Rose, L. (1994). Begging and Transfer of Coati Meat by White-Faced Capuchin Monkeys, *Cebus capucinus. Primates, 35*, 409-415.

Pika, S., Liebal, K., Call, J., & Tomasello, M. (2005). Gestural Communication of Apes. *Gesture, 5,* 41-56.

Pollick, A. S., Gouzoules, H., & de Waal, F. B. M. (2005). Audience Effects on Food Calls in Brown Capuchin Monkeys, *Cebus apella. Animal Behaviour, 70*, 1273–1281.

Povinelli, D. J., & Eddy, T. J. (1996). What Young Chimpanzees Know About Seeing. *Monographs of the Society for Research in Child Development, 61*, 1–152.

Rapaport, L. G. (1999). Provisioning of Young in Golden Lion Tamarins (Callitrichidae, *Leontopithecus rosalia*): A Test of the Information Hypothesis. *Ethology, 105,* 619–636.

Rapaport, L. G. (2006). Provisioning in Wild Golden Lion Tamarins (*Leontopithecus rosalia*): Benefits to Omnivorous Young. *Behavioural Ecology, 17*, 212–221.

Rapaport, L. G., & Ruiz-Miranda, C. R. (2006). Ontogeny of Provisioning in Two Populations of Wild Golden Lion Tamarins (*Leontopithecus rosalia*). *Behavioural Ecology and Sociobiology, 60*, 724–735.

Rekers, Y., Haun, D., & Tomasello, M. (2011). Children, but Not Chimpanzees, Prefer to Collaborate. *Current Biology, 21*, 1756-1758.

Rose, L. M. (1997). Vertebrate Predation and Food-Sharing in Cebus and Pan. *International Journal of Primatology, 18*, 727-765.

Roush, R. S., & Snowdon, C. T. (1994). Ontogeny of Food Associated Calls in Cotton-Top Tamarins. *Animal Behaviour, 47*, 263–273.

Roush, R. S., & Snowdon, C. T. (1999). The Effects of Social Status on Food Associated Calls in Captive Cotton-Top Tamarins. *Animal Behaviour, 58,* 1299–1305.

Roush, R. S., & Snowdon, C. T. (2000). Quality, Quantity, Distribution and Audience Effects on Food Calling in Cotton-Top Tamarins. *Ethology, 106*, 673–690.

Ruckstalis, M., Fite, J. E., & French, J. A. (2003). Social Change Affects Vocal Structure in a Callitrichid Primate (*Callithrix kuhlii*). *Ethology, 109*, 327–340.

Russon, A. E. (1997). Exploiting the Expertise of Others. In Whiten, A., & Byrne, R. W. (Eds.), *Machiavellian Intelligence II: Evaluations and Extensions* (pp. 174–206). Cambridge: Cambridge University Press.

Sabbatini, G., de Bortoli Vizioli, A., Visalberghi, E., & Schino, G. (2012). Food Transfers in Capuchin Monkeys: An Experiment on Partner Choice. *Biology Letters, 8,* 757-759.

van Schaik, C. P. (1983). Why Are Diurnal Primates Living in Groups? *Behaviour, 87*, 120–144. van Schaik, C. P., & Janson, C. (1988). Recognising the Many Faces of Primate Food Competition. *Behaviour, 105*, 165–186.

van Schaik, C.P., & Kappeler, P. (1997). Infanticide Risk and the Evolution of Permanent Male–Female Association in Nonhuman Primates: A New Hypothesis and Comparative Test. *Proceedings of the Royal Society of London, Series B, Biological Sciences, 264*, 1687–1694.

Schino, G., & Aureli, F. (2010). Primate Reciprocity and Its Cognitive Requirements. *Evolutionary Anthropology, 19,* 130-135.

Schino, G., Ventura, R., & Troisi, A. (2003). Grooming among Female Japanese Macaques: Distinguishing between Reciprocation and Interchange. *Behavioral Ecology, 14*, 887–891.

Shettleworth, S. (2001). Animal Cognition and Animal Behaviour. *Animal Behaviour, 61*, 277-286.

Silk, J. B. (2007). The Adaptive Value of Sociality in Mammalian Groups. *Philosophical Transactions of the Royal Society of London, Series B, 362,* 539-559.

Skerry, A. E., Sheskin, M., & Santos, L. R. (2011). Capuchin Monkeys Are Not Prosocial in an Instrumental Helping Task. *Animal Cognition, 14*, 647–654.

Slater, K., Schaffner, C. M., & Aureli, F. (2007). Embraces for Infant Handling in Spider Monkeys: Evidence for a Biological Market? *Animal Behaviour, 74*, 455-461.

Snowdon, C. T. (2001). Social Processes in Communication and Cognition in Callitrichid Monkeys: A Review. *Animal Cognition, 4,* 247–257.

Snowdon, C. T. (2009). Plasticity of Communication in Nonhuman Primates. *Advances in the Study of Behavior, 40*, 239-276.

Snowdon, C. T., & Boe, C.Y. (2003). Social Communication about Unpalatable Foods in Tamarins (*Saguinus oedipus*). *Journal of Comparative Psychology, 117*, 142–148.

Snowdon, C. T., & Cleveland, J. (1980). Individual Recognition of Contact Calls by Pygmy Marmosets. *Animal Behaviour, 28*, 717–727.

Snowdon, C. T., & Elowson, A. M. (2001). 'Babbling' in Pygmy Marmosets: Development after Infancy. *Behaviour, 138*, 1235–1248.

Snowdon, C. T., & Hodun, A. (1981). Acoustic Adaptations in Pygmy Marmoset Contact Calls: Locational Cues Vary with Distances between Conspecifics. *Behavioural Ecology and Sociobiology, 9*, 295–300.

Stevens, J. R. (2010). Donor Payoffs and Other-Regarding Preferences in Cotton-Top Tamarins (*Saguinus oedipus*). *Animal Cognition, 13,* 663–670.

Takimoto, A., Kuroshima, H., & Fujita, K. (2010). Capuchin Monkeys (*Cebus apella*) Are Sensitive to Others' Reward: An Experimental Analysis of Food-Choice for Conspecifics. *Animal Cognition, 13*, 249–261.

Tebbich, S., Seed, A. M., Emery, N. J., & Clayton, N. S. (2007). Non-Tool-Using Rooks, *Corvus Frugilegus*, Solve the Trap-Tube Problem. *Animal Cognition, 10*, 225-231.

Tiddi, B., Aureli, F., Polizzi di Sorrentino, E., Janson, C. H., & Schino, G. (2011). Grooming for Tolerance? Two Mechanisms of Exchange in Wild Tufted Capuchin Monkeys. *Behavioral Ecology, 22*, 663-669.

Tiddi, B., Aureli, F., & Schino, G. (2010) Grooming for Infant Handling in Tufted Capuchin Monkeys: A Reappraisal of the Primate Infant Market. *Animal Behaviour, 79,* 1115–1123.

Tomasello, M., & Call, J. (1997). *Primate Cognition*. New York, NY, USA: Oxford University Press. Tomasello, M., Call, J., & Hare, B. (1998). Five Primate Species Follow the Visual Gaze of Conspecifics. *Animal Behaviour, 55*, 1063–1069.

Ventura, R., Majolo, B., Koyama, N. F., Hardie, S., & Schino, G. (2006). Reciprocation and Interchange in Wild Japanese Macaques: Grooming, Cofeeding, and Agonistic Support. *American Journal of Primatology, 68*, 1138–1149.

Visalberghi, E. (1987). Acquisition of Nut-Cracking Behaviour by 2 Capuchin Monkeys (*Cebus apella*). *Folia Primatologica, 49,* 168–181.

Visalberghi, E. (1994). Learning Processes and Feeding Behaivour in Monkeys. In B. G. Galef, M. Mainardi, & P. Valsecchi (Eds.), *Behavioral Aspects of Feeding* (pp. 257–270). Chur, Switzerland: Harwood Academic.

Visalberghi, E., & Addessi, E. (2000). Response to Changes in Food Palatability in Tufted Capuchin Monkeys, *Cebus apella. Animal Behaviour, 59*, 231–238.

Visalberghi, E., & Fragaszy, D. (1990). Do Monkeys Ape? In S. Parker & K. Gibson (Eds.), *Language and Intelligence in Monkeys and Apes: Comparative Developmental Perspectives* (pp. 247–273). Cambridge, U.K: Cambridge University Press.

Visalberghi, E., Quarantotti, B. P., & Tranchida, F. (2000). Solving a Cooperation Task without Taking into Account the Partner's Behavior: The Case of Capuchin Monkeys (*Cebus apella*). *Journal of Comparative Psychology, 114*, 297-301.

de Waal, F. B. M. (2000). Attitudinal Reciprocity in Food Sharing among Brown Capuchin Monkeys. *Animal Behaviour, 60*, 253–261.

de Waal, F. B. M, & Berger, M. L. (2000). Payment for Labour in Monkeys. *Nature, 404*, 563.

de Waal, F. B. M., & Davis, J. M. (2003). Capuchin Cognitive Ecology: Cooperation Based on Projected Returns. *Neuropsychologia. 41*, 221-228.

de Waal, F. B. M., Leimgruber, K., & Greenberg, A. R. (2008). Giving Is Self-Rewarding for Monkeys. *Proceedings of the National Academy of Sciences, 105,* 13685–13689.

Wardenich, D, & Huber, L. (2002). Social Factors Determine Cooperation in Marmosets. *Animal Behaviour, 64*, 771-781.

Whiten, A. (1989). Transmission Processes in Primate Cultural Evolution. *Trends in Ecology and Evolution, 4*, 61–62.

Whiten, A. (2000). Primate Culture and Social Learning. *Cognitive Science, 24,* 477-508.

Whiten, A, & Byrne, R. W. (1988). *Machiavellian Intelligence. Social Expertise and the Evolution of Intellect in Monkeys, Apes and Humans.* New York: Oxford University Press.

Whiten, A., & Ham, R. (1992). On the Nature and Evolution of Imitation in the Animal Kingdom: Reappraisal of a Century of Research. In P. J. B. Slater, J. S. Rosenblatt, C. Beer, & M. Milinski (Eds.), *Advances in the Study of Behavior* (pp. 239–283). San Diego: Academic Press.

Wilson, E. O. (1975). *Sociobiology*. Cambridge/MA: Harvard University Press.

Yamamoto, S., & Takamoto, A. (2012). Empathy and Fairness: Psychological Mechanisms for Eliciting and Maintaining Prosociality and Cooperation in Primates. *Social Justice Research, 25*, 233–255.

Reviewed by Bonaventura Majolo, School of Psychology, University of Lincoln, Lincoln, U.K.

In: Monkeys
Editor: Hugo A. Barrera Saldaña
ISBN: 978-1-63117-851-1

Chapter 5

MONKEY PHYSICAL COGNITION: A COMPARATIVE REVIEW

Amici Federica [*] ***and Bräuer Juliane***
Department of Developmental and Comparative Psychology,
Max Planck Institute for Evolutionary Anthropology, Leipzig, Germany

ABSTRACT

In this chapter, we review the most important findings on physical cognition in New World monkeys. We first provide a theoretical background to frame the results of the studies that have been recently developed, by briefly introducing the different socio-ecological factors that might exert an important effect on the evolution of cognitive skills. Secondly, we discuss the different paradigms that have been used to test physical cognitive skills in captivity and partially in the wild. Thirdly, we put together the results of all the reported studies to obtain a more general picture of cognition across monkey species and therefore hint to the evolutionary factors that might have shaped their specific cognitive skills. Finally, we highlight the most controversial aspects of comparative cognition in monkeys and indicate possible directions for future research.

INTRODUCTION

In the last century, a large number of studies have experimentally investigated cognition in primates, providing us with a vast dataset on the distribution of specific cognitive skills across taxa (e.g., Tomasello & Call, 1997; Tomasello, 2000). Initially, the great majority of cognitive studies have been mainly conducted on very few species and mostly on chimpanzees, *Pan troglodytes* (Call, 2003). However, much effort has been recently made to also include other species which had been previously underrepresented in studies on comparative cognition, like New World monkeys. Today, the number of articles published in the area of animal cognition is so high, that it is hard to even keep up with all of them.

[*] Corresponding Author: abbepu@yahoo.it.

Through decades of experimental data and behavioural observations, monkeys have indeed unequivocally shown to be able to adjust their behaviour depending on the external situation and on the goal to be reached, and to use their previous knowledge before making complex decisions. In other words, monkeys are clearly able to behave flexibly, have mental representations and make complex decisions, using cognitive strategies to deal with the socio-ecological environment surrounding them (Tomasello & Call, 1997). Today much more information is available on the cognitive performance of different primate species. Thus, it is possible to try and reconstruct the evolutionary origins of specific cognitive skills and better understand their distribution across different taxa.

The socio-ecological environment in which all monkeys live is complex. Monkeys need to understand properties of the objects surrounding them in order to forage efficiently and they need to relate with their conspecifics, given that they are social animals. Therefore, it makes sense to predict that all monkey species will share some basic physical and social cognitive skills that allow them to understand spatial, causal and featural properties of objects, and to understand, predict and possibly manipulate their partners' behavior. The basic question is to what extent these cognitive skills are distributed across monkey species and whether this distribution can be better explained in terms of specific socio-ecological demands that the species are facing.

Evolutionary Hypotheses Predicting Inter-specific Differences in Cognitive Performance

There are three main kind of hypotheses that may explain inter specific differences in cognitive tasks.

1) The Ecological hypotheses consider specific characteristics of foraging strategies as the main selective pressure for the evolution of enhanced cognitive skills. Firstly, some authors have claimed that larger home ranges might require more sophisticated spatial memory and navigational ability (e.g., Clutton-Brock & Harvey, 1980; Milton, 1981).

Secondly, a higher degree of frugivory might be linked to enhanced cognitive skills, since sparsely and patchy distributed food items would call for an extra processing capacity to predict when and where to forage (e.g., Clutton-Brock & Harvey, 1980; Milton, 1981).

Thirdly, other authors have claimed that extractive foraging (i.e., the ability to process plants, open nuts, dig insects and use tools to access embedded food) is linked to higher hand-coordination, hierarchical coordination, stimulus enhancement, social facilitation and possibly social learning (e.g., Parker & Gibson, 1977). As a consequence, brain size and tool use ability would be causally linked (Reader & Laland, 2002).

Finally, a higher percentage of gum in the diet might be linked to the enhancement of specific cognitive skills, like inhibition, which would be needed to patiently wait for gum slowly exuding from tree barks (e.g., Stevens et al., 2005).

2) The Social hypotheses consider social complexity as the main selective pressure for the evolution of enhanced cognitive skills. When social life is complex, individuals need to use flexible cognitive strategies to recognize others, keep track of their relationships, predict, coordinate and manipulate their behaviours (e.g., Humphrey, 1976; Byrne & Whiten, 1988;

Dunbar & Shultz, 2007). Because there is no real agreement on how social complexity should be defined, different social measures have been used as indexes of social complexity.

Firstly, some authors have proposed social group size as an index of social complexity linked to enhanced cognitive skills (e.g. Dunbar, 1992). Social group size, however, does not necessarily correspond to the number of individuals within a group, but rather to the mean number of partners with whom individuals socially interact through time (i.e., to the "effective group size", see Pradhan et al., 2012). Nonetheless, both group size and effective group size might be a crude measure of social complexity and it has therefore been proposed that social complexity rather depends on the frequency and quality of social interactions (e.g., Kudo & Dunbar, 2001). More explicitly, it would not be species living in larger social groups showing enhanced cognitive skills, but species living in social groups in which grooming, social play and coalitions happen with a higher frequency. Although all these behaviours show some correlation with relative neocortex size, which is assumed to be one of the executive substrates for higher cognitive skills (e.g., Barton, 1999; Barton & Harvey, 2000), it is not yet clear how all these measures are linked to each other or how they differentially contribute to the complexity of social interactions. An attempt to combine different behaviours in uncorrelated measures through principle component analyses, for example, has shown that the quality of social interactions cannot simply be measured in terms of frequency of single behaviours, but can be estimated by using independent components (Fraser et al., 2008).

Furthermore, social complexity might be linked to dominance style, with more tolerant species interacting more freely and with a larger variation of social tactics (e.g. McFarland & Majolo, 2011; Amici et al., 2012a). In less tolerant species high within-group competition might favor niche construction, with high- and low-ranking individuals facing different environmental conditions and possibly developing different cognitive skills (Boesch, 2007; see also Hare et al., 2000; Bräuer et al., 2007). In particular, subordinates need to especially monitor and pay attention to the dominants (Chance, 1967; Emery, 2000). Subordinates gaze towards high-ranking individuals more often than vice versa to obtain information about the location of the dominant in order to avoid being displaced in agonistic encounters (McNelis & Boatright-Horowitz, 1998; Blois-Heulin & Girona, 1999). Low-ranking individuals can suffer from strong retaliations by higher-ranking group members during food or reproductive competition. They might therefore benefit from enhanced inhibitory skills in order not to engage in overtly competing behaviours with higher-ranking counterparts (e.g., Genty et al., 2004).

Cooperative breeding systems (i.e. those in which not only parents take care of infants) are not more socially complex than other systems, but they might be linked to psychological and motivational changes leading to greater prosociality and social tolerance, similarly to what happens in species with a more relaxed dominance style. Enhanced prosociality and social tolerance, in turn, would facilitate individuals' attention to conspecifics' behaviour, social learning and cooperative behaviour, resulting in individuals performing generally better in social (but not physical) cognition (e.g., Burkart et al., 2009). The higher the amount of time group members invest in caring for others' infants, the better would cognitive performance be, at least in the social field.

Other researchers have proposed that monogamy is linked to social complexity, in that species engaging in monogamous pair-bonding need to establish and coordinate complex behaviours with the partner (Dunbar, 2009). Although Dunbar's original hypothesis mainly referred to monogamy in terms of sexual behaviour, it seems reasonable to extend this hypothesis and predict that social complexity (and thus cognitive skills) will be higher in those species in which individuals preferentially involve in social interactions with only one or few partners, rather than in species in which social interactions are more evenly distributed among all group members. Monogamous individuals do indeed invest much in complexly coordinating behaviour with one extremely valuable partner.

Finally, social complexity has been defined in terms of fission-fusion dynamics: individuals living in groups frequently splitting in subgroups of varying size and composition might need specific enhanced cognitive skills to face specific cognitive challenges. These include enhanced memory to remember absent individuals, enhanced analogical skills to better deal with fragmentary information about group members and enhanced inhibitory skills to adapt to continuous changes in group size and composition (e.g., Barrett et al., 2003; Aureli et al., 2008; Amici et al., 2008, 2009, 2010).

3) The Phylogenetic hypothesis, finally, predicts that species sharing a long evolutionary history will show a similar cognitive performance, independently of their socio-ecological characteristics (e.g., Parker, 1991). Phylogenetic methods therefore take into account the species' evolutionary history when testing whether cognitive skills correlate with socio-ecological characteristics, and they provide information on the extent to which phylogenetic relatedness predicts the distribution of cognitive skills across species (MacLean et al., 2012).

Obviously, the list of hypotheses presented here is not exhaustive, and none of the socio-ecological hypotheses excludes the possibility that different selective pressures independently but simultaneously act on different cognitive skills (e.g., Dunbar, 1992).

New World monkeys allow testing the different evolutionary hypotheses, as the different species live in complex environments and dramatically vary in their socio-ecological characteristics. They inhabit different biomes, feed on different food items and present very different social structures (Smuts et al., 1987). While common marmosets (*Callithrix jacchus*) consume a high percentage of gum in their diets and are cooperative breeders, for instance, spider monkeys (*Ateles geoffroyi*) are mainly frugivorous and have very high levels of fission- fusion dynamics (Smuts et al., 1987). While white-headed capuchin monkeys (*Cebus capucinus*) live in large groups and mate promiscuously, common marmosets are monogamous and live in smaller groups. While black howler monkeys (*Alouatta pigra*) are largely folivorous and have smaller home ranges, brown capuchin monkeys (*Cebus apella*) are extractive foragers able to use tools to open nuts. Studying cognition in New World monkeys can therefore tell us a lot not only on the single species' cognitive abilities, but also on the species-specific socio-ecological demands which might have been the main selective pressure for the evolution of enhanced cognitive skills.

MONKEY PHYSICAL COGNITION

Studies on animal cognition can be roughly divided in those focusing on physical cognition and those focusing on social cognition. Physical cognition can be defined as knowledge of the physical world (see Tomasello & Call, 1997). The distinction is of course artificial, as animals may need to use physical skills also when solving problems in the social arena, or vice-versa, and the intertwinement of different cognitive skills being required to solve a task is not so easy to disentangle (e.g., Tebbich et al., 2007). However, it is generally accepted that animal cognitive skills are divided in the physical and the social domain, containing different cognitive skills more or less connected to each other (see Deaner et al., 2006; Amici et al., 2012b).

Below, we will report the most important studies that have been done on New World monkey physical cognition. The review does not aim to be exhaustive, as priority has been given to more recent, significant studies and to controlled experimental paradigms rather than behavioural observations. Moreover, studies using invasive paradigms have been excluded for ethical reasons.

UNDERSTANDING OBJECTS

One of the most basic cognitive skills is the ability to locate hidden objects and infer the location of objects that have moved out of view. Therefor animals have to understand object permanence, i.e., that objects continue to exist even when they cannot be observed. This skill is for instance necessary to forage, by allowing individuals to locate hidden food resources and keep track of their movements.

According to Piaget, who first investigated object permanence in human infants, there are six stages of object permanence (Piaget, 1954). Stage 4 of object permanence, for example, implies that individuals understand that objects are separate entities that continue to exist even out of sight. It is often tested by observing whether a subject retrieves the food that has been placed under one of multiple opaque containers while the subject observed the baiting procedure. Subjects completing a stage 5 are also able to solve visible displacements, finding the food that had been hidden under one opaque container and then moved to a new one in full view of the subject. Subjects completing a stage 6 are able to also solve invisible displacements, in which food is hidden under an opaque container and then the container is moved behind an opaque screen in full view of the subject. The food is then removed from the container when behind the opaque screen and out of the subject's view, and the subject is shown the empty container to infer the new location of the object (Piaget, 1954). Another task tracking the displacement of invisible objects is the transposition task, in which a food reward is placed under one of several opaque containers while the subjects is watching and then the containers are shuffled (Sophian, 1984).

Studies conducted on squirrel monkeys (*Saimiri sciureus*), brown capuchin monkeys (*Cebus apella*) and common marmosets (*Callithrix jacchus*) suggest that New World monkeys generally understand visible displacements typical of stage 4 of Piagetian object permanence (Dumas & Brunet, 1994; de Blois et al., 1998; Mendes & Huber, 2004). In more complex tasks typical of stage 5 or 6, New World monkeys do not excel and seem to use trial-and-error learning strategies or choose randomly, like 8-month-old children do and worse than apes (e.g., Call, 2001; Neiworth et al., 2003; Mendes & Huber, 2004; Anderson, 2012).

In particular, squirrel monkeys seem to only master stage 5 (Vaughter et al., 1972; de Blois et al., 1998), while brown capuchin monkeys, white-faced capuchins (*Cebus capucinus*), yellow-tailed woolly monkeys (*Lagothrica flavicauda*) and common marmosets would also complete a stage 6 (Mathieu et al., 1976; Schino et al., 1990; Mendes & Huber, 2004). However, subjects in these studies might have simply solved the task through associative learning, given the large amount of trials they received (Mathieu et al., 1976; Schino et al., 1990).

Interestingly, spider monkeys (*Ateles geoffroyi*) perform similarly to the great apes and better than brown capuchin monkeys in stage 5 tasks (Amici et al., 2008). However, tasks used to test stage 5 of object permanence also require inhibitory skills to be solved, since subjects need to refrain from choosing the container under which they previously found food but that has been recently emptied. As a consequence, the authors proposed that the different performance of spider monkeys and capuchin monkeys might be linked to inter-specific differences in terms of inhibition rather than a lack of understanding of object permanence *per se* (Amici et al., 2008).

Very little information is instead available on the performance of monkeys in transposition tasks, which seem harder to master than the traditional stage-6 invisible displacement task (Barth & Call, 2006; Rooijakkers et al., 2009). Brown capuchin monkeys perform above chance level only when the baited container does not switch its position with any of the other unbaited containers, while spider monkeys also perform above chance level, and better than brown capuchin monkeys, when the baited container switches its position only once with another unbaited container (Amici et al., 2010). However, it is worth noting that New World monkeys generally performed worse than apes in this task (Amici et al., 2010).

SPATIAL COGNITIVE SKILLS

Another important cognitive skill is the ability to represent and integrate spatial and ecological information, in order to develop a set of decision rules to for example locate productive feeding sites efficiently (Garber et al., 2009). Since food resources vary in their spatio-temporal distribution, individuals need to retain, integrate, and update information on the location and phenological patterns of potential food sources (Garber et al., 2009; Cheng, 2010; Deloache & Pickard, 2010; Porter & Garber, 2013). In general, all species can be expected to have spatial cognitive skills to a certain degree, as all species need to move in their environment and forage. However, it is also predictable that spatial cognitive skills vary depending on the socio- ecological requirements of a species. Highly frugivorous species, for example, need to remember the location of fruit trees with ripe fruit and need to locate them in the environment (Clutton- Brock & Harvey, 1980; Milton, 1981). Moreover, their home ranges are usually larger than folivorous species and require more sophisticated navigational skills (Clutton-Brock & Harvey, 1980; Milton, 1981).

So far, two general models have been proposed to explain how animals navigate between distant out of view sites in their environment. Firstly, using route-based mental maps implies that individuals habitually use familiar travel routes and landmarks as orientation or switch points to navigate, being able to generate novel routes but not to take the most direct route to reach their target (Garber, 2000; di Fiore & Suarez, 2007; Dolins, 2009; Dolins & Mitchell, 2010). Secondly, using coordinate-based mental maps implies that individuals encode and recall locations as coordinates, computing distances and directions in an accurate and more complex way, using direct routes and novel shortcuts like chimpanzees (Poucet, 1993; Normand et al., 2009).

The majority of New World monkeys seem to use a combination of these two strategies to navigate in different spatial scales. They do not forage randomly, but they decide where to forage based on the minimum distance to the nearest tree whose fruit they have not yet depleted (Garber & Hannon, 1993). Black-horned capuchins (*Cebus nigritus*), white-faced capuchins and spider monkeys use a combination of coordinate-based and route-based maps (Valero & Byrne, 2007; Urbani, 2009; Presotto & Izar, 2010; Porter & Garber, 2013). Tamarin species (*Saguinus mystax*, *S. fuscicollis nigrifrons* and *S. f. weddelli*), spider monkeys, black-horned capuchins and white-faced capuchins select feeding sites without travelling in straight lines between sequential feeding sites, but they reduce overall path length by maintaining some form of internal representation of the tree locations (Garber, 1989; Garber & Hannon, 1993; Valero & Byrne, 2007; Janson, 2007; Urbani, 2009; Porter & Garber, 2013). Tamarins, for example, navigate in large-scale space using a route-based spatial representation and integrate different types of spatial and non-spatial information in foraging decisions, including location and productivity of feeding sites and phenological state of future ones (Garber & Dolins, 1996; Garber, 2000; Deipolyi et al., 2001; Dolins, 2009; Garber et al., 2009; Porter & Garber, 2013; Garber et al., 2009).

Memory Skills

Remembering object locations can be useful when foraging, in order to know the locations in which food has been already exploited and those in which it has been recently seen (Milton, 1988; Call, 2000). Moreover, memory skills can be useful to locate conspecifics and to maintain social relationships, retain knowledge about them and refrain from continuously renegotiating long-term relationships (Fiset & Doré, 2006; Aureli et al., 2008; Amici et al., 2010).

The classical way to assess whether subjects remember the location of objects consists of hiding a food reward under one of several opaque containers and let the subject retrieve it after a varying delay of time. Numerous studies have shown that primates can remember the location of food rewards after different delays of time, but most of these studies did not include New World monkeys or only tested species with different paradigms, allowing no comparative conclusions (e.g. MacDonald et al., 1994; Barth & Call, 2006).

In a recent study, however, Amici and colleagues (2010) tested brown capuchin monkeys and spider monkeys by hiding food rewards under one of three opaque containers. Subjects were allowed to retrieve the food after 30 seconds or 30 minutes. Brown capuchin monkeys performed above chance level only with a delay of 30 seconds, while spider monkeys performed above chance level in both conditions and also performed significantly better than brown capuchin monkeys in the 30-second condition. The authors hypothesized that inter-specific differences in performance reflected the distribution of fission-fusion dynamics among monkey species (Amici et al., 2010).

Working memory, instead, refers to a system enabling the temporary storage and manipulation of information. That information can be continuously, repeatedly reactivated, protecting it not only from passive, temporal decay, but also from the retroactive interference produced by other distracting stimuli (Baddeley, 2003; Saults & Cowan, 2007; Scott et al., 2012). Since working memory is required for the active maintenance and manipulation of information in the face of concurrent processing and distraction (Conway & Getz, 2010), non-invasive tasks in primates can be done by simultaneously presenting them with a memory task and a distracting task.

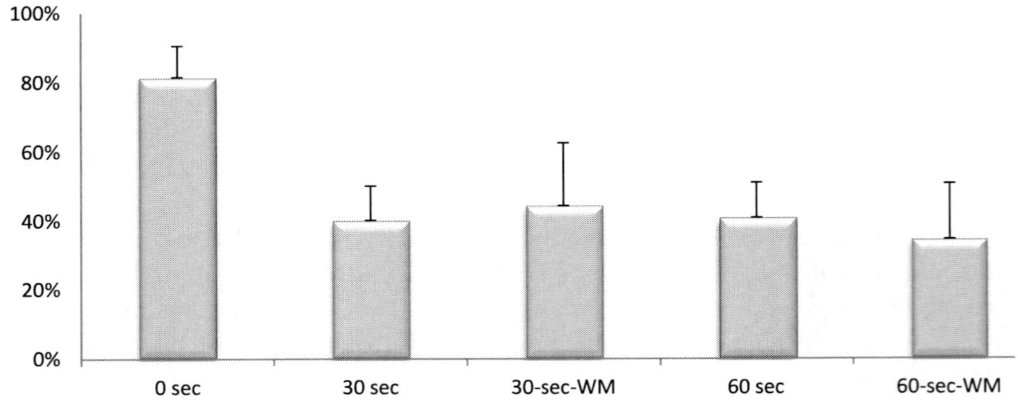

Figure 1. For spider monkeys, mean percentage (+SE) of correct trials in each condition of the working memory task (no delay, 30-second delay with no distracting task, 30-second delay with distracting task, 60-second delay with no distracting task and 60-second delay with distracting task).

Spider monkeys, for example, have been tested by presenting them with five different opaque containers, one of which had been baited in full view of the subject (Amici, unpublished data). After a varying delay of time (30 seconds or 60 seconds), spider monkeys were given the chance to retrieve the baited container. In the working memory condition (but not in the memory condition), subjects were also presented with a distracting task after the baiting procedure before they could retrieve the food. With this set-up, spider monkeys performed above chance level (20%) in the working memory condition, retrieving the right container in 44% and 34% of the trials when the delay was respectively 30 and 60 seconds (Figure 1). Their performance did not significantly differ when the distracting task was administered (i.e. between memory and working memory conditions), but spider monkeys' performance decreased to chance levels in the distracting task.

Spider monkeys have also been tested with a different paradigm (Hanus et al., 2003; Abramson et al., 2011), in which subjects have been presented with two different food quantities (Amici, unpublished data). The food was prepared by dropping pieces of food one by one into two opaque containers, so that the subjects had to keep track of the amount of food placed in each container, store this memory and subsequently compare the two food quantities mentally, in order to select the larger one (Hanus et al., 2003; Abramson et al., 2011). Preliminary analyses show that spider monkeys performed above chance level (50%), choosing the larger food quantity in 59% of the trials.

Tsujimoto & Sawaguchi (2002) have tested squirrel monkeys and common marmosets with another paradigm. In this task, the experimenter baited one of two cups out of the subject's view and then allowed the subject to select one of the two cups and retrieve the food in case of a correct choice. The subject waited a varying delay of time while the experimenter baited once again the same cup out of the subject's view, and finally was allowed to choose once more one of the two cups. With this paradigm, squirrel monkeys were better than marmosets in making their second choice based on the memory of their previous action (i.e. choosing the same location again if the first choice was right, shifting if it was wrong). The authors hypothesized that the observed difference depended on squirrel monkeys having a larger relative size of the neocortex, probably as a result of their more frugivorous diet

(Tsujimoto & Sawaguchi, 2002). However, because only two species have been tested with the same paradigm, it is difficult to draw reliable conclusions on inter-specific differences of the working memory in New World monkeys.

Some authors argue that the same mechanisms that are responsible for certain memory skills might be also responsible for future planning abilities (Tulving, 1993; Suddendorf & Corballis, 1997; Suddendorf & Busby, 2005). Squirrel monkeys do not only perform well in memory tasks, but they may also anticipate future needs independent from current motivational states. In a study by Naqshbandi & Roberts (2006) a non-thirsty animal had its water bottle removed and could then choose between a smaller and larger quantity of food. Consumption of the food in question induced thirst. When subjects selected the smaller quantity the water bottle was returned sooner than when they selected the larger quantity. Monkeys reversed their baseline preference for the larger quantity of food when the experimental contingencies were introduced. The authors could exclude that monkeys' behaviour was controlled by a food aversion induced by thirst that followed of large amounts of food as subjects preferred the large quantity again when water was available (Naqshbandi & Roberts, 2006). However, this paradigm has been criticized because trial and error learning cannot be ruled out (Shettleworth, 2007; Suddendorf & Corballis, 2008).

SKILLS INHIBITORY

Inhibitory skills allow the suppression of prepotent but ineffective responses (Santos et al., 1999). In this respect, inhibition is a basic aspect of intelligence, by allowing subjects a certain degree of flexibility in their behavioural patterns.

Inhibitory skills have been tested with a variety of paradigms. In most of the tasks, the basic procedure consists of presenting two or more alternatives for the subjects to choose from. In some cases, this entails subjects' retrieving hidden rewards from under three cups, avoiding the cup under which they previously found food but that has been recently emptied (A not B task: Piaget, 1954), or avoiding the empty cup close to a cup from which they have just retrieved food (Middle cup task: Call, 2001). In other cases, it entails subjects refraining from reaching directly for a visible reward to instead take a detour movement through a lateral side (Plexiglas hole task: Amici et al., 2008; Swing door task: Vlamings et al., 2010). Primates have also been tested with different versions of the Delay of gratification task, in which subjects can choose between two different food quantities, a smaller immediately available one, and a larger one which they can only obtain after waiting a varying delay of time (Rosati et al., 2007; Amici et al., 2008). Another task that is used to test inhibitory skills is the Reverse-reward contingency task, in which subjects are presented with two different food quantities and they have to select the smaller one in order to obtain the larger reward (Boysen & Berntson, 1995).

Brown capuchin monkeys and spider monkeys have been tested with a battery of tasks including the A not B, the Middle cup, the Plexiglas hole, the Swing door and one version of the Delay of gratification tasks (Amici et al., 2008). In this study, spider monkeys performed significantly better than brown capuchin monkeys in three of the five inhibition tasks (the Middle cup, the Plexiglas hole and the Delay of gratification tasks). Moreover, spider monkeys generally performed similarly to the three great apes sharing a similar social system, so that the authors explained the results in terms of higher fission-fusion dynamics being linked to a general enhancement of inhibitory skills (Amici et al., 2008). When tested with a similar version of the Swing door task, cotton-top tamarins failed to perform above chance level even after 24 trials, and only mastered the task after a longer training phase (Santos et al., 1999).

With a slightly different version of the Delay of gratification task, Addessi and colleagues (2011) tested some of the same brown capuchin monkeys tested by Amici and colleagues (2008), using 2 versus 6 food rewards (instead of 1 versus 3) and administering only one session a day to each subject (instead of up to 9 sessions). Brown capuchin monkeys tested with the new procedure waited significantly longer, reaching a mean indifference point (the time when monkeys equally valued the two different food amounts) of 81 seconds versus 22 seconds (Addessi et al., 2011). The authors suggested that brown capuchin monkeys' better performance correlated to their proficiency in tool use, which would require individuals to show foresight and deliberate control over one's action, therefore leading them to postpone immediate smaller benefits for larger delayed rewards (Addessi et al., 2011; Evans & Westergaard, 2006). Similarly, some brown capuchin monkeys were successful at exchanging less preferred food rewards for more preferred ones, and the most successful subjects were the ones with more tool use experience (Evans & Westergaard, 2006). However, brown capuchin monkeys do not always succeed on inhibitory tasks. Other studies have reported that they perform worse than other species and show consistent impulsivity (e.g., Pelé et al., 2011).

Stevens et al., (2005a) tested common marmosets and cotton-top tamarins with a version of the Delay of gratification task, and showed that marmosets waited significantly longer to obtain larger food rewards than tamarins. The authors explained this inter-specific difference in terms of feeding ecology, as marmosets extensively feed on gum, which is acquired by waiting for exudate to flow from trees, whereas tamarins mainly feed on insects, which can be better caught through impulsive action (Stevens et al., 2005a). Similarly, marmosets - but not tamarins - inhibited reaching for a small, nearby reward and rather travelled for a larger, distant one (Stevens et al., 2005b).

In another version of the Delay of gratification task, brown capuchin monkeys and to a slightly extent squirrel monkeys stopped the accumulation of food rewards by reaching for them, suggesting difficulties in inhibiting the selection of smaller immediate rewards (Anderson et al., 2010). However, it is hard to understand whether the differences in that study depend on inter- specific or inter-individual differences, as the number of subjects tested was very low. Indeed, important inter-individual differences seem to explain also brown capuchin monkeys' performance in a task in which they could exchange smaller or less preferred food rewards for a larger or better one (Drapier et al., 2005).

Inter-individual differences in performance are relatively common, especially in tasks testing inhibitory skills (e.g. Anderson et al., 2000; Kralik et al., 2002). Several authors have hypothesized that they might depend on subjects' age, rank, personality or previous experience (see Albiach-Serrano et al., 2007). When it comes to inhibitory skills, therefore, it seems that both individual life histories and different socio-ecological factors (tool-using, gum consumption, fission-fusion dynamics) simultaneously affect subjects' performance, and a lot of caution is needed when extending comparisons to species tested with different paradigms (see Addessi et al., 2013). Especially puzzling is the performance of brown capuchin monkeys, which seem to perform very differently when procedures are changed, leading some authors to propose that the paradigms used in most inhibition tasks might simply be not intuitive enough (Bramlett et al., 2012).

The Reverse-reward contingency task has also been often used to test inhibitory skills. However, some authors have proposed that failing to solve this task may not necessarily indicate lack of inhibitory skills, but rather reflect an inability to evaluate and generate alternative innovative responses (Kralik, 2005; Kralik et al., 2002). For example, subjects might inhibit the prepotent response of reaching for the larger food quantity, but still fail to understand that selecting the smaller food quantity would provide them with the larger one. As a consequence, inter-specific differences in performance might not necessarily reflect inter-specific difference in inhibitory skills.

In the original version of the Reverse-reward contingency task, most New World monkeys failed to master the task after many trials (Anderson et al., 2000; Kralik et al., 2002). Consequently, the task was often presented with correction procedures that presumably made the task easier to master. For example the large-or-none condition was implemented in which subjects received the larger quantity in case of choosing the smaller one, but received no food at all in case of choosing the larger quantity (e.g., Silberberg & Fujita, 1996). Other studies used correction trials in which the position of food quantities remained the same until the trial was correctly solved (e.g., Anderson et al., 2000), or they used qualitatively but not quantitatively different food rewards (e.g., Anderson et al., 2008) or symbols instead of food (to minimize the intrinsic incentive of the larger quantity's perceptual features: e.g., Boysen et al., 1999). When these correction procedures were used, several subjects mastered the task or improved their performance, including cotton-top tamarins (Kralik, 2005; Kralik et al., 2002), squirrel monkeys (Anderson et al., 2000; Anderson et al., 2004) and brown capuchin monkeys (Anderson et al., 2008; Addessi & Rossi, 2011). However, the use of correction procedures in some studies makes it difficult to compare the results obtained in the Reverse-reward contingency task by different species, with no evolutionary hypothesis clearly explaining the inter-specific differences found so far and with inter-individual differences being often larger than inter-specific ones (e.g., Kralik, 2005).

UNDERSTANDING OF CAUSALITY AND TOOL USE

Once monkeys have found their food, they need to obtain it and process it. Several primates show morphological adaptations that allow them to obtain and process the food, like the elongated finger used by aye-ayes (*Daubentonia madagascariensis*) to extract insects from tree holes (e.g., Erickson, 1994). Other primates, however, show behavioural adaptations, whose complexity can greatly vary (see Tomasello & Call, 1997). Some species, for instance, habitually use tools to obtain food that would otherwise be unattainable. Investigating the behavioural strategies that primates use to obtain and process food can tell us a lot on their intuitive theories about the physical world. Moreover, tool use and manufacture are also informative on other aspects of cognition (see Seed & Byrne, 2010, for a review). Making tools, for example, requires delaying gratification until the tool is made, while transporting tools to the food site requires some sort of planning (see Emery & Clayton, 2009).

Tool use has been documented in a variety of primates and not simply for foraging purposes. Wild spider monkeys, for example, use tools to scratch (Lindshield & Rodrigues, 2009), while captive golden lion tamarins (*Leontopithecus rosalia*) use tools for grooming (Stoinski & Beck, 2001). In the wild, however, tools are habitually used by only few species, and the only New World monkey which is known to regularly use tools in the wild is brown capuchin monkey (see Tomasello & Call, 1997). Wild bearded capuchin monkeys (*Sapajus libidinosus*), for example, use anvils and stones as pounding tools to crack nuts opened (Fragaszy et al., 2004). Interestingly, capuchin monkeys use these tools in a very complex way, selecting hammer tools on the basis of stone mass and distance from the anvil, selecting stones and nuts in order to minimize the strikes required to open the nuts, and placing the nuts in anvils selectively (Fragaszy et al., 2010; Massaro et al., 2010; Liu et al., 2011). Recently, wild bearded capuchin monkeys have also shown the ability to strategically place palm nuts in a stable position on anvils during nut-cracking (Fragaszy et al., 2013). Monkeys minimized their efforts by placing the nuts flat side down more frequently than expected by chance, and knocking them a few times to check for stability before putting them in place. When compared to blindfolded humans, capuchin monkeys used identical methods to position the nuts, following tactile cues to perform more efficiently (Fragaszy et al., 2013). Although wild brown capuchin monkeys only carry food to the stone sites (Jalles-Filho et al., 2001), captive brown capuchin monkeys carry stones and tools where the food is located and even if it is out of view, suggesting that in captivity they show more complex forms of tool use involving some sort of planning and self-control (Cleveland et al., 2004). In captivity, for example, brown capuchin monkeys also use less valuable food as a tool to obtain more valuable food, suggesting enhanced inhibitory skills in this context (Evans & Westergaard, 2006). Moreover, captive capuchin monkeys (*Cebus apella*) learn to select the most effective tools to open nuts by attending to their physical properties (Schrauf et al., 2008). Such a complex way of using tools suggests that capuchin monkeys are paying simultaneous attention to the various properties of these objects in an attempt to optimize the opening process, adjusting their behaviour to the existing conditions.

Using tools, however, does not necessarily imply that subjects understand the causal properties of the tools, as tool-using can simply be a reflection of a species' specialized feeding mechanism (e.g. Visalberghi & Tomasello, 1998; Emery & Clayton, 2009; Seed & Byrne, 2010). Researchers have therefore used different paradigms to test how tool using and non-tool using species understand the causal properties of different objects.

Some studies, for example, have shown that cotton-top tamarins can select tools based on their functionality, suggesting that they are able to distinguish causally relevant from causally irrelevant properties of tools (Hauser, 1997; Hauser et al., 2002; but see Penn & Povinelli, 2007 for a critical review). One of the easiest set-ups to test monkeys' understanding of causality is the cloth-pulling experiment, in which subjects are presented with two food rewards, one lying on a cloth and the other one closed to another cloth, both out of the subjects' reach. If subjects understand the support relation, they should pull the cloth with the food reward on its top.

When tested with this set-up, cotton-top tamarins succeed at selecting the right cloth after extensive training and then generalize performance to novel combinations of tools, suggesting that they understand the properties that are relevant to a functional pulling tool (Hauser et al., 1999; Santos et al., 2005, 2006). Moreover, tamarins with tool use experience can also overcome the irrelevant aspects of the task (Hauser et al., 2002).

However, no relation between performance at this task and the use of tools in the wild has been found by other studies (Herrmann et al., 2008). Common marmosets tested with a similar paradigm showed that they do not differ from tamarins in understanding the physical causality of contact/support (Spaulding & Hauser, 2005). Moreover, marmosets could learn to select the further retrievable food reward over the closer one, but they failed to do so when the cloth pattern was made more complex (e.g. by dividing one cloth into two pieces so that pulling that cloth would still not allow subjects to retrieve the food positioned on top of one piece; Yamazaki et al., 2011). The authors suggested that common marmosets were influenced by irrelevant perceptual features in solving this task, paying attention to the food rather than to the presence of a gap in some conditions (Yamazaki et al., 2011).

Similarly, Penn & Povinelli (2007) have suggested that primates solving the cloth-pulling experiment are only sensitive to certain perceptual relations, like the degree of surface contact between cloth and reward, but not to the structural relation casually relevant to obtaining the reward. Common marmosets have also been tested in the wild in a basic version of the cloth-pulling task, but their performance was worse than the one by captive conspecifics, probably because of the added difficulties imposed by the natural setting of the task, implying an increase in observed vigilance behaviour by subjects (Halsey et al., 2006).

When tested with a cloth-pulling paradigm, capuchin monkeys could also solve this task only after some training (Fujita et al., 2003). When tested with the same procedure but different cloth patterns and no training, spider monkeys but not brown capuchin monkeys performed above chance level (50%) in all conditions, significantly outperforming brown capuchin monkeys and partially great apes (Amici et al., 2010). The authors did not explain these results in terms of spider monkeys' better understanding of causal relations, but they suggested that spider monkeys might have enhanced analogical skills (see below), allowing them to better understand the physical relation between a functional tool and the food in one condition, and extend this knowledge to other conditions (Amici et al., 2010).

A more complex set-up to test monkeys' understanding of causality is the trap-tube task (Visalberghi & Limongelli, 1994). In the original version of this task, brown capuchin monkeys were presented with a transparent tube containing a food reward, which they had to retrieve using a straight stick (Visalberghi & Limongelli, 1994). Since the tube contained a highly visible hole with a small transparent cup attached (i.e. the trap), subjects had to avoid the trap while retrieving the food, which would have otherwise fallen into the trap and become inaccessible. Subjects had two different ways to solve the task. They could learn to solve it through trial and error via a simple associative rule based on the perceptual features of the task (i.e., always push the food form the furthest side); or they could use causal reasoning, avoiding the trap only when it was causally relevant (i.e., when the trap was positioned downward, but not when the trap was rotated upward and the food could thus also be accessed from that side). Only one brown capuchin monkey learned to solve the task, but when the tube was rotated upward and became causally irrelevant the brown capuchin monkey persisted in avoiding the trap, suggesting that she was simply solving the task through associative learning, without causal understanding.

Several other studies have been done on a variety of primates and birds and produced different results (see Emery & Clayton, 2009, for a review), also depending on the procedure (Mulcahy & Call, 2006; Seed et al., 2009). The few studies conducted on New World monkeys have basically confirmed the results of the first study. Monkeys learn to solve the trap-tube task but showing no understanding of causality even if regularly using tools in the wild (Visalberghi & Trinca, 1989; Visalberghi et al., 1995; Visalberghi & Tomasello, 1998). Does that mean that New World monkeys have no understanding what properties a tool should have in order to be useful?

Santos and colleagues (2006) explored causal understanding in two non-tool using species (cotton-top tamarins and the Old World vervet monkeys) and found that tamarins were worse than vervet monkeys at selecting tools based on their functionally relevant properties, lacking a general understanding of unobservable causal properties like gravity and support. The authors concluded that monkeys' understanding of causality is more limited and that tool using does not need to be correlated with understanding of causality, tool use probably deriving from domain-general rather than domain-specific cognitive skills evolved for tool use (Santos et al., 2006).

However, capuchin monkeys treat the same visual feature differently depending on task-relevant physical properties (Visalberghi et al., 2009). Wild capuchins presented with stones at some distance from a nut-cracking anvil selected large, heavy, robust stones to transport. Even when the stones were artificial and differed in density and weight, the capuchins selected the heavy, effective tool.

UNDERSTANDING OF NUMEROSITY AND QUANTITIES

The ability to count and judge quantities is ecologically relevant in several contexts, including predation, foraging and breeding (Dehaene, 1999; Schmitt & Fischer, 2011). Numerical skills have indeed been found in a variety of species, including mosquitofish (*Gambusia holbrooki*: Agrillo et al., 2007), suggesting that the system for representing numbers is evolutionarily ancient and widely spread across several species.

In order to test quantity discrimination in primates, two different paradigms have often been used. The first paradigm consists in presenting subjects with two different food quantities and letting them select and thus retrieve the larger one (e.g., Hanus et al., 2003; Evans et al., 2009). The second paradigm consists in training subjects to associate symbols with food quantities and then testing their ability to discriminate quantities by presenting them with symbols (e.g., Olthof et al., 1997; Addessi et al., 2008). Since symbols help primates overcome the saliency of food, primates usually perform much better when they are presented with symbols instead of food quantities (but see Schmitt & Fischer, 2011, for a more complex hypothesis on food saliency in Old World monkeys).

The number of studies conducted in New World monkeys is unfortunately very low and drawing conclusions on the distribution of numerical skills across these species is therefore impossible. One study on cotton-top tamarins showed that they look significantly longer at inconsistent outcomes of easy sums, suggesting that numerical representations are spontaneously available to several species (Uller et al., 2001). Squirrel monkeys are able to use symbols to represent quantities and learn to choose the largest number within a set, approximating the counting mechanisms used by humans (Olthof et al., 1997).

Also brown capuchin monkeys are able to reliably select the larger of two visible quantities of 1–5 food items (Addessi et al., 2008) and make accurate judgments between food sets after extra items are added to one of the sets (Beran, 2008). Moreover, they accurately select the larger food quantity when presented with sets of food items in which items are provided one by one, suggesting that they have sophisticated skills to form representations of food quantities, but not as enhanced as in chimpanzees (van Marle et al., 2006; Evans et al., 2009).

Moreover, brown capuchin monkeys can be trained to associate specific quantities with specific tokens and sum the value of several different tokens to make accurate quantity judgments (Evans et al., 2010). They can also successfully compare individual tokens to one another or to sets of visible food items. Like chimpanzees, however, only few individuals managed to master the task by flexibly combining symbols to judge quantity (Evans et al., 2010).

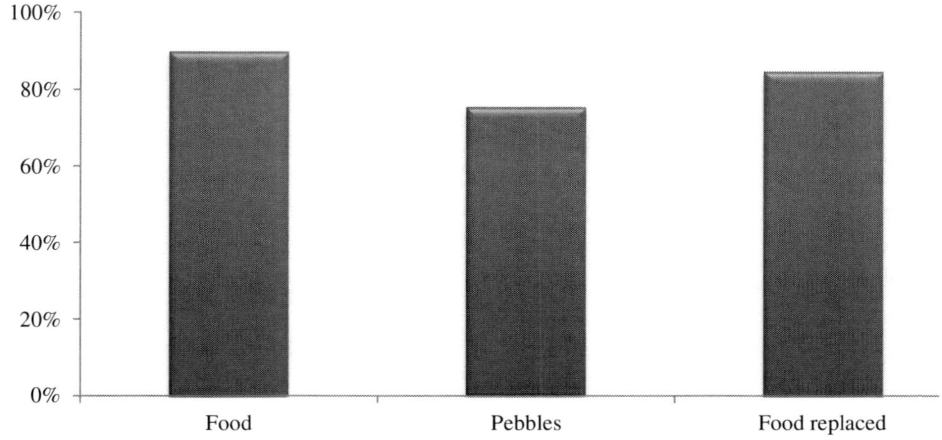

Figure 2. For spider monkeys, mean percentage of correct trials in each condition of the numerosity task (when they can choose between different food quantities that they receive, when they can choose between different pebble quantities that are subsequently replaced by food and when they can choose between different food quantities that are subsequently replaced by food).

Preliminary analyses show that spider monkeys can discriminate between different pairs of food quantities comparably to Old World monkeys, successfully retrieving the larger food quantity in more than 90% of the trials (Amici, unpublished data). Following the procedure by (Schmitt & Fischer, 2011), spider monkeys have also been tested using pebbles instead of food rewards and symbolic food rewards (i.e. food rewards were shown to the subject, but after selection food rewards were replaced by an equivalent quantity of food). In contrast to Old World monkeys, spider monkeys' performance decreased when pebbles were used instead of food rewards and partially also when they were provided with symbolic food rewards (Figure 2). These results suggest that spider monkeys have good numerical skills which are not enhanced through the use of symbolic stimuli, in contrast to what seems to happen in Old World monkeys (Schmitt & Fischer, 2011).

Abstract Concepts and Analogical Skills

Using abstract concepts improves the ability to sort objects, relations and events into common classes with similar perceptual, associative and relational properties (analogical reasoning; e.g. Truppa et al., 2011). When individuals form abstract concepts, they can more easily transfer knowledge to new stimuli or contexts, reducing the cognitive load of retaining much "off-line" information (see Aureli et al., 2008).

Several studies have tested primates on their ability to learn abstract concepts, and the paradigm that has been mostly used is the matching to sample task, in which individuals are presented with two stimuli and they have to choose which one is perceptually equivalent to the sample stimulus shown before. This paradigm is especially interesting as it allows the use of multiple levels of abstraction varying in relational complexity (e.g. Truppa et al., 2011). Instead of first-order relations (i.e., in which subjects assess which stimulus is equivalent to the sample), for example, it is also possible to test second-order relations (i.e., relations-between-relations, in which subjects assess whether the relationship among attributes of one set of objects is equivalent to the relationship among attributes of another set of objects). Acquiring abstract concepts implies that subjects solve the task not only with familiar stimuli, but also extending the acquired criteria to novel stimuli through relational learning (e.g., D'Amato & Colombo, 1989).

Brown capuchin monkeys trained with a large number of stimuli show a good transfer performance to novel stimuli in matching to sample tasks, suggesting that they can acquire the identity concept and transfer it to novel visual dimensions (Truppa et al., 2010). They are also able to solve the matching to sample task when less realistic images are used as stimuli (Truppa et al., 2009). Moreover, after a long training they can successfully acquire the identity and difference abstract concepts to solve a matching to sample task with second-order relations, suggesting that analogical reasoning can be elicited in brown capuchin monkeys under certain training conditions (Truppa et al., 2011).

Finally, brown capuchin monkeys can also solve other matching to sample tasks based on the above/below concept (Spinozzi et al., 2004) and on the relative size of a set of 3-dimensional objects (Kennedy & Fragaszy, 2008). The authors suggested that acquiring abstract concepts for spatial relations is ecologically relevant to gather information about the location of objects and move through the physical environment (Spinozzi et al., 2004). These recent studies contrast with previous findings suggesting that brown capuchin monkeys can master matching to sample tasks by acquiring an identity concept with a very limited number of stimuli (e.g., D'Amato et al., 1985; Barros et al., 2002). However, these studies repeatedly used not-novel stimuli as transfer stimuli, and could not therefore rule out the possibility that monkeys simply solved the task through associative learning processes (see Katz et al., 2007).

CONCLUSION: CURRENT PROBLEMS AND FUTURE RESEARCH

New World monkeys have a complex understanding of the physical world surrounding them. They usually understand visible displacements typical of stage 4 of Piagetian object permanence, but probably use alternative strategies to solve more complex tasks typical of stage 5 or 6. Partially, inter-specific differences in object permanence can be probably better explained in terms of differences in other cognitive skills which are also needed to solve the tasks (e.g., inhibition). When taken together, however, these studies also suggest that there might be some differences between New World monkeys and other primates in terms of object permanence (see also Deaner et al., 2006).

New World monkeys also possess complex navigational skills to move through their environment, often using a combination of coordinate-based and route-based maps. To some extent foraging habits seem to predict travelling habits, with frugivorous species using more complex navigational skills to move through the environment. However, the most consistent inter-specific difference is the one between chimpanzees and the other primate species, as chimpanzees seem so far the only species extensively using coordinate-based mental maps.

Memory skills have been tested across different species by using very different paradigms. The few studies having included more New World monkey species suggest that memory differences exist and can be partially explained in terms of fission-fusion dynamics and degree of frugivory in the diet. Inhibitory skills also show a great variability across and within species. In particular, new studies seem to suggest that fission-fusion dynamics, tool use and consumption of gum might be all linked to a general enhancement of inhibitory skills. However, because inter-individual variability is also very high, studies with larger sample sizes are needed to confirm these results.

Studies on tool use and understanding of causality are abundant and also include non-tool users. These studies are consistent in showing that New World monkeys can learn to master the cloth-pulling and the trap-tube task. However, most of these studies cast doubt on the fact that subjects really understand causality in the majority of tasks, suggesting that New World monkeys possibly use alternative strategies to solve causal tasks. In general, monkeys' understanding of causality seems to be more limited than usually thought and it is probably not correlated with the ability to use tools in the wild.

The number of studies conducted on numerosity in New World monkeys is unfortunately very low and drawing conclusions on the distribution of numerical skills across these species is therefore too risky. All tested species seem able to discriminate food quantities and represent numbers, but using symbols instead of food seems to differentially affect subjects' performance across different species. Similarly, studies on the understanding of abstract concepts have been done with different paradigms and have only included few species, so that so far no conclusions can be drawn on the possible distribution of analogical reasoning across New World monkeys.

Given these premises, it seems hard to hint to specific socio-ecological factors that might be linked to an enhancement of specific cognitive skills in New World monkeys. In general, evidence has been collected on the influence of both social and ecological factors on cognition. Fission-fusion dynamics, tool use, degree of frugivory and consumption of gum in the diet have all been linked to an enhancement of some specific cognitive skills. It makes sense to think that none of these factors can alone predict the distribution of all cognitive skills, but that their complex inter-play rather affects the emergence of enhanced independent cognitive skills across different primate species, in a complex relationship of social, ecological and phylogenetic factors (e.g. Dunbar, 1992; Amici et al., 2012b).

In order to disentangle this complex relationships, several problematic aspects should be addressed in future research in comparative cognition. First of all, more species should be tested in studies on animal cognition, including species notoriously underrepresented. The inclusion of species with very different socio-ecological characteristics and phylogenetic relatedness would ensure the possibility to reliably contrast the different socio-ecological hypotheses. Secondly, more subjects should be included in these studies, to ensure that differences in personality, age or rank cannot alone explain the results obtained. Moreover, the inclusion of both wild and captive individuals would ideally allow controlling for the effects of ontogenetic experience on cognition. Thirdly, experimental paradigms should be used consistently across species, to ensure that differences in performance do not simply reflect differences in experimental paradigms. Moreover, simple paradigms should be implemented to ensure that specific cognitive skills are tested and that subjects do not require mastery of other complex cognitive skills to be successful. Finally, socio-ecological factors can dramatically vary also across conspecific populations, so that simultaneously collecting ecological, social and cognitive data from the same population seem necessary in order to reliably contrast the different evolutionary hypotheses and understand to what extent cognitive skills are linked to the socio-ecological factors experienced during ontogeny or evolution. Addressing these controversial aspects will provide further strength to the study of comparative cognition and allow us better understanding the evolutionary factors explaining the distribution of cognitive skills in New World monkeys.

Acknowledgments

This chapter was written while the first author held a Humboldt Research Fellowship for Postdoctoral Researchers (Humboldt ID number 1138999). Thanks to Bino Majolo for revising a previous draft of the manuscript.

References

Abramson, J. Z., Hernández-Lloreda, V., Call, J., & Colmenares, F. (2011). Relative Quantity Judgments in South American Sea Lions (*Otaria flavescens*). *Animal Cognition, 14*, 695-706.

Addessi, E., Crescimbene, L., & Visalberghi, E. (2008). Food and Token Quantity Discrimination in Capuchin Monkeys (*Cebus apella*). *Animal Cognition, 11*, 275-282.

Addessi, E., Paglieri, F., Beran, M. J., Evans, T. A., Macchitella, L., De Petrillo, F., & Focaroli, V. (2013). Delay Choice Versus Delay Maintenance: Different Measures of Delayed Gratification in Capuchin Monkeys (*Cebus apella*). *Journal of Comparative Psychology*

Addessi, E., Paglieri, F., & Focaroli, V. (2011). The Ecological Rationality of Delay Tolerance: Insights from Capuchin Monkeys. *Cognition, 119*, 142 – 147.

Addessi, E., & Rossi, S. (2011). Tokens Improve Capuchin Performance in the Reverse–Reward Contingency Task. *Proceedings of the Royal Society of London B Biological Sciences, 278*, 849-854.

Agrillo, C., Dadda, M., & Bisazza, A. (2007). Quantity Discrimination in Female Mosquitofish. *Animal Cognition, 10*, 63-70.

Albiach-Serrano, A., Guillén-Salazar, F., & Call, J. (2007). Mangabeys (*Cercocebus torquatus lunulatus*) Solve the Reverse Contingency Task without a Modified Procedure. *Animal Cognition, 10*, 387- 396.

Amici, F., Aureli, F., & Call, J. (2008). Fission-Fusion Dynamics, Behavioral Flexibility, and Inhibitory Control in Primates. *Current Biology, 18*, 1415-1419.

Amici, F., Aureli, F., & Call, J. (2010). Monkeys and Apes: Are Their Cognitive Skills Really So Different? *American Journal of Physical Anthropology, 143,* 188–197.

Amici, F., Barney, B., Johnson, V. E., Call, J., & Aureli, F. (2012b). A Modular Mind? A Test Using Individual Data from Seven Primate Species. *PLoS ONE, 7*(12), e51918.

Amici, F., Call, J., & Aureli, F. (2009). Variation in Withholding of Information in Three Monkey Species. *Proceedings of the Royal Society of London B Biological Sciences, 276*, 3311-3318. Amici, F., Call, J., & Aureli, F. (2012a). Aversion to Violation of Expectations of Food Distribution: The Role of Social Tolerance and Relative Dominance in Seven Primate Species. *Behaviour, 149*, 345–368.

Anderson, J. R., Awazu, S., & Fujita, K. (2000). Can Squirrel Monkeys (*Saimiri sciureus*) Learn Self- Control? A Study Using Food Array Selection Tests and Reverse-Reward Contingency. *Journal of Experimental Psychology Animal Behavior Processes, 26*, 87-97.

Anderson, J. R., Awazu, S., & Fujita, K. (2004). Squirrel Monkeys (*Saimiri sciureus*) Choose Smaller Food Arrays: Long-Term Retention, Choice with Nonpreferred Food, and Transposition. *Journal of Comparative Psychology, 118*, 58–64.

Anderson, J. R., Hattori, Y., & Fujita, K. (2008). Quality Before Quantity: Rapid Learning of Reverse- Reward Contingency by Capuchin Monkeys (*Cebus apella*). *Journal of Comparative Psychology, 122*, 445-448.

Anderson, J. R., Kuroshima, H., & Fujita, K. (2010). Delay of Gratification in Capuchin Monkeys (*Cebus apella*) and Squirrel Monkeys (*Saimiri sciureus*). *Journal of Comparative Psychology, 124*, 205-210.

Anderson, M. R. (2012). Comprehension of Object Permanence and Single Transposition in Gibbons. *Behaviour, 149*, 441-459.

Aureli, F., Schaffner, Colleen M., Boesch, C., Bearder, Simon K., Call, J., Chapman, Connor, R., di Fiore, A., Dunbar, R. I. M., Henzi, S. P., Holekamp, K., Korstjens, A. H., Layton, R., Lee, P., Lehmann, J., Manson, J. H., Ramos-Fernandez, G., Strier, K. B., van Schaik, C. P. (2008). Fission-Fusion Dynamics: New Research Frameworks. *Current Anthropology, 49*, 627-654.

Baddeley, A. (2003). Working Memory: Looking Back and Looking Forward. *National Review of Neuroscience, 4*, 829-839.

Barrett, L., Henzi, S. P., & Dunbar, R. (2003). Primate Cognition: From 'What Now?' to 'What If ?'. *Trends in Cognitive Sciences, 7*, 494-497.

Barros, R., Galvão, O., & McIlvane, W. (2002). Generalized Identity Matching-to-Sample in *Cebus apella*. *Psychological Record, 52*, 441–460.

Barth, J., & Call, J. (2006). Tracking the Displacement of Objects: A Series of Tasks with Great Apes (*Pan troglodytes*, *Pan paniscus*, *Gorilla gorilla*, and *Pongo pygmaeus*) and Young Children (*Homo sapiens*). *Journal of Experimental Psychology Animal Behavior Processes, 32*, 239-252.

Barton, R. A. (1999). The Evolutionary Ecology of the Primate Brain. In P. C. Lee (Ed.), *Comparative Primate Socioecology* (pp. 167-203). Cambridge: Cambridge University Press.

Barton, R. A., & Harvey, P. H. (2000). Mosaic Evolution of Brain Structure in Mammals. *Nature, 405*, 1055-1058.

Beran, M. J. (2008). Capuchin Monkeys (*Cebus apella*) Succeed in a Test of Quantity Conservation. *Animal Cognition, 11*, 109-116.

Blois-Heulin, C., & Girona, B. (1999). Patterns of Social Visual Attention in the Red-Capped Mangabey (*Cercocebus torquatus torquatus*) in the Context of Cood Competition. *Folia Primatologica, 70*, 180-184.

Boesch, C. (2007). What Makes Us Human (*Homo sapiens*)? The Challenge of Cognitive Cross-Species Comparison. *Journal of Comparative Psychology, 121*, 227-240.

Boysen, S. T., & Berntson, G. G. (1995). Responses to Quantity: Perceptual Versus Cognitive Mechanisms in Chimpanzees (*Pan troglodytes*). *Journal of Experimental Psychology Animal Behavior Processes, 21*, 82-86.

Boysen, S. T., Mukobi, K. L., & Berntson, G. G. (1999). Overcoming Response Bias Using Symbolic Representations of Number by Chimpanzees (*Pan troglodytes*). *Animal Learning & Behavior, 27*, 229-235.

Bramlett, J. L., Perdue, B. M., Evans, T. A., & Beran, M. J. (2012). Capuchin Monkeys (*Cebus apella*) Let Lesser Rewards Pass Them by to Get Better Rewards. *Animal Cognition, 15*, 963-969.

Bräuer, J., Call, J., & Tomasello, M. (2007). Chimpanzees Really Know What Others Can See in a Competitive Situation. *Animal Cognition, 10*, 439-448.

Burkart, J. M., Hrdy, S. B., & Van Schaik, C. P. (2009). Cooperative Breeding and Human Cognitive Evolution. *Evolutionary Anthropology: Issues, News, and Reviews, 18*, 175-186.

Byrne, R. W., & Whiten, A. (1988). *Machiavellian Intelligence: Social Expertise and the Evolution of Intellect in Monkeys, Apes and Humans*. Oxford: Clarendon Press.

Call, J. (2000). Representing Space and Objects in Monkeys and Apes. *Cognitive Science, 24*, 397-422. Call, J. (2001). Object Permanence in Orangutans (*Pongo pygmaeus*), Chimpanzees (*Pan troglodytes*), and Children (*Homo sapiens*). *Journal of Comparative Psychology, 115*, 159-171.

Call, J. (2003). Comparative Ape Cognition in Leipzig: The Wolfgang Köhler Primate Research Centre. *Folia Primatologica, 74*, 179.

Chance, M. R. A. (1967). Attention Structure as the Basis of Primate Rank Orders. *Man, 2*, 503-518. Cheng, K. (2010). Common Principles Shared by Spatial and Other Kinds of Cognition. In F. L. Dolins & R. W. Mitchell (Eds.), *Spatial Cognition, Spatial Perception: Mapping the Self and Space* (pp. 54–74). Cambridge, UK: Cambridge University Press.

Cleveland, A., Rocca, A. M., Wendt, E. L., & Westergaard, G. C. (2004). Transport of Tools to Food Sites in Tufted Capuchin Monkeys (*Cebus apella*). *Animal Cognition, 7*, 193-198. Clutton-Brock, T. H., & Harvey, P. H. (1980). Primates, Brain and Ecology. *Journal of Zoology (London), 190*, 309-323.

Conway, A. R. A., & Getz, S. J. (2010). Cognitive Ability: Does Working Memory Training Enhance Intelligence? *Current Biology, 20*, 362-364.

D'Amato, M. R., & Colombo, M. (1989). Serial Learning With Wild Card Items by Monkeys (*Cebus apella*): Implications for Knowledge of Ordinal Position. *Journal of Comparative Psychology, 103*, 252-261.

D'Amato, M. R., Salmon, D. P., & Colombo, M. (1985). Extent and Limits of the Matching Concept in Monkeys (*Cebus apella*). *Journal of Experimental Psychology Animal Behavior Processes, 11*, 35-51.

de Blois, S. T., Novak, M. A., & Bond, M. (1998). Object Permanence in Orangutans (*Pongo pygmaeus*) and Squirrel Monkeys (*Saimiri sciureus*). *Journal of Comparative*

Psychology, 112, 137-152. Deaner, R. O., van Schaik, C. P., & Johnson, V. (2006). Do Some Taxa Have Better Domain-General Cognition Than Others? A Meta-Analysis of Nonhuman Primate Studies. *Evolutionary Psychology, 4*, 149-196.

Dehaene, S. (1999). *The Number Sense: How the Mind Creates Mathematics*: Oxford University Press.

Deipolyi, A., Santos, L., & Hauser, M. D. (2001). The Role of Landmarks in Cotton-Top Tamarin Spatial Foraging: Evidence for Geometric and Non-Geometric Features. *Animal Cognition, 4*, 99-108. Deloache, J. S., & Pickard, M. B. (2010). Spatial Cognition, Spatial Perception: Mapping the Self and Space. In F. L. Dolins & R. W. Mitchell (Eds.), *Spatial Cognition, Spatial Perception: Mapping the Self and Space* (pp. 486–501). Cambridge, UK: Cambridge University Press.

Di Fiore, A., & Suarez, S. A. (2007). Route-Based Travel and Shared Routes in Sympatric Spider and Woolly Monkeys: Cognitive and Evolutionary Implications. *Animal Cognition, 10*, 317-329. Dolins, F. L. (2009). Captive Cotton-Top Tamarins' (*Saguinus oedipus oedipus*) Use of Landmarks to Localize Hidden Food Items. *American Journal of Primatology, 71*, 316-323.

Dolins, F. L., & Mitchell, R. W. (2010). Linking Spatial Cognition and Spatial Perception. In F. L. Dolins & R. W. Mitchell (Eds.), *Spatial Cognition, Spatial Perception: Mapping the Self and Space* (pp. 1–31). Camebridge, UK: Cambridge University Press.

Drapier, M., Chauvin, C., Dufour, V., Uhlrich, P., & Thierry, B. (2005). Food-Exchange with Humans in Brown Capuchin Monkeys. *Primates, 46*, 241–248.

Dumas, C., & Brunet, C. (1994). Object Permanence in the Capuchin Monkey (*Cebus apella*): individual displacement tasks. *Canadian Journal of Experimental Psychology, 48*, 341-357.

Dunbar, R. I. M. (1992). Neocortex Size as a Constraint on Group Size in Primates. *Journal of Human Evolution, 22*, 469-493.

Dunbar, R. I. M. (2009). The Social Brain Hypothesis and Its Implications for Social Evolution. *Annals of Human Biology, 36*, 562-572.

Dunbar, R. I. M., & Shultz, S. (2007). Evolution in the Social Brain. *Science, 317*, 1344-1347. Emery, N. J. (2000). The Eyes Have It: The Neuroethology, Function and Evolution of Social Gaze. *Neuroscience & Biobehavioral Reviews, 24*, 581-604.

Emery, N. J., & Clayton, N. S. (2009). Tool Use and Physical Cognition in Birds and Mammals. *Current Opinion in Neurobiology, 19*, 27-33.

Erickson, C. J. (1994). Tap-Scanning and Extractive Foraging in Aye-Ayes, *Daubentonia madagascariensis*. *Folia Primatologica, 62*, 125-135.

Evans, T. A., Beran, M., Harris, E. H., & Rice, D. F. (2009). Quantity Judgments of Sequentially Presented Food Items by Capuchin Monkeys (*Cebus apella*). *Animal Cognition, 12*, 97-105. Evans, T. A., Beran, M. J., & Addessi, E. (2010). Can Nonhuman Primates Use Tokens to Represent and Sum Quantities? *Journal of Comparative Psychology, 124*, 369-380.

Evans, T. A., & Westergaard, G. C. (2006). Self-Control and Tool Use in Tufted Capuchin Monkeys (*Cebus apella*). *Journal of Comparative Psychology, 120*, 163-166.

Fiset, S., & Doré, F. Y. (2006). Duration of Cats' (*Felis catus*) Working Memory for Disappearing Objects. *Animal Cognition, 9*, 62-70.

Fragaszy, D. M., Greenberg, R., Visalberghi, E., Ottoni, E. B., Izar, P., & Liu, Q. (2010). How Wild Bearded Capuchin Monkeys Select Stones and Nuts to Minimize the Number of Strikes per Nut Cracked. *Animal Behaviour, 80*, 205-214.

Fragaszy, D., Izar, P., Visalberghi, E., & Ottoni, E. B. (2004). Wild Capuchin Monkeys (*Cebus libidinosus*) Use Anvils and Stone Pounding Tools. *American Journal of Primatology, 64,* 359–366.

Fragaszy, D. M., Liu, Q., Wright, B. W., Allen, A., Welch Brown, C., & Visalberghi, E. (2013). Wild Bearded Capuchin Monkeys (*Sapajus libidinosus*) Strategically Place Nuts in a Stable Position During Nut-Cracking. *Plos ONE, 8 (2)*: e56182.

Fraser, O. N., Schino, G., & Aureli, F. (2008). Components of Relationship Quality in Chimpanzees. *Ethology, 114*, 834-843.

Fujita, K., Kuroshima, H., & Asai, S. (2003). How Do Tufted Capuchin Monkeys (*Cebus apella*) Understand Causality Involved in Tool Use? *Journal of Experimental Psychology Animal Behavior Processes, 29*, 233-242.

Garber, P., & Hannon, B. (1993). Modeling Monkeys: A Comparison of Computer-Generated and Naturally Occurring Foraging Patterns in Two Species of Neotropical Primates. *International Journal of Primatology, 14*, 827-852.

Garber, P. A. (1989). Role of Spatial Memory in Primate Foraging Patterns: *Saguinus mystax* and *Saguinus fuscicollis*. *American Journal of Primatology, 19*, 203-216.

Garber, P. A. (2000). The Ecology of Group Movement: Evidence for the Use of Spatial, Temporal, and Social Information in Some Primate Foragers. In S. Boinski & P. A. Garber (Eds.), *On the Move: How and Why Animals Travel in Groups* (pp. 261–298). Chicago: University of Chicago Press.

Garber, P. A., Bicca-Marques, J. C., & Azevedo-Lopes, M. A. O. (2009). Primate Cognition: Integrating Social and Ecological Information in Decision-Making. In P. A. Garber, A. Estrada, J. C. Bicca- Marques, E. Heymann & K. B. Strier (Eds.), *South American Primates: Comparative Perspectives in the Study of Behavior, Ecology, and Conservation* (pp. 365–385). New York: Springer.

Garber, P. A., & Dolins, F. L. (1996). Testing Learning Paradigms in the Field: Evidence for Foraging in Wild Moustached Tamarins. In M. Norconk, A. L. Rosenberger & P. A. Garber (Eds.), *Adaptive Radiation of Neotropical Primates* (pp. 201–216). New York: Plenum Press.

Genty, E., Palmier, C., & Roeder, J.-J. (2004). Learning to Suppress Responses to the Larger of Two Rewards in Two Species of Lemurs, *Eulemur fulvus* and *E. macaco*. *Animal Behaviour, 67*, 925-932.

Halsey, L. G., Bezerra, B. M., & Souto, A. S. (2006). Can Wild Common Marmosets (*Callithrix jacchus*) Solve the Parallel Strings Task? *Animal Cognition, 9*, 229-233.

Hanus, D., Call, J., & Tomasello, M. (2003). Quantity-Based Judgments by Orangutans (*Pongo pygmaeus*), Gorillas (*Gorilla gorilla*) and Bonobos (*Pan paniscus*). *Folia Primatologica, 74*, 196-197.

Hare, B., Call, J., Agnetta, B., & Tomasello, M. (2000). Chimpanzees Know What Conspecifics Do and Do Not See. *Animal Behaviour, 59*, 771-785.

Hauser, M. D. (1997). Artifactual Kinds and Functional Design Features: What a Primate Understands Without Language. *Cognition, 64*, 285-308.

Hauser, M. D., Kralik, J., & Botto-Mahan, C. (1999). Problem Solving and Functional Design Features: Experiments on Cotton-Top Tamarins, *Saguinus oedipus oedipus*. *Animal Behaviour, 57*, 565-582.

Hauser, M. D., Pearson, H., & Seelig, D. (2002). Ontogeny of Tool Use in Cottontop Tamarins, *Saguinus oedipus*: Innate Recognition of Functionally Relevant Features. *Animal Behaviour, 64*, 299-311.

Herrmann, E., Wobber, V., & Call, J. (2008). Great Apes' (*Pan troglodytes*, *Pan paniscus*, *Gorilla gorilla*, *Pongo pygmaeus*) Understanding of Tool Functional Properties after Limited Experience. *Journal of Comparative Psychology, 122*, 220-230.

Humphrey, N. K. (1976). The Social Function of Intellect. In P. P. Bates & R. A. Hinde (Eds.), *Growing Points in Ethology*. Oxford: Cambridge University Press.

Jalles-Filho, E., Grassetto, R., da Cunha, T., & Salm, R. A. (2001). Transport of Tools and Mental Representation: Is Capuchin Monkey Tool Behaviour a Useful Model of Plio-Pleistocene Hominid Technology? *Journal of Human Evolution, 40*, 365-377.

Janson, C. (2007). Experimental Evidence for Route Integration and Strategic Planning in Wild Capuchin Monkeys. *Animal Cognition, 10*, 341-356.

Katz, J. S., Wright, A. A., & Bodily, K. D. (2007). Issues in the Comparative Cognition of Abstract- Concept Learning. *Comparative Cognition & Behavior Review, 2*, 79-92.

Kennedy, E. H., & Fragaszy, D. M. (2008). Analogical Reasoning in a Capuchin Monkey (*Cebus apella*). *Journal of Comparative Psychology, 122*, 167-175.

Kralik, J. D. (2005). Inhibitory Control and Response Selection in Problem Solving: How Cotton-Top Tamarins (*Saguinus oedipus*) Overcome a Bias for Selecting the Larger Quantity of Food. *Journal of Comparative Psychology, 119*, 78-89.

Kralik, J. D., Hauser, M. D., & Zimlicki, R. (2002). The Relationship Between Problem Solving and Inhibitory Control: Cotton-Top Tamarin (*Saguinus oedipus*) Performance on a Reversed Contingency Task. *Journal of Comparative Psychology, 116*, 39-50.

Kudo, H., & Dunbar, R. I. M. (2001). Neocortex Size and Social Network Size in Primates. *Animal Behaviour, 62*, 711-722.

Lindshield, S., & Rodrigues, M. (2009). Tool Use in Wild Spider Monkeys (*Ateles geoffroyi*). *Primates, 50*, 269-272.

Liu, Q., Fragaszy, D., Wright, B., Wright, K., Izar, P., & Visalberghi, E. (2011). Wild Bearded Capuchin

Monkeys (*Cebus libidinosus*) Place Nuts in Anvils Selectively. *Animal Behaviour, 81*, 297-305. MacDonald, S. E., Pang, J. C., & Gibeault, S. (1994). Marmoset (*Callithrix jacchus jacchus*) Spatial

Memory in a Foraging Task: Win-Stay Versus Win-Shift Strategies. *Journal of Comparative Psychology, 108*, 328-334.

MacLean, E. L., Matthews, L. J., Hare, B., Nunn, C. L., & Anderson, R. C. (2012). How Does Cognition Evolve? Phylogenetic Comparative Psychology. *Animal Cognition, 15*, 223-238.

van Marle, K., Aw, J., McCrink, K., & Santos, L. R. (2006). How Capuchin Monkeys (*Cebus apella*) Quantify Objects and Substances. *Journal of Comparative Psychology, 120*, 416-426.

Massaro, L., Liu, Q., Visalberghi, E., & Fragaszy, D. (2012). Wild Bearded Capuchin (*Sapajus libidinosus*) Select Hammer Tools on the Basis of Both Stone Mass and Distance from the Anvil. *Animal Cognition, 15*, 1065-1074.

Mathieu, M., Bouchard, M.-A., Granger, L., & Herscovitch, J. (1976). Piagetian Object-Permanence in *Cebus capucinus*, *Lagothrica flavicauda* and *Pan troglodytes*. *Animal Behaviour, 24*, 585-588. McFarland, R., & Majolo, B. (2011). Reconciliation and the Costs of Aggression in Wild Barbary

Macaques (*Macaca sylvanus*): A Test of the Integrated Hypothesis. *Ethology, 117*, 928-937. McNelis, N. L., & Boatright-Horowitz, S. L. (1998). Social Monitoring in a Primate Group: The

Relationship Between Visual Attention and Hierarchical Ranks. *Animal Cognition, 1*, 65-69. Mendes, N., & Huber, L. (2004). Object Permanence in Common Marmosets (*Callithrix jacchus*). *Journal of Comparative Psychology, 118*, 103-112.

Milton, K. (1981). Distribution Patterns of Tropical Plant Foods as an Evolutionary Stimulus to Primate Mental Development. *American Anthropologist, 83*, 534-548.

Milton, K. (1988). Foraging Behaviour and the Evolution of Primate Intelligence. In R. W. Byrne & A. Whiten (Eds.), *Machiavellian Intelligence: Social Expertise and the Evolution of Intellect in Monkeys, Apes, and Humans* (pp. 285-305). New York, NY: Clarendon Press/Oxford University Press.

Moura, A. C. d. A., & Lee, P. C. (2004). Capuchin Stone Tool Use in Caatinga Dry Forest. *Science, 306*, 1909.

Mulcahy, N. J., & Call, J. (2006). How Great Apes Perform on a Modified Trap-Tube Task. *Animal Cognition, 9*, 193-199.

Naqshbandi, M., & Roberts, W. A. (2006). Anticipation of Future Events in Squirrel Monkeys (*Saimiri sciureus*) and Rats (*Rattus norvegicus*): Tests of the Bischof-Kohler Hypothesis. *Journal of Comparative Psychology, 120*, 345-357.

Neiworth, J. J., Steinmark, E., Basile, B. M., Wonders, R., Steely, F., & DeHart, C. (2003). A Test of Object Permanence in a New-World Monkey Species, Cotton Top Tamarins (*Saguinus oedipus*). *Animal Cognition, 6*, 27-37.

Normand, E., Ban, S. D., & Boesch, C. (2009). Forest Chimpanzees (*Pan troglodytes verus*) Remember the Location of Numerous Fruit Trees. *Animal Cognition, 12*, 797-807.

Olthof, A., Iden, C. M., & Roberts, W. A. (1997). Judgments of Ordinality and Summation of Number Symbols by Squirrel Monkeys (*Saimiri sciureus*). *Journal of Experimental Psychology Animal Behavior Processes, 23*, 325-339.

Ottoni, E. B., & Izar, P. (2008). Capuchin Monkey Tool Use: Overview and Implications. *Evolutionary Anthropology, 17*, 171-178.

Ottoni, E. B., & Mannu, M. (2001). Semifree-Ranging Tufted Capuchins (*Cebus apella*) Spontaneously Use Tools to Crack Open Nuts. *International Journal of Primatology, 22*, 347-358.

Parker, S. T. (1991). A Developmental Approach to the Origins of Self-Recognition in Great Apes. *Human Evolution, 6*, 435-449.

Parker, S. T., & Gibson, K. R. (1977). Object Manipulation, Tool Use and Sensorimotor Intelligence as Feeding Adaptations in Cebus Monkeys and Great Apes. *Journal of Human Evolution, 6*, 623- 641.

Pelé, M., Micheletta, J., Uhlrich, P., Thierry, B., & Dufour, V. (2011). Delay Maintenance in Tonkean Macaques (*Macaca tonkeana*) and Brown Capuchin Monkeys (*Cebus apella*). *International Journal of Primatology, 32*, 149-166.

Penn, D. C., & Povinelli, D. J. (2007). Causal Cognition in Human and Nonhuman Animals: A Comparative, Critical Review. *Annual Review of Psychology, 58*, 97-118.

Piaget, J. (1954). *The construction of reality in the child.* NY: Basic Books.

Porter, L. M., & Garber, P. (2013). Foraging and Spatial Memory in Wild Weddell's Saddleback Tamarins (*Saguinus fuscicollis weddelli*) When Moving Between Distant and Out-of-Sight Goals. *International Journal of Primatology, 34*, 30-48.

Poucet, B. (1993). Spatial Cognitive Maps In Animals: New Hypotheses on Their Structure and Neural Mechanism. *Psychological Review, 100*, 163–182.

Pradhan, G. R., Tennie, C., & van Schaik, C. P. (2012). Social Organization and the Evolution of Cumulative Technology in Apes and Hominins. *Journal of Human Evolution, 63*, 180-190. Presotto, A., & Izar, P. (2010). Spatial Reference of Black Capuchin Monkeys in Brazilian Atlantic Forest: Egocentric or Allocentric? *Animal Behaviour, 80*, 125–132

Reader, S. M., & Laland, K. N. (2002). Social Intelligence, Innovation, and Enhanced Brain Size in Primates. *Proceedings of the National Academy of Sciences of the United States of America, 99*, 4436-4441.

Rooijakkers, E., Kaminski, J., & Call, J. (2009). Comparing Dogs and Great Apes in Their Ability to Visually Track Object Transpositions. *Animal Cognition, 12*, 789-796.

Rosati, A.G., Stevens, J. R., Hare, B., & Hauser, M. D. (2007). The Evolutionary Origins of Human

Patience: Temporal Preferences in Chimpanzees, Bonobos, and Human Adults. *Current Biology, 17,* 1663–1668.

Santos, L. R., Ericson, B. N., & Hauser, M. D. (1999). Constraints on Problem Solving and Inhibition: Object Retrieval in Cotton-Top Tamarins (*Saguinus oedipus oedipus*). *Journal of Comparative Psychology, 113*, 186-193.

Santos, L. R., Pearson, H. M., Spaepen, G. M., Tsao, F., & Hauser, M. D. (2006). Probing the Limits of Tool Competence: Experiments With Two Non-Tool-Using Species (*Cercopithecus aethiops* and *Saguinus oedipus*). *Animal Cognition, 9*, 94-109.

Santos, L. R., Rosati, A., Sproul, C., Spaulding, B., & Hauser, M. D. (2005). Means-Means-End Tool Choice in Cotton-Top Tamarins (*Saguinus oedipus*): Finding the Limits on Primates' Knowledge of Tools. *Animal Cognition, 8*, 236-246.

Saults, J. S., & Cowan, N. A. (2007). A Central Capacity Limit to the Simultaneous Storage of Visual and Auditory Arrays in Working Memory. *Journal of Experimental Psychology, 136*, 2663-2684. Schino, G., Spinozzi, G., & Berlinguer, L. (1990). Object Concept and Mental Representation in *Cebus apella* and *Macaca fascicularis*. *Primates, 31*, 537-544.

Schmitt, V., & Fischer, J. (2011). Representational Format Determines Numerical Competence in Monkeys. *Nature Communications, 2*, 1-5.

Schrauf, C., Huber, L., & Visalberghi, E. (2008). Capuchin Monkeys Learn to Use Weight to Select the Most Effective Tool to Crack Open Nuts. *Animal Cognition, 11*, 413-422.

Scott, B. H., Mishkin, M., & Yin, P. (2012). Monkeys Have a Limited Form of Short-Term Memory in Audition. *PNAS, 109*, 12237-12241.

Seed, A. M., & Byrne, R. (2010). Animal Tool-Use. *Current Biology, 20*, R1032-R1039.

Seed, A. M., Call, J., Emery, N. J., & Clayton, N. S. (2009). Chimpanzees Solve the Trap Problem When the Confound of Tool Use Is Removed. *Journal of Experimental Psychology Animal Behavior Processes, 35*, 23-34.

Shettleworth, S. J. (2007). Planning for Breakfast. *Nature, 445*, 825-826.

Silberberg, A., & Fujita, K. (1996). Pointing at Smaller Food Amounts in an Analogue of Boysen and Berntson's (1995) Procedure. *Journal of the Experimental Analysis of*

Behavior, 66, 143-147. Smuts, B. B., Cheney, D. L., Seyfarth, R. M., Wrangham, R. W., & Struhsaker, T. T. (1987). *Primate Societies*. Chicago and London: The University of Chicago Press.

Sophian, C. (1984). Spatial Transpositions and the Early Development of Search. *Developmental Psychology, 20*, 21-28.

Spaulding, B., & Hauser, M. D. (2005). What Experience Is Required for Acquiring Tool Competence? Experiments With Two Callitrichids. *Animal Behaviour, 70*, 517-526.

Spinozzi, G., Lubrano, G., & Truppa, V. (2004). Categorization of Above and Below Spatial Relations by

Tufted Capuchin Monkeys (*Cebus apella*). *Journal of Comparative Psychology, 118*, 403-412. Stevens, J. R., Hallinan, E. V., & Hauser, M. D. (2005a). The Ecology and Evolution of Patience in Two New World Monkeys. *Biology Letters, 1*, 223-226.

Stevens, J. R., Rosati, A. G., Ross, K. R., & Hauser, M. D. (2005b). Will Travel for Food: Spatial Discounting in Two New World Monkeys. *Current Biology, 15*, 1855-1860.

Stoinski, T., & Beck, B. (2001). Spontaneous Tool Use in Captive, Free-ranging Golden Lion Tamarins (*Leontopithecus rosalia rosalia*). *Primates, 42*, 319–326.

Suddendorf, T., & Busby, J. (2005). Making Decisions With the Future in Mind: Developmental and Comparative Identification of Mental Time Travel. *Learning and Motivation, 36*, 110-125. Suddendorf, T., & Corballis, M. C. (1997). Mental Time Travel and the Evolution of the Human Mind. *Genetic, Social, and General Psychology Monographs, 123*, 133-167.

Suddendorf, T., & Corballis, M. C. (2008). New Evidence for Animal Foresight? *Animal Behaviour, 75*, e1-e3.

Tebbich, S., Seed, A. M., Emery, N. J., & Clayton, N. S. (2007). Non-Tool-Using Rooks, *Corvus frugilegus*, Solve the Trap-Tube Problem. *Animal Cognition, 10*, 225-231.

Tomasello, M. (2000). Two Hypotheses about Primate Cognition. In C. E. Source Heyes & L. E. Huber (Eds.), *The Evolution of Cognition* (pp. 165-183). Cambridge, US: The Mit Press.

Tomasello, M., & Call, J. (1997). *Primate Cognition*. New York, NY, USA: Oxford University Press. Truppa, V., E., P. M., Garofoli, D., Privitera, S., & Visalberghi, E. (2011). Same/Different Concept Learning by Capuchin Monkeys in Matching-to-Sample Tasks. *PLoS ONE, 6*(8), e23809.Truppa, V., Garofoli, D., Castorina, G., Piano Mortari, E., Natale, F., & Visalberghi, E. (2010). Identity Concept Learning in Matching-to-Sample Tasks by Tufted Capuchin Monkeys (*Cebus apella*). *Animal Cognition, 13*, 835-848.

Truppa, V., Spinozzi, G., Stegagno, T., & Fagot, J. (2009). Picture Processing in Tufted Capuchin Monkeys (*Cebus apella*). *Behavioural Processes, 82*, 140-152.

Tsujimoto, S., & Sawaguchi, T. (2002). Working Memory of Action: A Comparative Study of Ability to Selecting Response Based on Previous Action in New World Monkeys (*Saimiri sciureus* and *Callithrix jacchus*). *Behavioral Processes, 58*, 149-155.

Tulving, E. (1993). What is Episodic Memory? *Current Directions in Psychological Science, 2*, 67-70. Uller, C., Hauser, M., & Carey, S. (2001). Spontaneous Representation of Number in Cotton-Top Tamarins (*Saguinus oedipus*). *Journal of Comparative Psychology, 115*, 248-257. Urbani, B. (2009). *Spatial Mapping in Wild White-Faced Capuchin Monkeys (Cebus capucinus).* University of Illinois, Urbana, IL.

Valero, A., & Byrne, R. W. (2007). Spider Monkey Ranging Patterns in Mexican Subtropical Forest: Do Travel Routes Reflect Planning? *Animal Cognition, 10*, 305-315.

Vaughter, R. M., Smotherman, W., & Ordy, J. M. (1972). Development of Object Permanence in the Infant Squirrel Monkey. *Developmental Psychology, 7*, 34-38.

Visalberghi, E., Addessi, E., Truppa, V., Spagnoletti, N., Ottoni, E., Izar, P., & Fragaszy, D. (2009).

Selection of Effective Stone Tools by Wild Bearded Capuchin Monkeys. *Current Biology, 19*, 213-217.

Visalberghi, E., Fragaszy, D. M., & Savage-Rumbaugh, S. (1995). Performance in a Tool-Using Task by Common Chimpanzees (*Pan troglodytes*), Bonobos (*Pan paniscus*), an Orangutan (*Pongo pygmaeus*), and Capuchin Monkeys (*Cebus apella*). *Journal of Comparative Psychology, 109*, 52-60.

Visalberghi, E., & Limongelli, L. (1994). Lack of Comprehension of Cause-Effect Relations in Tool- Using Capuchin Monkeys (*Cebus apella*). *Journal of Comparative Psychology, 108*, 15-22.

Visalberghi, E., & Tomasello, M. (1998). Primate Causal Understanding in the Physical and Psychological Domains. *Behavioural Processes, 42*, 189-203.

Visalberghi, E., & Trinca, L. (1989). Tool Use in Capuchin Monkeys: Distinguishing Between Performing and Understanding. *Primates, 30*, 511-521.

Vlamings, P., Barth, J., & Call, J. (2003). Inhibition in Great Apes: Performance on a Detour-Reaching Task. *Folia Primatologica, 74*, 227.

Vlamings, P., Hare, B., & Call, J. (2010). Reaching Around Barriers: The Performance of the Great Apes and 3–5-Year-Old Children. *Animal Cognition, 13*, 273-285.

Yamazaki, Y., Iriki, A., & Watanabe, S. (2011). Modulation of Physical Understanding by Common Marmosets (*Callithrix jacchus*). *Animal Cognition, 14*, 175-186.

Reviewed by Bonaventura Majolo, School of Psychology, University of Lincoln, Lincoln, U.K.

ZOONOTIC DISEASES

In: Monkeys
Editor: Hugo A. Barrera Saldaña

ISBN: 978-1-63117-851-1

Chapter 6

ZOONOSES AND ANTHROPONOSES IN NON-HUMAN PRIMATES OF COSTA RICA

A. Chaves[*] and G. Gutiérrez-Espeleta
Escuela de Biología, Universidad de Costa Rica, Costa Rica

ABSTRACT

The majority of emerging infectious diseases are zoonotics. Animal wildlife represents a large and unknown reservoir of new and emerging zoonotic diseases, or source for the reemergence of zoonosis that were believed under control. The group of animals that represents the greatest threat for transmitting diseases to humans are non-human primates because they are susceptible to a variety of similar or identical infectious agents that affect humans.

We have decided to illustrate those pathogens that over the last few years have been recognized as disease-causing infectious agents specifically in the Neotropical region: West Nile Virus (WNV), Dengue Virus (DENV), Malaria, Tuberculosis and *Toxoplasma gondii.*

Keywords: Non-human primates, zonoses, antrhoponoses, West Nile Virus, Dengue Virus, Malaria, Tuberculosis, T. gondii, Costa Rica

INTRODUCTION

Status of Nonhuman Primates in Costa Rica

Non-human primates are animals that have always attracted the interest of human beings due to their characteristics and close evolutionary relationship (Kennedy et al., 1997), or cultural and scientific meaning (Flynn et al., 2003).

---

[*] Corresponding author: A. Chaves. Escuela de Biología, Universidad de Costa Rica. Teléfonos (506) 25118652 y (506) 25115955. E-mail: andrea.chavesramirez@ucr.ac.cr.

But human activities have caused habitat loss and vulnerability of these species. Currently in the world of the 394 species evaluated, 114 species of non-human primates are classified as in danger of extinction according to the IUCN Red List (IUCN, 2013).

Costa Rica has four non-human primate species in different regions of the country. All four species have suffered a drastic reduction both in distribution and abundance. Currently, according to IUCN 2013, the red spider monkey *(Atelesgeoffroyi)* is endangered and has a decreasing population trend. The squirrel monkey *(Saimirioerstedii)* is categorized as vulnerable with a decreasing population trend. The white-faced monkey (*Cebuscapucinus*) and the howler monkey *(Alouattapalliata)* are classified as of least concern with an unknown population trend (IUCN, 2013).

Habitat loss, due to the disappearance of tropical forests for agriculture and illegal logging continues to be the most important factor in reducing the number of non-human primates (IUCN, 2013). However, the impact of infectious diseases on wild animal populations has been recognized as an important factor influencing the distribution and density of species (Anderson, 1979). Likewise, parasitic infections have been identified as critical components to be considered in conservation biology of the species (May, 1988; Scott, 1988).

As the natural habitat of nonhuman primates is reduced in size and fragmented by human activity, populations are concentrated and reduced to narrow areas, favoring the opportunity for disease increase and transmission (Gilbert and Dodds, 1987). The fragmented landscapes result in a decrease in the availability of adequate nutritional resources, reduction in habitat size and repeated use of aerial routes. The inability to increase the size of their home ranges or the magnitude of their movements has enhanced disease transmission due to the possibility of humans and their proximity to their pets (Scott, 1988). These risks have a negative impact on the health of individuals and their potential for development, impacting directly on their conservation status and an obvious risk to human populations near these fragments.

Deforestation in the tropics, with the resulting loss of plant and animal species, has been well documented (Sader and Joyce, 1988). In Costa Rica, despite a much-publicized effort to devote extensive areas to national parks and reserves, large forest areas have been destroyed at a very fast rate and the pace of deforestation has not diminished much. Deforestation, along with hunting and illegal trade, threatens the survival of wild primates living in tropical regions (Thorington and Heltne, 1976). The extent of habitat disturbance and destruction in Costa Rica led the Costa Rican Wildlife Service to elaborate a list of endangered species. Most monkeys in Costa Rica, who are obligate herbivores that utilize much of the plant species present in the forest and depend on these for their survival (Glander, 1978; Crockett and Eisenberg, 1987), are included in the Costa Rican Wildlife Service list of endangered species.

Conservation efforts in Costa Rica have been mostly directed to the creation of national parks and protected areas. However, this strategy may not guarantee the medium- and long-term conservation of primate species within the protected areas. Habitat destruction adjacent to the protected zones and, hence, the disappearance of primate populations from those areas, isolates protected populations and limits the exchange of genes with other groups. Moreover, because of limitations in the size of areas destined for conservation, primate populations within them have relatively small sizes. Populations living outside protected areas have even smaller sizes. In addition, among some of these species (i.e. *A. palliata*), only the dominant male is believed to father the offspring produced within the group (Glander, 1980).

This further reduces the effective size of the population and promotes genetic drift. In summary, both the lack of migration and the small size of the populations result in the loss of genetic variation. Although the exact relationship between genetic variation and species conservation cannot be easily modeled, there is a positive relationship between both variables; i.e., the greater the amount of genetic variation the greater the chances of medium- and long-term conservation (Soule, 1986).

Non-human primates may illustrate the impact of deforestation and the effectiveness of conservation efforts in the preservation of biodiversity, particularly of animal populations.

Zoonosis and Anthroponosis

The risk that an animal may be carrying zoonotic pathogens depends on the taxonomic position and the region of origin of the species to which it belongs. It can be considered that the order of increased risk in non-human primates is the following: prosimians, marmosets, tamarins, New World monkeys, Old World monkeys and apes. This risk is also greater in non-human primates captured in the wild than in captive nonhuman primates bred in captivity and maintained in a defined environment under veterinary supervision (OIE - Código Sanitario para los Animales Terrestres, 2011).

It is important to point out that animal wildlife health can provide an indication of health in an ecosystem, as the health of wildlife integrating it is a reflection of the state of the ecosystem, since in the species present it is observed or determined that alterations such as the presence of pathogens, may reflect changes that occur in the dynamics of the ecosystem.

Therefore, interactions that occur in a specific environment, determine serious consequences on the ecosystem, such as, increased competition for habitat or physical spaces, greater predation, and increased disease transmission between species.

Interrelationships between human beings, and domestic and wild animals, as well as the impact of these interactions in ecosystems, impact the occurrence of emerging diseases or resurgence of other already controlled reemerging diseases.

Non-human primates are an example of animal populations that represent many wild reservoirs for zoonotic diseases. Environmental degradation, pollution, climate changes and other anthropogenic factors can cause changes in the dynamics of these populations. Changes in migration patterns and population density increase may be anthropogenic effects of these actions, which in turn can affect patterns of infection and parasite load of wild populations. Therefore, the occurrence of zoonosis and public health problems obey well-established and potentially preventable causes (Wallis and Rick Lee, 1999).

More than half (62%) of human pathogens are zoonotic (Taylor, 2001) and constitute the bulk of emerging and reemerging diseases.

Many of these diseases are recognized as being of global significance and of worldwide distribution, but generally, they are more common in tropical areas where the conditions for transmission are particularly favorable.

In most zoonosis the actual incidence is unknown due to the lack of knowledge of the disease. A large proportion of subclinical infection may go unnoticed, and diagnostic methods are not available in endemic areas (Haydon et al., 2002).

In this case, it is essential to determine the presence of zoonotic pathogens of interest in the region.

We have decided to illustrate those pathogens that over the last few years have been recognized as disease-causing agents specifically in the neotropical region, and the possible relation between these pathogens in human beings and in the primates of the New World: West Nile Virus (WNV), Dengue Virus (DENV), Malaria, Tuberculosis and *Toxoplasma gondii.* Although there is no clear understanding of the role played by non-human primates in the cycles of some of the pathogens in the Neotropics (WNV and DENV), it was decided to include them because of their importance to the region and the close social and biological relationship between humans and non-human primates.

West Nile Virus

West Nile Virus (WNV) is a member of the genus *Flavivirus* (Flaviviridae) that forms part of the Japanese Encephalitis Virus serocomplex (Murphy et al., 1995). WNV surfaced as an emerging infectious disease in the northeastern United States in 1999 and it gradually spread across the continent. Countries of Latin America and the Caribbean that reported activity for WNV between 2001 and 2004 are: Mexico, Belize, Guatemala, El Salvador, Cuba, Bahamas, Cayman Islands, Jamaica, Dominican Republic, Puerto Rico (United States), Guadeloupe (French West Indies), Trinidad and Tobago, Colombia, Venezuela, and Argentina (Komar and Clark, 2006; Bosch et al., 2007). In 2003, the first case for Central America was reported (El Salvador) (Cruz, et al., 2005) and in 2005, for South America (Colombia) (Mattar, et al., 2005). In Costa Rica, according to the National Service for Animal Health (SENASA), WNV was first documented in 2009, and is currently considered endemic across the country. WNV circulates in natural transmission cycles involving primarily *Culex* species mosquitoes and birds, and humans are incidental hosts. Although human infections in endemic areas are common, such infections are usually mild or subclinical. Severe disease is commonly associated with the elderly (Hayes, 1989). WNV infection can cause mortality among equines as well as among certain domestic and wild birds (Komar, 2003). To date, most of the studies of WNV ecology have focused on humans, equines and a few susceptible bird species. However, WNV has an extremely wide host range among vertebrates (CDC, 2002; Endy and Nisalak, 2002). The role of some of these other species in the amplification and maintenance of WNV in North America remains to be determined. WNV activity was demonstrated in surveillance at the nonhuman primate-breeding colony and in the neighboring community during this same period. The high infection rate in this captive nonhuman primate population illustrates the intensity of WNV transmission that can occur silently in nature among other susceptible vertebrates during epidemic periods (Ratterree, et al., 2003).

Dengue Virus

Dengue virus (DENV) is a positive-sense, single-stranded RNA virus of the genus *Flavivirus* (Flaviviridae) that uses mosquitoes of the genus *Aedes* as transmission vectors among primates. DENV occurs as four antigenetically distinct but genetically related serotypes (DENV-1 to DENV-4) within the dengue antigenic complex (Halstead, 1988).

Each of the four DENV serotypes is maintained in two ecologically and evolutionarily distinct transmission cycles: The transmission of DENV presents two distinct cycles: (i) endemic / epidemic or urban that involves humans as hosts and transmission by *Ae. aegypti* with *Ae. Albopictus* and other *Aedes* mosquitoes serving as secondary vectors, and (ii) a zoonotic or sylvatic cycle in forest habitats of Central Africa, East Africa and Southeast Asia, involving nonhuman primates as reservoir hosts and various canopy mosquitos from the genus *Aedes* (Feng and Velasco-Hernández, 1997; Weaver and Vasilakis, 2009).

The sylvatic transmission cycles have been little studied. However, phylogenetic studies suggest that this jungle virus form is ancestral, which means that the origin of urban DENV is the jungle (Weaver and Vasilakis, 2009). In addition, entomological studies agree with this finding by proposing that the vector linking the sylvatic cycle and human communities located in or near the woods, could be *Ae. furcifer,* a forest mosquito with easy dispersion in peridomestic habitats (Weaver and Vasilakis, 2009). Studies suggest that adaptation to new vectors and vertebrate hosts may have occurred repeatedly due to selective pressures that might lead to genetic changes in the pathogen (Gubler, 2001; Weaver and Vasilakis, 2009).

DENV has a close relationship with humans from about 2000 years ago, which has intensified in recent decades due to altered ecological conditions and human behavior (global trade, urbanization, large population movements) and unsustainable vector control programs (Gubler, 2001; Weaver and Vasilakis, 2009). Since *Ae. aegypti* did not exist in Asia and Oceania before the establishment of trade ships, *Ae. albopictus* was probably the original vector in humans. The role of *Ae. aegypti* as a vector may have occurred only during the last few centuries, when mosquitoes were distributed in the tropics via trading vessels (Weaver and Vasilakis, 2009).

In the New World, unlike the Old World, it is believed that dengue infection is absent in wildlife since the viral transmission cycle was endemically introduced, with the emergence of *Ae. aegypti* in the tropics of the world (Weaver and Vasilakis, 2009). However, several studies have documented the presence of the virus in mammalian wildlife in New World countries (Platt et al., 2000; de Thoisy et al., 2004; Setién Aguilar et al., 2008; Lavergne et al., 2009; de Thoisy et al. 2009). However, it is unknown what might be the potential mosquito vectors in the forest, as *Ae. aegypti* is usually associated with urban sites (Lavergne et al., 2009). Changes in tropical land use, particularly the insertion of human habitations and agriculture into previously undeveloped areas that are enzootic for arboviruses, could promote interactions among DENV vectors and hosts (Vasilaskis et al., 2012), increasing the likelyhood of sylvatic-DENV re-emergence. Alternatively, deforestation could limit enzootic DENV circulation by the reduction of the habitat for mosquito vectors and monkey hosts.

MALARIA

Malaria is a parasitic infection caused by protozoa of the genus *Plasmodium*, family Plasmodiidae, suborder Haemosporidae, order Coccidia (Siden et al., 2002). Approximately 200 species of mammals, reptiles and birds are potential hosts to the parasite.

Primates represent more than half of the host species (Perkins et al., 2009). The protozoa is present in multiple species of prosimians, New World monkeys, Old World monkeys, African apes, and Asian apes (Coatney et al., 1971; Krief et al., 2010).

Most malarias are transmitted through mosquito vectors and all human malarias are transmitted by *Anopheles* mosquitoes (Sindenand Gilles, 2002).

This group of malaria parasites includes at least 172 known species, parasites of birds, reptiles, rodents and primates (Leclerc et al., 2004). But only four *(P. falciparum, P. vivax, P. ovale and* P. malariae) infect humans (Collins and Aikawa, 1977).

The two most common species are: *P. falciparum*, which has a global distribution, but is most common in Africa and is the most aggressive, causing mainly death, coma or anemia. *P. vivax* has a global distribution and may cause debilitating and recurrent infections, but rarely death (Tramputz et al., 2003; Batista et al., 2009).

Currently, 22 species are recognized as nonhuman primate *Plasmodium* (Cornier, 2011): Two species (*P. brasilianum and P. simium*) occur in New World, monkeys of the Cebidae family; 11 (*P. coatneyi, P. cynomolgi, P. fieldi, P. fragile, P. gonderi, P. georgesi, P. inui, P. knowlesi, P. petersi, P. shortti* and *P. simiovale)* in Old World monkeys of the Cercopithecidae family; 4 (*P. eylesi, P. hylobati, P. jefferyi, P. youngi*) in gibbons of the Hylobatidae family (localized in South East Asia), and 5 in the great apes, *Pan troglodytes* and *Gorilla gorilla* in West and Central Africa (*P. reichenowi, P. rodhaini, P. schwetzi*), and *Pongo pygmaeus* in South Asia (*P. pitheci* and *P. silvaticum*).

Under experimental conditions, numerous New World monkey species have been infected with human *P. falciparum* and *P. vivax* including howler monkeys (*Alouatta* spp.), owl monkeys *(Aotus* spp.), spider monkeys (*Ateles* spp.), squirrel monkeys (*Saimiri* spp.), and tamarins (*Saguinus* sp.) (Coatney et al., 1971, Collins, 1994).

New World monkeys, unlike Old World monkeys are susceptible to *P. falciparum* (Martin et al., 2005). Only two species of simian malaria have been described in Brazil: *P. brasilianum*, first described in bald uakari (*Cacajao calvus*) in the upper Amazon (northern Brazil) and *Plasmodium simium*, a tertian malaria, in a howler monkey (*Alouatta fusca*) in the State of São Paulo (southern Brazil) (Araujo et al., 2013).

P. brasilianum has the widest host range. It has been described as infecting 35 species of monkeys, mostly Cebidae and Atelidae, and rarely in members of the family Callithricidae (Lorenco and Deane, 1995).

In contrast, *P. simium* has been described in only three primate species: *Alouatta guariba, Brachyteles arachnoids*and *Alouattacaraya*. Geographic distribution of hosts infected by *P. simium* is restricted to the Atlantic forest (southern Brazil) while *P. brasilianum* infected hosts have been described in Brazil, Panama, Venezuela, Peru and Colombia. *P. simium* was never described in other countries or even in northern Brazil. However, the lists of both susceptible monkey species and malaria parasite species themselves in the Amazon forest might be incomplete (Araujo, et al., 2013).

Costa Rica has been a country of low malaria incidence. The incidence of malaria has declined in the country since 1999, but in 2005, there was a spike which could be dangerous if not controlled. Chinchilla and colleagues (2006) studied 104 animals and three specimens from Cahuita, Limón presented very immature forms of the parasite.

In the Central Pacific (La Uvita), three *A. palliata* were positive with a fair amount of forms of trophozoites, gametocytes, and schizonts, easily identifiable and appropriate to carry out classification according to *P. brasilanum*.

TUBERCULOSIS

Tuberculosis is an infectious disease caused by a subspecies of the *Mycobacterium tuberculosis* complex (MTBC) (*M. tuberculosis*, *M. africanum*, *M. bovis*, *M. canettii*, *M. microti*, and others, together known as the MTBC). Some traditionally endemic diseases, including diseases with control programs, can, in certain circumstances, increase its incidence. This has happened with tuberculosis, which is of great importance in the fields of public and animal health. In this regard, it is estimated that between 2002 and 2020 about one billion people will be infected, more than 150 million will develop the disease and 36 million will die from tuberculosis. This disease, caused by *Mycobacterium tuberculosis*, has probably affected humans since its origin, but 40 years ago it was hoped that people would soon be free of it. Now that its incidence is on the rise, we can consider it a re-emerging disease (Dye et al., 1999).

The MTBC, which includes *M. tuberculosis*, *M. bovis* and *M. africanum*, has been implicated as the cause of tuberculosis in several genera of captive and free-ranging non human primates. Tuberculosis (TB) is globally distributed, infecting an estimated 2 billion humans and causing some 2 million deaths per year (WHO, 2009). Concern about diseases like tuberculosis, plays an increasing role in the conservation and management of non human primates. However, little is known about the prevalence of mycobacterial infection in non human primates and in the New World few studies of non-human primates kept in captivity have been done. Monkeys acquire classic TB by contact with other sick monkeys or humans via inhalation or the digestive route (Moreland, 1970), becoming either reservoirs or disease sources.

TB has been mainly reported in captive animals, but there are some studies describing this disease in free-ranging wild animals. The degree of susceptibility to TB varies in different primate genera; the New World species *Callithrix*, *Cebus* and *Saimiri* have been observed to be the most susceptible (Leathers and Hamm, 1976), whilst monkeys from the *Aotus* genera may have a certain degree of natural resistance to the disease (Bone and Soave, 1970; Moreland, 1970). Old World primates are thought to be more susceptible to TB than New World monkeys; disseminated TB in captive baboons has been reported (Fourie and Odendaal, 1983; Thorel et al., 1998) and macaques are considered to be disease reservoirs. The presence of mycobacteria as disease-causing agents in New World primates has been little studied and only a few reports mention isolated cases of TB that have appeared in monkeys kept in captivity (Moreland, 1970; Snyder et al., 1970). After colonizing the respiratory tract via inhalation or the digestive tract via ingestion, mycobacteria can be detected in bronchial lavage, gastric lavage, feces and blood (Good, 1984).

TOXOPLASMOSIS

Toxoplasma gondii is the only protozoan responsible for toxoplasmosis. It presents oocysts with two sporocysts, each of which have four sporozoites (Dubey et al., 1998). Heteroxenous parasites are either optional or mandatory. Toxoplasmosis is a disease of worldwide distribution in a wide variety of mammals and birds, including humans and all domestic mammals (Nicolle and Manceaux, 1908, Dubey and Frenkel, 1972).

Toxoplasma is eliminated from infected felines, through urine, saliva, nasal, ocular and vaginal secretions, milk and feces (Elmore et al., in *press*). The definitive hosts confirmed in America are the domestic cat *(Felis catus),* jaguarundi *(Herpailurus jaguaroundi),* ocelot *(Leopardus pardalis)* and mountain lion *(Puma concolor).*

This acute fatal infection affects New World monkeys, marsupials and some other animals. New World monkeys are particularly susceptible to clinical toxoplasmosis, while Old World monkeys are resistant. Experimental infection was performed with *T. gondii* in New World monkeys and owl monkeys *(Aotus* spp) and two species of Callitrichids *(Saguinus* spp) that resulted in fatal infection.

Natural infection has happened in Cebids, including the genera A*llouatta, Aotus, Ateles, Cacajao, Callicebus, Cebus, Lagothrix, Pithecia* and *Saimiri*. Many of these cases are reported from zoos where the source of infection is the diet. This association between feeding and contaminated food results in fatal toxoplasmosis in New World monkeys. The sudden and strong characteristics of the disease may be due to the fact that these primates are highly susceptible to infection by *Toxoplasma* and are deficient in resistance to this parasite (Furuta et al., 2001; Carme et al., 2009; Dubey, 2010).

Reports (Salles et al., 1998) indicate that *T. gondii* can cause latent infection in New World monkeys and therefore potentially cause abortions and congenital infection in primate females of the *Leontopithecus rosalia* species that may be reintroduced to the wild.

Cunningham et al. (1992) reported that toxoplasmosis naturally causes 30% mortality and 100% morbidity, suggesting immune compromise prior to the clinical presentation of the disease and death.

Infection in some animals and humans may be prevalent, but it is usually asymptomatic; however, in New World primates, because of their susceptibility, the disease does not permit a clear clinical manifestation and they perish in a short interval.

COSTA RICA – CURRENT INVESTIGATION

In Costa Rica, the presence of Dengue Virus, West Nile Virus, *Plasmodium sp*, *Mycobacterium* sp and *Toxoplasma gondii* was evaluated. We sampled a total of 598 NHP all over the country between the years 1993 to 1995 and 2001 to 2013.

The species sampled were *Alouatta palliata* (n= 354), *Cebus capucinus* (n= 92), *Ateles geoffroyi* (n=61), and *Saimiri oerstedii* (n=54), including samples of free ranging NHP (n=491) from national parks, and other protected and non-protected areas, as well as individuals kept in captivity (n=92). For detection of infectious agents the Polymerase Chain Reaction (PCR) was used with the following protocols: *Mycobacterium* and more specifically *M. Tuberculosis*, Del Portillo et al. (1991) and Matsuo et al. (1990); Lanciotti et al. (1992) for Dengue virus, types 1 to 4; Bretagne et al. (1993) for Toxoplasma *gondii*; Lanciotti et al. (2000) for West Nile Virus; and Rubio et al. (2002) for *Plasmodium.*

Thirty-two individuals positive to *Plasmodium* were detected from a total of 367 blood samples belonging to four species of monkeys in Costa Rica. These were caught in the provinces of Alajuela, Puntarenas, Guanacaste, Heredia and Limón. Both in fragmented forests, as well as unfragmented forests, the presence of positive primates were detected (Chaves et al., 2011; Chaves et al., 2013a).

This discovery had been previously reported by Chinchilla et al. (2006) using one portion of the samples in 6 individuals of the species *A. palliate* from Puntarenas (3) and Limon (3), performing the diagnosis by culture and also through extended Giemsa stained slides for subsequent morphometric analysis to determine the presence of *P. malariae*.

Of 207 individuals tested for DENV, eight were positive, four for *A. palliata*, one for *C. capucinus* type 2, one for *C. capucinus* type 3, and two for *C. capucinus* type 4. Four positive cases were found in a group of animals taken from the retrospective study from the same site (Palo Verde), in the province of Guanacaste, in an unfragmented forest classified as wetland. The other four positive individuals were from the province of Puntarenas, from individuals caught in the Osa peninsula in protected areas of rainforest near wetlands (Chaves et al., 2011; Jiménez et al., 2013; Dolz, et al., 2013). According to Tran et al. (2008), wetlands are characterized by high density of mosquitoes and thus possible vectors of disease.

Four of 267 individuals tested were positive for West Nile virus, one for *C. capucinus* and three for *A. palliata*. Three positive monkeys (*A. palliata*) were captured in Guanacaste (Nosara and Ostional sites), sites with characteristics of fragmented forest, close to human settlements.

Meanwhile one *C. capucinus* held captive was found positive in the province of Guanacaste (Cañas) on a site with similar environmental characteristics to the site for wildlife positives. This represents the first report of molecular detection of West Nile virus in non human primates in the wild (Chaves et al., 2011; Jiménez et al., 2013; Dolz et al., 2013).

The presence of *T. gondii* was determined in blood and plasma samples of 45 *S. oerstedii* and 76 *A. mantled* for a total of 16 positives, 10 positive individuals were determined of *S. oerstedii* and six of *A. palliata*. The positive individuals belong to Puntarenas and Limon in protected sites (Cahuita National Park and Manuel Antonio National Park), but heavily influenced by human activities due to high tourist visitation and nearby villages (Chaves et al., 2013b).

For *Mycobacterium sp* a total of 598 serum and plasma samples were analyzed by PCR amplification of the Alpha Antigen, present in all reported mycobacteria, and the *Mycobacterium tuberculosis* Mtp40 species-specific gene; no positive individuals were detected in either the wild or in captivity (Chaves et al., 2013c).

Although blood samples are not ideal for the detection of this agent, it has been reported in positive individuals in captivity using blood samples and these samples are ideal for working with individuals captured in the field (Wilbur et al., 2012). Meanwhile Costa Rica is a country with a very low incidence of human tuberculosis (WHO, 2009).

With regard to environmental characteristics, all sampled sites had temperatures of 25°C to 30°C, high rainfall throughout the year, except for Ostional and Nosara, Guanacaste (where individuals were detected positive for WNV) and Palo Verde, Guanacaste (where individuals were detected positive for DENV), which had a marked dry season of 4 to 5 months (Dolz, et al., 2013). The distribution of the sampling sites was at an altitude less than 500 meters above sea level, which is also the highest reported for the presence of non human primates in Costa Rica (Zaldivar, et al., 2004).

With this investigation, the presence of microbial agents reported as a situation that could impact populations of non human primates in natural habitats. The impact of these agents on animal populations in the wild has been recognized as an important factor influencing the distribution and abundance of species (Anderson, 1979).

REFERENCES

Aguilar-Setién, A., M. L. Romero-Almaraz, C. Sánchez-Hernández, R. Figueroa, L. P. Juárez-Palma, M. M. García-Flores, C. Vázquez-Salinas, M. Salas-Rojas, A. C. Hidalgo-Martínez, S. Pierlé-Aguilar, C. García-Estrada, C. Ramos. 2008. Dengue virus in mexican bats. *Epidemiology and infection.* 136 (12): 1678-83.

Araújo, M., Messias, M. R., Figueiró, M. R., Gil, L. H., Probst, C. M., Vidal, N., Katsuragawa, T. H., Krieger, M. A., Pereira da Silva, L., Ozaki, L. S., 2013. Natural Plasmodium infection in monkeys in the state of Rondônia (Brazilian Western Amazon). *Malaria Journal* 12:180.

Anderson, R. M. 1979, Parasite pathogenicity and the depression of host population equilibria. *Nature* 279:150-152.

Batista, R., A. Silva and A. Braga de Oliveria. 2009. Plantderived antimalarial agents: New leads and efficient phytomedicines. Part II. Non-Alkaloidal Natural Products. *Molecules* 14: 3037-3072.

Bosch, I., Herrera, F., Navarro, J., Lentino, M., Dupuis, A., Maffei, J., Jones, M., Fernández, E., Pérez, N., Pérez-Emán, J., Guimarães, A. E., Barrera, R., Valero, N., Ruiz, J., Velásquez, G., Martinez, J., Comach, G., Komar, N., Spielman, A., Kramer, L. 2007. West Nile virus, Venezuela. *Emerg. Infect. Dis.* 13: 651-653.

Carme, B., Demar, M., Ajzenberg, D., Darde, M. L., 2009b. Severe acquired toxoplasmosis caused by wild cycle of Toxoplasma gondii, French Guiana. *Emerg. Infect. Dis.* 15 (4), 656-658.

CDC, Centers for Disease Control and Prevention. Provisional surveillance summary of the West Nile virus epidemic-United States, January–November 2002. *MMWR Morb. Mortal. Wkly Rep.* 2002;51:1129–33.

Chaves, A., Ortíz, E., Jiménez, A., Jiménez, M., Gutiérrez-Espeleta, G., Corrales, R., Avendaño, R., Hernández, J., Bernal, S., Hagnauer, I., Valverde, N., DiMare, M. L., Sánchez, R., Wong, G., Dolz, G. 2011. Presencia de agentes infecciosos en primates no humanos de costa rica y caracterización ambiental, resultados preliminares. *Congreso de Medicina de la Conservación.* Querétaro, México.

Chaves, A., Gutiérrez-Espeleta, G., Ortiz, E., Jiménez, M., Bernal, S., Dolz, G. 2013a. *Distribution of different Plasmodium species in non-human primates from Costa Rica.* Unpublished.

Chaves, A., Ardón, C., Gutiérrez-Espeleta, G. 2013b. *Toxoplasma gondii in non-human primates from Costa Rica.* Unpublished.

Chaves, A., Santos, A., Sotomayor, J. J., Gutiérrez-Espeleta, G. 2013c. *Molecular detection of Mycobacterium sp. in free-ranging and captivity non-human primates from Costa Rica.* Unpublished.

Chinchilla, M., Guerrero, O. M., Gutiérrez, G., Sánchez, R. 2006. Presencia de *Plasmodium brasilianum* (Apicomplexa, Plasmodidae) en el mono congo (*Alouatta palliata*, Primates: Cebidae) de Costa Rica. Importancia epidemiológica en relación con el ser humano. *Parasitol. Latinoam.* 61: 192 – 196.

Chomel, B. B., Belotto, A., Meslin, F. X. 2007. Wildlife, exotic pets, and emerging zoonoses. *Emerging Infectious Diseases*, 13: 6–11.

Coatney, G. R., Collins, W. E., Warren, M., Contacos, P. G. 1971. *The primate malarias*. US Department of Health, Education, and Bethesda, MD, US: Welfare.

Cormier, L. A. 2011 *The Ten-Thousand Year Fever: Rethinking Human and Wild Primate Malaria*. Left Coast Press, CA, US.

Collins, W. E. 1994. The owl monkey as a model for malaria. In: *Aotus: The Owl Monkey*. Baer, J., Weller, R. E., Kakoma, I., Eds.; Academic Press: San Diego, CA, US, pp. 217-244.

Collins, C. 2000. The bovine tubercle bacillus. *Br. J. Biomed.*, 57, 234-240.

Collins, W. E. and Aikawa, M. 1977. Plasmodia in non-human primates. *Parasitic Protozoa*. Vol. 3. New York, Academic Press.

Crockett, C. M. and Eisenberg, J. F. 1987. Howlers: Variations in group size and demography. Pages 54-69 In: B. B. Smuts, D. L. Cheney, R. M. Seyfarth, R. W. Wrangham and T. T. Struhsaker, editors. *Primate Societies*. University of Chicago Press, Chicago.

Cruz, L., Cardenas, V. M., Abarca, M., Rodriguez, T., Reyna, R. F., Serpas, M. V. 2005. Serological evidence of West Nile virus activity in El Salvador. *Am. J. Trop. Med. Hyg*. 72 (5):612–5.

Cunningham, A. A., Buxton, D. and Thomson, K. M. 1992. An epidemic of toxoplasmosis in a captive colony of squirrel monkeys (Saimirisciureus). *J. Comp. Pathol*. 107:207-219.

Deane, L. M., Ferreira-Neto, J. A. 1973. Malária de macacos no Estado de Goiás, Brasil: Encontro de guaribas, Alouattacaraya infectados comPlasmodiumbrasilianum. *Rev. Inst. Med. Trop. Sao Paulo*. 15:107–111.

De Thoisy, B., Dussart, P. and Kazanji, M. 2004. Wild terrestrial rainforest mammals as potential reservoirs for flaviviruses (yellow fever, dengue 2 and St. Louis encephalitis viruses) in French Guiana. *Transactions of the Royal Society of Tropical Medicine and Hygiene*. 98 (7): 409-12.

De Thoisy, B., Lacoste, V., Germain, A., Muñoz-Jordán, J., Colón, C., Mauffrey, J. F., Delaval, M., Catzeflis, F., Kazanji, M., Matheus, S., Dussart, P., Morvan, J., Aguilar-Setién, A., Deparis, X., and Lavergne, A. 2009. Dengue infection in neotropical forest mammals. *Vector borne and zoonotic diseases*. 9 (2): 157-70.

Dolz, G., Chaves, A., Gutiérrez-Espeleta, G., Ortiz, E., Jiménez, M., Bernal, S. 2013. *Serological and molecular detection of flavivirus in wild non-human primates from Costa Rica*. Unpublished.

Dubey, J. P. and Frenkel, J. K. 1972. Cyst-induced toxoplasmosis in cats. *J. Protozool*. 19: 155–177.

Dubey, J. P., Lindsay, D. S. and Speer, C. A. 1998.Structures of *Toxoplasma gondii* Tachyzoites, Bradyzoites, and Sporozoites and Biology and Development of Tissue Cysts. *Clin. Microbiol. Rev*. 11(2): 267–299.

Dubey, J. P., 2010. *Toxoplasmosis of Animals and Humans*. CRC Press, Boca Raton, FL, US, 313 pp.

Elmore, S., Jones, J., Conrad, P., Patton, S., Lindsay, D. S., and Dubey, J. P. Toxoplasma gondii: epidemiology, feline clinical aspects, and prevention. *Cell*. In press.

Dye, C., Scheele, S., Dolin, P., Pathania, V., Raviglione, M. C. 1999. Consensus statement. Global burden of tuberculosis: estimated incidence, prevalence, and mortality by country. WHO Global Surveillance and Monitoring Project. *J. Am. Med. Assoc*. 282: 677–686.

Endy, T. P. and Nisalak, A. 2002. Japanese encephalitis virus: ecology and epidemiology. *Curr. Top Microbiol. Immunol*. 267:11–48.

Feng, Z. and Velasco-Hernández, J. X. 1997. Competitive exclusion in a vector-host model for the dengue fever. *Journal of Mathematical Biology*. 35: 523–544.

Fiennes, R. 1967. *Zoonoses of primates: the epidemiology and ecology of simian diseases in relation to man.* Weidenfeld and Nicholson, London.

Flynn, J., Capuano, S., Croix, D., Pawar, S., Myers, A., Zinovik, A., Klein, E. 2003. Nonhuman primates: a model of tuberculosis research. *Tuberculosis* 83: 116-118.

Fourie, P. B. and Odendaal, M. W. 1983. *Mycobacterium tuberculosis* in a closed colony of baboons (*Papioursinus*), *Lab. Anim.* 17: 125 – 128.

Furuta, T., Une, Y., Omura, M., Matsutani, N., Nomura, Y., Kikuchi, T., Hattori, S., Yoshikawa, Y., 2001. Horizontal transmission of Toxoplasma gondii in squirrel monkeys (Saimirisciureus). *Exp. Anim.* 50 (4), 299-306.

Gilbert and Dodds. 1987, *The philosophy and practice of wildlife management.* Robert E. Krieger Publishing, Malabar, Florida.

Glander, K. E. 1978. Howling monkey feeding behavior and plant secondary compounds: A study of strategies. Pages 561-574 In: G. G. Montgomery, (editor). *The Ecology of Arboreal Folivores.* Smithsonian Institution Press, Washington, DC.

Glander, K. E. 1980. Reproduction and population growth in free-ranging mantled howling monkeys. *American Journal of Primatology* 53: 25-36.

Good, R. C., 1984. Diseases in nonhuman primates, In: Kubica, G. P., Wayne, L. G. (Eds.), *The Mycobacteria: A Sourcebook.* Dekker, New York, pp. 903–924.

Gubler, D. J. 2001. Human arbovirus infections worldwide. *Annals of the New York Academy of Sciences.* 951: 13-24.

Halstead, S. B. 1988. Pathogenesis of dengue: challenges to molecular biology. *Science* 239, 476-481.

Haydon, D., Cleaveland, S., Taylor, L., Laurenson, M. K. 2002. Identifying Reservoirs of Infection: A Conceptual and Practical Challenge. *Emerging Infectious Diseases* 8 12: 1468-1473.

Hayes, C. G. 1989. West Nile fever, p. 59–88. In: T. P. Monath (ed.), *The arboviruses: epidemiology and ecology,* vol. V. CRC Press, Inc., Boca Raton, Fla.

IUCN 2013. *IUCN Red List of Threatened Species.* Version 2013.2.<www.iucnredlist.org>. Downloaded on 25 November 2013.

Jiménez, A. E., Urbina, A., Chaves, A., Bernal, S., Calderon, A., Ortiz, E., Jiménez, M., Dolz, G. 2013. *Parasitological, mycological and virological findings in wild non-human primates from Costa Rica.* Unpublished.

Kennedy, R., Shearer, M., Hilderbrand, W. 1997. Nonhuman primate models to evaluate vaccine safety and immunogenicity. *Vaccine* 5 (8) 903-908.

Komar, N. 2003. Epidemiology and ecology in North America. *Adv. Vir. Res.* 61:185–234.

Komar, N. and Clark, G. G. 2006. West Nile virus activity in Latin America and the Caribbean. *Rev. Panam. Salud. Publica* vol. 19 no. 2 Washington Feb. 2006.

Krief, S., Escalante, A. A., Pacheco, M. A., Mugisha. L., André, C., Halbwax, M., Fischer, A., Krief, J. M., Kasenene, J. M., Crandfield, M., Cornejo, O. E., Chavatte, J. M., Lin, C., Letourneur, F., Grüner, A. C., McCutchan, T. F., Rénia, L., Snounou, G. 2010. On the diversity of malaria parasites in African apes and the origin of Plasmodium falciparum from Bonobos. *PLoS Pathog.*, 6:e1000765.

Lanciotti, R. S., Calisher, C. H., Gubler, D. J., Chang, G. J., Vorndam, A. V. 1992. Rapid detection and typing of dengue viruses from clinical samples by using reverse transcriptase-polymerase chain reaction. *Journal of Clinical Microbiology*. 30(3): 545–551.

Lanciotti, R. S., Kerst, A. J., Nasci, R. S., Godsey, M. S., Mitchell, C. J., Savage, H. M., Komar, N., Panella, N. A., Allen, B. C., Volpe, K. B., Davis, B. S., Roehrig, J. T., 2000. Rapid Detection of West Nile Virus from Human Clinical Specimens, Field-Collected Mosquitoes, and Avian Samples by a TaqMan Reverse Transcriptase-PCR Assay. *Journal of Clinical Microbiology* 38: 4066–4071.

Lavergne, A., Lacoste, V., Germain, A., Matheus, S., Dussart, P., Deparis, X., de Thoisy, B. 2009. Infection par le virus de la dengue de mammifères sauvages en région néotropicale : hôtes accidentels ou réservoirs potentiels ? *Médecine Tropicale*. 69 (4): 345-350.

Leclerc, M. C., Hugot, J. P., Durand, P., Renaud, F. 2004. Evolutionary relationships between 15 Plasmodium species from New and Old World primates (including humans): a 18S rDNAcladistic analysis. *Parasitology*, 129, pp. 677-684.

Leathers, C. W. and Hamm, T. E. 1976.Naturally occurring tuberculosis in a squirrel monkey and a cebus monkey. *J. Am. Vet. Med. Assoc*. 169(9): 909 – 911.

Lourenço-de-Oliveira, R., Deane, L. M. 1995. Simian Malaria at Two Sites in the Brazilian Amazon. *Mem. Inst. Oswaldo Cruz,* 90:331–339.

Martin, M. J., Rayner, J., Gagneux, P., Barnwell, J. W., Varki, A. 2005. Evolution of human-chimpanzee differences in malaria susceptibility: Relationship to human genetic loss of *N*-glycolylneuraminic acid. *Proc. Natl. Acad. Sci. US.* 102: 12819-12824.

Mattar, S., Edwards, E., Laguado, J., González, M., Alvarez, J., Komar, N. 2005. West Nile virus antibodies in Colombian horses. *Emerg. Infect. Dis.* 9: 1497-1498.

May, 1988. Conseravtion and Disease. *Conservation Biology* 2:28-30.

Moreland, A. F. 1970. Tuberculosis in New World primates. *Lab. Anim. Care* 20: 262-264.

Murphy, F. A., Fauquet, C. M., Bishop, D. H. L., Ghabrial, S. A., Jarvis, A. W., Martelli, G. P., Mayo, M. A., Summers, M. D. 1995. Virus taxonomy, classification and nomenclature of viruses. *Arch. Virol.* 10(Suppl.):1–586.

Nicolle, C. and Manceaux, L. 1908. Sur une infection à corps de Leishman (ouorganismes-voisins) du gondi. C. R. Hebd. *Seances Acad. Sci.* 147:763–766.

[OIE] Organización Mundial de Sanidad Animal. 2011. París: OIE. Código sanitario para los animales terrestres 1 (Cap. 1, 3). 144 p.

Perkins, S. L. and Austin, C. C. 2009. Four new species of Plasmodium from New Guinea lizards: integrating morphology and molecules. *J. Parasitol.* 95:424–433.

Platt, K. B., Mangiafico, J. A., Rocha, O. J., Zaldivar, M. E., Mora, J., Trueba, G., Rowley, W. A. 2000. Detection of dengue virus neutralizing antibodies in bats from Costa Rica and Ecuador. *Journal of medical entomology*. 37 (6): 965-7.

Ratterree, M. S., da Rosa, A. P., Bohm Jr., R. P., Cogswell, F. B., Phillippi, K. M., Caillouet, K., Schwanberger, S., Shope, R. E., Tesh, R. B. 2003. West Nile virus infection in nonhuman primate breeding colony, concurrent with human epidemic, southern Louisiana. *Emerging Infect. Dis*., 9 (11): 1388–1394.

Rubio, J. M., Post, R. J., Docters van Leeuwen, W. M., Henry, M. C., Lindergard, G., Hommel, M. 2002. Alternative polymerase chain reaction method to indentify *Plasmodium* species in human blood samples: semi-nested multiplex malaria PCR (SnM-PCR). *Transactions of the Royal Society of Tropical Medicine and Hygiene* 96: 199-204.

Sader, S. A. and Joyce, T. 1988. Deforestation rates and trends in Costa Rica, 1940 to 1983. *Biotropica* 20: 11-19.

Salles, C., Prats, N., Marco, A. J., 1998, Fatal acute toxoplasmosis in three golden lion tamarins (*Leontopithecusrosalia*). *J. Zoo Wild. Med.* 29:55–60.

Scott, 1988. The impact of infection and disease on animal populations: implications for conservation biology. *Conservation Biology* 2:40-56.

Sinden, R. E. 2002. Molecular Interactions between Plasmodium and its insect vectors. *Cell. Microbiol.* 4, 713-724.

Sinden, R. E. and Gilles, H. M. 2002. The malarial parasites. In: *Essential Malariology,* 4th ed.; Warrell, D. A., Gilles, H. M., Eds.; Oxford University Press: New York, NY, US, pp. 8-34.

Snyder, S., Peace, T., Soave, O., Lund, J. 1970. Tuberculosis in an owl monkey (*Aotustrivir-gatus*), *J. Am. Vet. Med. Assoc.* 157, 712 – 713.

Soulé, M. E. (editor). 1986. *Conservation Biology: The science of scarcity and diversity.* Sinauer Associates, Sunderland, MA.

Taylor, L. H., Latham, S. M., Woolhouse, M. E. J. Risk factors for human disease emergence. *Philos. Trans. R Soc. Lond. B Biol. Sci.* 2001;356:983–9.

Thorington, R. W. and Heltne, P. G. (editors). 1976. *Neotropical primates field studies and conservation.* National Academy of Sciences, Washington DC, US.

Thorel, M. F., Karoui, C., Varnerot, A., Fleury, C., Vincent, V. 1998. Isolation of *Mycobacteriumbovis* from baboons, leopards and a sea-lion, *Vet. Res.* 29, 207 – 212.

Tramputz, A., Jereb, M., Muslovic, I., Prabhu, R. M. 2003. Severe Malaria. *J. Cardioth. Surg.* 7: 315-323.

Vasilakis, N., Cardosa, J., Hanley, K. A., Holmes, E. C., Weaver, S. C. 2011. Fever from de forest: prospect for the continued emergence for sylvatic dengue virus and its impact on public health. *Nat. Rev. Microbiol.* 9 (7): 532-541.

Wallis, J. and Rick Lee, D. 1999. Primate Conservation: The Prevention of Disease Transmission. *International Journal of Primatology*. 20 (6): 803-826.

Weaver, S. C. and Vasilakis, N. 2009. Molecular evolution of dengue viruses: contributions of phylogenetics to understanding the history and epidemiology of the preeminent arboviral disease. *Infection, genetics and evolution: journal of molecular epidemiology and evolutionary genetics in infectious diseases*. 9 (4): 523-40.

WHO. Global tuberculosis control-- epidemiology, strategy, financing. 2009. p. 1-28.

Wilbur, A. K., Engel, G. A., Rompis, A., Putra, I. G., Lee, B. P., Aggimarangsee, N., Chalise, M., Shaw, E., Oh, G., Schillaci, M. A., Jones-Engel, L. 2012. From the mouths of monkeys: detection of Mycobacterium tuberculosis complex DNA from buccal swabs of synanthropic macaques. *Am. J. Primatol.* 74(7):676-86.

In: Monkeys
Editor: Hugo A. Barrera Saldaña

ISBN: 978-1-63117-851-1

Chapter 7

THE ROLE OF NATURAL OCCURRING INFECTIONS IN EXPERIMENTAL STUDIES IN NON-HUMAN PRIMATES

***Natalia Schlabritz-Loutsevitch*[1], *Edward J. Dick Jr.*[2] *and Gene B. Hubbard*[3]**

[1]Department of Obstetrics and Gynecology,
University of Tennessee Health Science Center, Memphis, Tennessee, US
[2]Southwest National Primate Research Center,
Texas Biomedical Research Institute, San Antonio, Texas, US
[3]Department of Pathology, University of Texas Health Science Center at San Antonio,
San Antonio, Texas, US

ABSTRACT

Non-human primates (NHP) have been an indispensable part of experimental medicine for centuries. Baboons (*Papio* spp.) have been extensively used in reproductive research and are a well characterized model of premature labor, endometriosis and pregnancy perturbations. However, the role of naturally occurring infectious agents in reproductive research using these non-human primate species has received less attention. Microbial participation is essential for host functions such as defense, metabolism, and reproduction. The possible role of turbulences in the placental microbial communities has been linked to preterm deliveries [1]. Changes in the fecal microbiome have been recently recognized as an essential part of both inflammation [2] and pregnancy [3]. This chapter will describe the most common infectious pathologies in baboons and their influence on reproductive research.

INTRODUCTION

The use of Old World non-human primates, specifically baboons (*Papio* spp.), in reproductive research is indispensable due to their metabolism, endocrinology and morphology of the reproductive organs and feto-placental unit being very similar to human [4-6]. The use and interpretation of the data retrieved from non-human primates requires an understanding of confounds influencing the parameters studied. Such confounds include, among others, seasonality [7], indoor vs. outdoor husbandry [8, 9], natural infections and parasitic diseases [10]. The urgent need to understand the microbial background in animal studies is underlined by the recently emerging field of microbiomes, including the human microbiome, which plays an important role in body metabolism and function [11]. The possible role of turbulences in the microbial communities in the placenta has been linked to preterm delivery [12] and changes in fecal microbiome are recognized as an essential part of both obesity-related inflammation and pregnancy development [3]. The fecal microbiome is connected to the tissue microbiomes though the mechanism of bacterial drift to the mesenteric lymph nodes and subsequently to other tissues, e.g., cardiovascular [13]. The peripheral mononuclear cells (PBMC) are believed to be part of the mechanism of bacterial translocation, especially in the perinatal period [14]. Interestingly PBMCs demonstrate inflammatory responses in preterm labor [15] and pre-eclampsia [16]. Resident bacteria could harbor viruses and change the maternal immune response [17, 18]. The recent discovery of inflammasomes (infection-associated inflammatory complexes) in placental tissues [19] and increased inflammasome expression in the placentas of obese patients [20] and patients with pre-eclampsia links the microbial environment to pregnancy. The importance of viral and microbial priming for reproduction has been shown by Budda et al. [21] in baboons, where adverse reproductive events (abortions, stillbirths) strongly correlated with colony type (specific pathogen-free (SPF) vs. conventional housing), when corrected for parity. Interestingly, the duration of gestation was significantly shorter in SPF animals [21]. Thus, knowledge of how naturally occurring and experimental conditions influence the microbial environment, and how microbial environment influences study results and the data interpretations is extremely important.

1. Cytomegalovirus (CMV) and Herpes Virus 2 (HVS2)

Human Cytomegalovirus (HCMV) and herpes virus 2 (HVS2) are the leading causes of congenital viral infection, affecting about 1% of newborns in the United States [22] and can cause abortions and stillbirths. These types of viruses are widespread among captive baboons [23-25, 47-48].

1.1. Pathophysiology of HCMV in Human Reproduction

HCMV is a major cause of congenital disease, the leading cause of prenatal viral infection, and a known viral cause of mental retardation [26-30]. There are an estimated 30,000-40,000 cases of congenital HCMV infections per year in the United States. Five percent of human infants with symptomatic congenital HCMV die and 10% of those who are asymptomatic at birth develop neurological sequelae [29, 30].

Analysis of spontaneously aborted conceptuses shows that HCMV infects the placenta before the embryo or fetus [31, 32]. HCMV targets endothelial cells [33, 34] and these infected endothelial cells may trigger direct vascular injury, leading to microvascular endothelial damage, vasculitis, thrombotic occlusion, and hemorrhage [32, 35]. Human cytotrophoblasts are permissive for HCMV infection and play a central role in viral transmission to the fetus [36, 37]. HCMV influences inflammatory cytokine release from the placenta [38].

Sinzger et al. [33] and Satosar et al. [39] found that HCMV are localized in Hofbauer cells and trophoblasts. Viral HCMV proteins were found in the decidua, IGFBP1-positive resident decidual cells and endovascular cytotrophoblasts in remodeled uterine blood vessels [40]. Cytomegalovirus infection has been described to limit extravillus trophoblast invasion in an *in vitro* system [41, 42]. HCMV infection in the decidua illustrates how this virus utilizes the host immunity [43]. In the human hemochorial placenta, maternal blood directly contacts syncytiotrophoblasts that cover chorionic villi and cytotrophoblasts that invade uterine vessels, suggesting possible routes for HCMV transmission [41].

Infection during first half of pregnancy appears to cause the most pathology [44, 45]. HCMV infection may result in impaired placental physiological function as the changes described above influence placental structure, tissue tropism, tissue remodeling and perfusion [33, 46].

1.2. CMV in Baboons (Papio spp)

Baboon Cytomegalovirus (BaCMV) is a common virus in the baboon population. Data by Blewett et al. [47, 48] indicated that 95% of adult baboons are infected with BaCMV. Susceptibility of baboons to CMV was shown decades ago by Deitchman et al. [49]. Extensive reviews of the pathology of non-human primates [10, 50, 51] demonstrate that the most common system affected by congenital anomaly in the baboons was the nervous system, the most common organ affected was brain, and the third most common pathologic finding was hydrocephalus. Hydrocephalus is one typical HCMV lesion (Figure 1).

1.3. Herpes Simplex Virus (HSV-1 and HSV-2) Infections in Humans

HSV infection during pregnancy causes abnormalities in the fetal central nervous system [52], and is vertically transmitted [53]. In humans syncytiotrophoblasts are a barrier to maternal/fetal transmission of herpes simplex virus, however infection of these cells could prevent the normal process of placentation [54, 55]. Herpes virus–8 and -6 infections have been reported to be associated with a higher risk of human stillbirth [56] and neurological disorders in newborns having mother-child transmissions [57].

1.4. Papilineherpesvirus 2 in Baboons (Papio spp.)

Papilineherpesvirus 2 (HPV-2) is a major disease entity of the baboon with a reported antibody prevalence of up to 85% in some groups of captive baboons [23-25]. Infection in baboons with HVP-2 is very similar to Herpes simplex virus (HSV-1 and HSV-2) infections in humans [58].

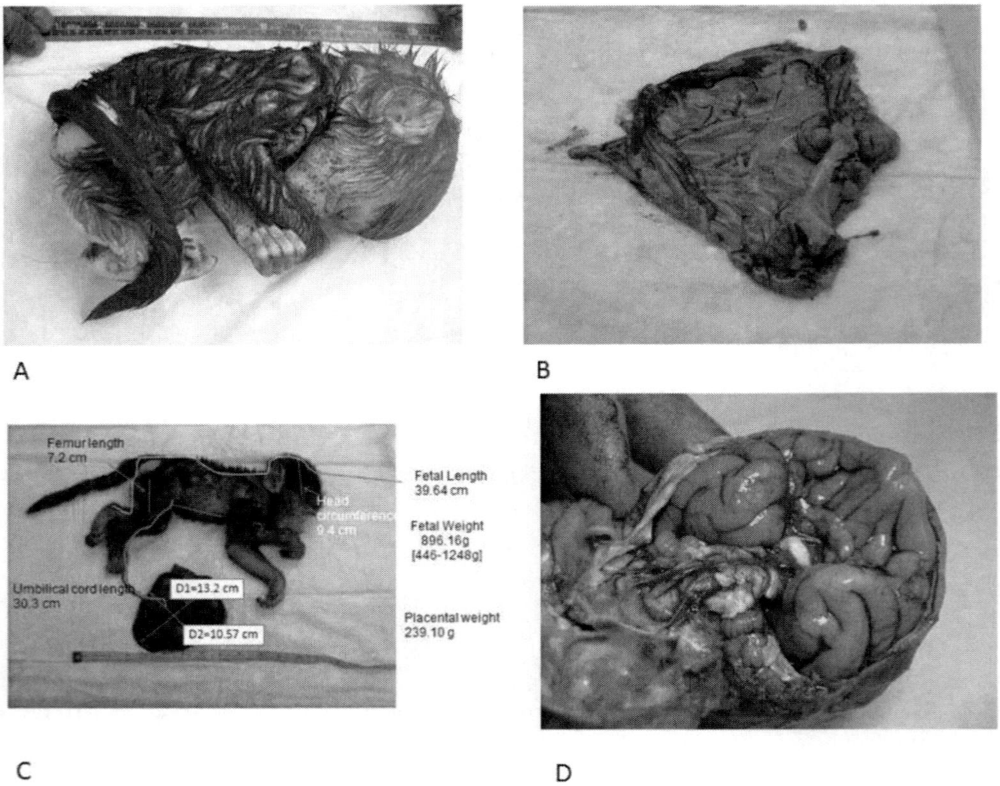

Figure 1. (A) Stillborn baboon at term with hydrocephalus. (B) Brain from a stillborn baboon with hydrocephalus. (C) Stillborn baboon with normal brain structure. (D) Brain from a fetal baboon at term (stillbirth).

2. BRUCELLA

Brucellosis is a recognized problem in both veterinary and human medicine [59, 60]. Hundreds of thousands of new cases are recorded annually, new foci of disease endemicity have emerged in recent years, and a significant percentage of cases are being imported into the developed, officially brucellosis-free, world through international travel and trade. Moreover, *Brucella* remains the most common cause of laboratory-acquired infections [61] and is abundant in wildlife (including novel species) [62]. *Brucella* is able to survive for prolonged periods in humans by residing predominantly in *Brucella*-containing vacuoles (BCV) that progressively evolve through the pathogen's interaction with macrophage and non-professional phagocytes' organelles [63], including HELA cells and trophoblasts [64]. The virulent *B. abortus*, for example, transits from autophagosomes to the endoplasmic reticulum of the trophoblasts where actual bacterial multiplication occurs [63, 65]. Although several primate species have shown antibodies to *Brucella*, only a single *Brucella melitensis* positive culture in *Papio* spp. has been reported [66]. Published data of a novel *Brucella* isolate identified in a female baboon [62] suggest that the genetic difference separating the baboon species of *Brucella* from any other species is equal or greater than that which separates other extant species such as *B. neotomae* or *B. ovis*. (Figure 2).

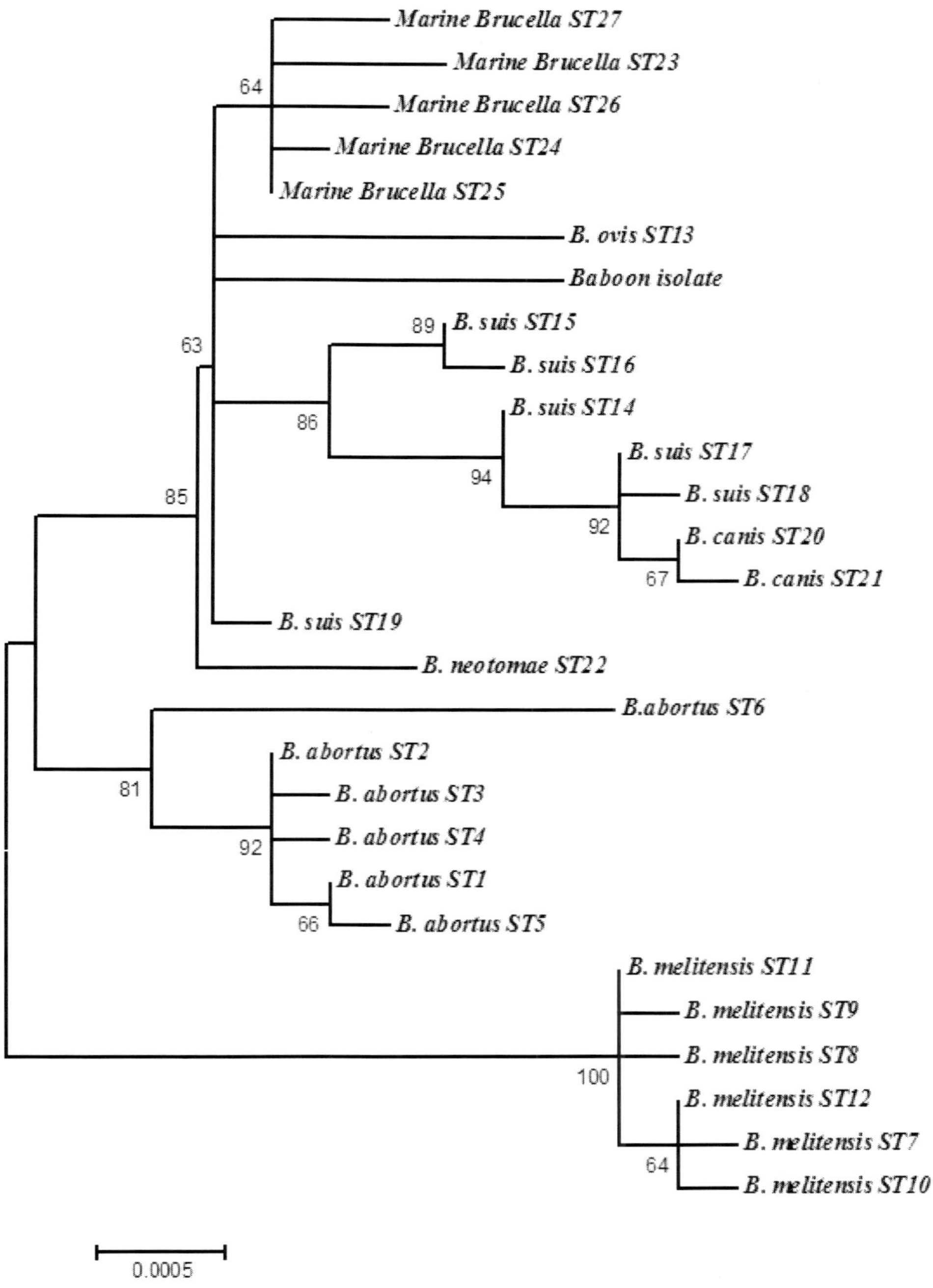

Figure 2. Unrooted phylogenetic reconstruction of the relationship between the novel baboon isolate and *Brucella* Sequence Types previously described. The tree was constructed with concatenated data of nine genetic loci (4,396 bp) using the neighbor-joining approach implemented in the MEGA3.1 package. The Jukes-Cantor model was used to determine genetic distances. The percentage bootstrap confidence levels of internal branches were calculated from 1,000 resamplings of the original data [62].

Seropositive animals might be a chronic reservoir of *Brucella* infection in macrophages and mononucleated cells [67]. Since macrophages and mononucleated cells are also key players in metabolic disorders and pregnancy, the possible presence of *Brucella* should be taken into consideration during data interpretation.

3. TRYPANOSOMA CRUZI

Chagas' disease affects 16-18 million people in Central and South America (World Health Organization). The cause of this disease, the protozoan parasite *Trypanosoma cruzi* (*T. cruzi*), is traditionally transmitted by the triatomine insect vector, but blood transfusion has also been a significant source of infection. These routes of transmission have become less important as insect control and blood screening programs have been implemented in endemic countries. In fact, congenital transmission is now more common in Latin America than disease caused by vector transmission [68-72]. Chagas' disease is characterized by an acute and a chronic stage. Transmission of *T. cruzi* from mother to fetus can occur during either stage of the disease [71-74]. Congenital Chagas' disease is associated with premature labor, stillbirth, abortion and placentitis [68, 75]. Materno-fetal transmission cannot currently be prevented because the most common therapeutic agents used to treat *T. cruzi* infection, nifurtimox and benznidazole, have not been used in pregnant women due to high toxicity [76]. The seroprevalence in pregnant women in endemic areas has been shown to be as high as 81% and congenital transmission can occur in up to 21% of seropositive cases [77-80].

Chagas' disease in a colony baboon was first observed at the Texas Biomedical Research Institute in 1986 [81] and has been well characterized since then [82-86]. The outdoor housing of animals results in natural infection though exposure to the *T. cruzi* triatomine vector. In primates it is presumed that transmission is mainly through ingestion of infected insects [87] or possibly though bloodsucking lice [82]. An analyses of the stillbirths in the baboon colony at Texas Biomedical Research Institute revealed presence of *T. cruzi* parasite in the placenta of one animal (Figure 3) with no association between the rate of stillbirth and seropositivity for *T. cruzi* [85]. However, the presence of this infection should be taken into consideration when interpreting experimental studies, since maternal infection could independently influence the development of fetal immune responses and susceptibility to postnatal infections.

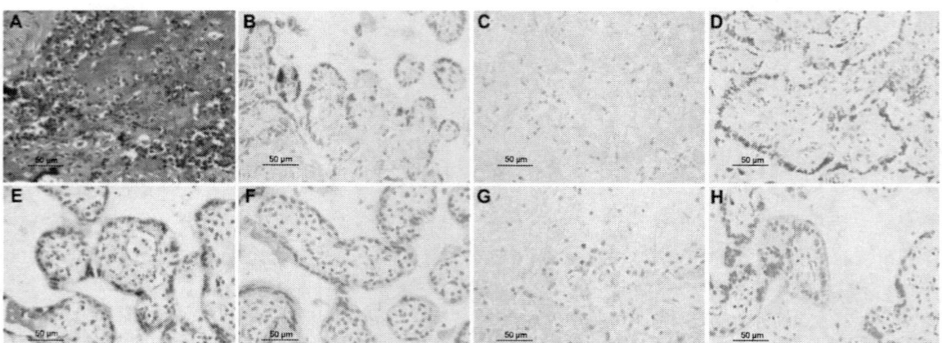

Figure 3. Photomicrographs of placentas from baboons seronegative (A, B, C, D) and seropositive (E, F, G, H) for *T. cruzi*. Hematoxylin and eosin (A and E); CD45RO (B and F), CD57 (C and G), and CD68 (D and H) immunohistochemistry. Note the absence of immunostaining of CD45RO (F) and CD57 (G) in the placenta from the seropositive dam [85].

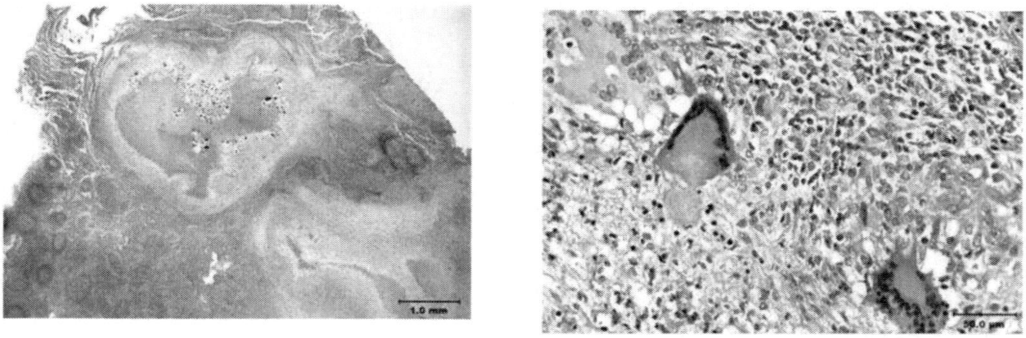

A B

Figure 4. (A). TB-positive baboon tracheobronchial lymph node (tubercle). Lower-power magnification reveals two foci composed of caseonecrotic debris surrounded by abundant pyogranulomatous inflammation. Several mineralized foci are noted (bar = 1.0 mm). (B). Baboon tracheobronchial lymph node. Large multinucleated foreign body type giant cells are surrounded by abundant extracellular debris and macrophages. On the left, many degenerative neutrophils and nuclear debris are present. To the right, rows of lymphocytes separated by fibrous connective tissue and fibroblasts complete the criteria for a pyogranuloma (tubercle) (bar = 50 μm) [90].

This influence has been shown to be independent of in-utero transmission of the pathogen [88].

4. Tuberculosis

Tuberculosis transmission to the human fetus via the placenta or amnion has been reported in less than 300 patients in the English language literature [89]. The diagnosis of tuberculosis in pregnancy is challenging, since it could be masked by other pregnancy-related pathology (such as pre-eclampsia, for example).

Tuberculosis occurs most commonly in captive primates exposed to humans, and appears to be a larger threat to the primate population than it is to the human handlers [90, 91]. Disease onset in the nonhuman primate is usually more rapid and progressively fatal. Martino et al. [90] described the first case of tuberculosis at a large captive baboon colony in more than 25 years; shortly after that, tuberculosis was reported in another large baboon colony [91].

One of these cases involved a pregnant baboon [90] but was not associated with any placental changes. The recent resurgence of tuberculosis, including multi-drug resistant strains, suggests that tuberculosis may be more likely to be introduced into a colony.

In Summary

Experimental conditions are influenced by the microbial environment and vice versa. Deeper knowledge of the NHP microbiome is important for data interpretation and analyses in pregnancy-related experimental work with Baboons (*Papio spp*).

ACKNOWLEDGMENT

Authors are thankful Ms. Malinda Mann and Ms. Maria Messenger for their editing help.

DISCLOSURE STATEMENT

The authors report no conflict of interest.

GRANT SUPPORT

Southwest National Primate Research Center grant P51 RR013986 from the National Center for Research Resources, National Institutes of Health and which are currently supported by the Office of Research Infrastructure Programs through P51 OD011133; Office of Research Infrastructure Programs (ORIP) of the National Institutes of Health through Grant Numbers 1 C06 RR014578 and 1 C06 RR015456.

REFERENCES

[1] Kuehn, B. M. Center takes broader look at preterm birth. *JAMA*, 2012 Dec. 19, 308:2448.

[2] Turnbaugh, P. J.; Hamady, M.; Yatsunenko, T.; Cantarel, B. L.; Duncan, A.; Ley, R. E.; Sogin, M. L.; Jones, W. J.; Roe, B. A.; Affourtit, J. P.; Egholm, M.; Henrissat, B.; Heath, A. C.; Knight, R.; Gordon, JI. A core gut microbiome in obese and lean twins. *Nature*, 2009 Jan. 22, 457:480-484.

[3] Koren, O.; Goodrich, J. K.; Cullender, T. C.; Spor, A.; Laitinen, K.; Bäckhed, H. K.; Gonzalez, A.; Werner, J. J.; Angenent, L. T.; Knight, R.; Bäckhed, F.; Isolauri, E.; Salminen, S.; Ley, R. E. Host remodeling of the gut microbiome and metabolic changes during pregnancy. *Cell*, 2012 Aug. 3,150:470-480.

[4] Houston M. L. The development of the baboon (*Papio* sp.) placenta during the fetal period of gestation. *Am. J. Anat.*, 1969 Sep., 126:17-29.

[5] Enders, A. C.; Lantz, K. C.; Peterson, P. E.; Hendrickx, A. G. From blastocyst to placenta: the morphology of implantation in the baboon. *Hum. Reprod. Update*, 1997 Nov.-Dec., 3:561-573.

[6] Castracane, V. D.; Hendrickx, A. G.; Henson, M. C. Serum leptin in nonpregnant and pregnant women and in old and new world nonhuman primates. *Exp. Biol. Med. (Maywood)*, 2005 Apr., 230:251-254.

[7] Alberts, S. C.; Hollister-Smith, J.; Mututua, R. S.; Sayialel, S. N.; Muruthi, P. M.; Warutere, J. K.; Altmann, J. Seasonality and long-term change in a savanna environment. In: Brockman D. K., van Schaik C. P. editors. Title: Seasonality in primates: studies of living and extinct human and non-human primates. New York: Cambridge University Press; 2005; 157-195.

[8] Cary, M. E.; Valentine, B.; White, G. L. The effects of confinement environment on reproductive efficiency in the baboon. *Contemp. Top Lab. Anim. Sci.*, 2003 Jan., 42: 35-39.

[9] Caperton, L.; Eddy, C.; Leland, M. M.; Carey, K. D.; McCarrey, J. R. Alteration of the menstrual cycle in baboons placed on tethering devices and moved to individual housing--a stress model for a follicular phase defect. *J. Med. Primatol.*, 2006 Dec., 35:341-345.

[10] Dick E. J. Jr., Owston M. A., David J. M., Sharp R. M., Rouse S., Hubbard G. B. Mortality in Captive Baboons (*Papio* spp.): A 23 Year Study. *J. Med. Primatol.*, 2014 Feb. 1. [Epub ahead of print]

[11] Human Microbiome Project Consortium. Structure, function and diversity of the healthy human microbiome. *Nature*, 2012 Jun. 13, 486:207-214.

[12] Cardenas, I.; Means, R. E.; Aldo, P.; Koga, K.; Lang, S. M.; Booth, C. J.; Manzur, A.; Oyarzun, E.; Romero, R.; Mor, G. Viral infection of the placenta leads to fetal inflammation and sensitization to bacterial products predisposing to preterm labor. *J. Immunol.*, 2010 Jul. 15, 185:1248-1257.

[13] Macpherson, A. J.; Smith, K. Mesenteric lymph nodes at the center of immune anatomy. *J. Exp. Med.*, 2006 Mar. 20, 203:497-500.

[14] Donnet-Hughes, A.; Perez, P. F.; Dore, J.; Leclerc, M.; Levenez, F.; Benyacoub, J.; Serrant, P.; Segura-Roggero, I.; Schiffrin, E. J. Potential role of the intestinal microbiota of the mother in neonatal immune education. *Proc. Nutr. Soc.*, 2010 Aug., 69:407-415.

[15] Andrews, W. W.; Goldenberg, R. L.; Faye-Petersen, O.; Cliver, S.; Goepfert, A. R.; Hauth, J. C. The Alabama Preterm Birth study: polymorphonuclear and mononuclear cell placental infiltrations, other markers of inflammation, and outcomes in 23- to 32-week preterm newborn infants. *Am. J. Obstet. Gynecol.*, 2006 Sep.,195:803-808.

[16] Uszynski, W.; Uszynski, M. Two hypotheses about preeclampsia: endothelium injury and inflammation. *Ginekol. Pol.*, 2003 Mar., 74:237-245.

[17] Pennisi, E. European Society for Evolutionary Biology meeting. Mothering beetles suppress microbes to protect food for their young. *Science*, 2011 Sep. 16, 333:1562.

[18] Pennisi, E. Mysteries of development. How do microbes shape animal development? *Science*, 2013 Jun. 7, 340:1159-1160.

[19] Mulla, M. J.; Myrtolli, K.; Potter, J.; Boeras, C.; Kavathas, P. B.; Sfakianaki, A. K.; Tadesse, S.; Norwitz, E. R.; Guller, S.; Abrahams, V. M. Uric acid induces trophoblast IL-1β production via the inflammasome: implications for the pathogenesis of preeclampsia. *Am. J. Reprod. Immunol.*, 2011 Jun., 65:542-548.

[20] Aye, I. L. M. H.; Ramirez, V. I.; Gaccioli, F.; Lager, S.; Jansson, T.; Powell, T. L. Activation of Placental Inflammasomes in Pregnant Women with High BMI. *Reprod. Sci.*, 2013, 20(S3)73A.

[21] Budda, M. L.; Ely, J. J.; Doan, S.; Chavez-Suarez, M.; White, G. L.; Wolf, R. F. Evaluation of reproduction and raising offspring in a nursery-reared SPF baboon (*Papio hamadryas anubis*) colony. *Am. J. Primatol.*, 2013 Aug., 75:798-806.

[22] Britt, W. Virus entry into host, establishment of infection, spread in host, mechanisms of tissue damage. In: Arvin A., Campadelli-Fiume G., Mocarski E., Moore P. S., Roizman B., Whitley R., Yamanishi K., editors. Title: Human Herpesviruses: Biology,

Therapy, and Immunoprophylaxis. Cambridge: Cambridge University Press; 2007; Chapter 41.

[23] Eberle, R.; Black, D. H.; Blewett, E. L.; White, G. L. Prevalence of Herpesvirus papio 2 in baboons and identification of immunogenic viral peptides. *Lab. Anim. Sci.*, 1997 Jun., 47:256-262.

[24] Eberle, R.; Hilliard, J. The simian herpesviruses. *Infect. Agents Dis.*, 1995 Jun., 4: 55-70.

[25] Martino, M. A.; Hubbard, G. B.; Butler, T. M.; Hilliard, J. K. Clinical disease associated with simian agent 8 infection in the baboon. *Lab. Anim. Sci.*, 1998 Feb., 48:18-22.

[26] Stagno, S. Cytomegalovirus. 3rd Edition. In: Remington J. S., Klein J. O., editors. Title: Infectious diseases of fetus and new born infant. Philadelphia: Saunders; 1990;240-281.

[27] Demmler, G J. Infectious Diseases Society of America and Centers for Disease Control. Summary of a workshop on surveillance for congenital cytomegalovirus disease. *Rev. Infect. Dis.*, 1991 Mar.-Apr., 13:315-329.

[28] Stratton, K. R.; Howe, C. J.; Johnston, R. B. Jr. Adverse events associated with childhood vaccines other than pertussis and rubella. Summary of a report from the Institute of Medicine. *JAMA*, 1994 May 25, 271:1602-1605.

[29] Boppana, S. B.; Fowler, K. B.; Britt, W. J.; Stagno, S.; Pass, R. F. Symptomatic congenital cytomegalovirus infection in infants born to mothers with preexisting immunity to cytomegalovirus. *Pediatrics*, 1999 Jul., 1104:55-60.

[30] Dar, L.; Pati, S. K.; Patro, A. R.; Deorari, A. K.; Rai, S.; Kant, S.; Broor, S.; Fowler, K. B.; Britt, W. J.; Boppana, S. B. Congenital cytomegalovirus infection in a highly seropositive semi-urban population in India. *Pediatr. Infect. Dis. J.*, 2008 Sep., 27:841-843.

[31] Terauchi, M.; Koi, H.; Hayano, C.; Toyama-Sorimachi, N.; Karasuyama, H.; Yamanashi, Y.; Aso, T.; Shirakata, M. Placental extravillous cytotrophoblasts persistently express class I major histocompatibility complex molecules after human cytomegalovirus infection. *J. Virol.*, 2003 Aug., 77:8187-8195.

[32] Benirschke, K.; Kauffman, P. *Pathology of Human Placenta*, 2nd Edition. New York: Springer, Berlin, Heidelberg; 1990;542-635.

[33] Sinzger, C.; Muntefering, H.; Loning, T.; Stoss, H.; Plachter, B.; Jahn, G. Cell types infected in human cytomegalovirus placentitis identified by immunohistochemical double staining. *Virchows Arch. A. Pathol. Anat. Histopathol.*, 1993, 423:249-256.

[34] Maidji, E.; Percivalle, E.; Gerna, G.; Fisher, S.; Pereira, L. Transmission of human cytomegalovirus from infected uterine microvascular endothelial cells to differentiating/invasive placental cytotrophoblasts. *Virology*, 2002 Dec. 5, 304:53-69.

[35] Persoons, M. C.; Stals, F. S.; van dam Mieras, M. C.; Bruggeman, C. A. Multiple organ involvement during experimental cytomegalovirus infection is associated with disseminated vascular pathology. *J. Pathol.*, 1998 Jan., 184:103-109.

[36] Halwachs-Baumann, G.; Wilders-Truschnig, M.; Desoye, G.; Hahn, T.; Kiesel, L.; Klingel, K.; Rieger, P.; Jahn, G.; Sinzger C. Human trophoblast cells are permissive to the complete replicative cycle of human cytomegalovirus. *J. Virol.*, 1998 Sep., 72:7598-7602.

[37] Hemmings, D. G.; Guilbert, L. J. Polarized release of human cytomegalovirus from placental trophoblasts. *J. Virol.*, 2002 Jul., 76:6710-6717.

[38] Hamilton, S. T.; Scott, G.; Naing, Z.; Iwasenko, J.; Hall, B.; Graf, N.; Arbuckle, S.; Craig, M. E.; Rawlinson, W. D. Human cytomegalovirus-induces cytokine changes in the placenta with implications for adverse pregnancy outcomes. *PLoS One,* 2012, 7:e52899.

[39] Satosar, A.; Ramirez, N. C.; Bartholomew, D.; Davis, J.; Nuovo, G. J. Histologic correlates of viral and bacterial infection of the placenta associated with severe morbidity and mortality in the newborn. *Hum. Pathol.*, 2004 May, 35:536-545.

[40] Pereira, L.; Maidji, E.; McDonagh, S.; Genbacev, O.; Fisher, S. Human cytomegalovirus transmission from the uterus to the placenta correlates with the presence of pathogenic bacteria and maternal immunity. *J. Virol.*, 2003 Dec., 77:13301-13314.

[41] Fisher, S.; Genbacev, O.; Maidji, E.; Pereira, L. Human cytomegalovirus infection of placental cytotrophoblasts in vitro and in utero: implications for transmission and pathogenesis. *J. Virol.*, 2000 Aug., 24:6808-6820.

[42] Yamamoto-Tabata, T.; McDonagh, S.; Chang, H. T.; Fisher, S.; Pereira, L. Human cytomegalovirus interleukin-10 downregulates metalloproteinase activity and impairs endothelial cell migration and placental cytotrophoblast invasiveness in vitro. *J. Virol.*, 2004 Mar., 78:2831-2840.

[43] Mocarski, E. S. Jr. Immunomodulation by cytomegaloviruses: manipulative strategies beyond evasion. *Trends Microbiol.*, 2002 Jul., 10:332-339.

[44] Yow, M. D.; Williamson, D. W.; Leeds, L. J.; Thompson, P.; Woodward, R. M.; Walmus, B. F.; Lester, J. W.; Six, H. R.; Griffiths, P. D. Epidemiologic characteristics of cytomegalovirus infection in mothers and their infants. *Am. J. Obstet. Gynecol.*, 1988 May, 158:1189-1195.

[45] Ahlfors, K.; Forsgren, M.; Ivarsson, S. A.; Harris, S.; Svanberg, L. Congenital cytomegalovirus infection: on the relation between type and time of maternal infection and infant's symptoms. *Scand. J. Infect. Dis.*, 1983, 15:129-138.

[46] Yamamoto, R.; Ishii, K.; Shimada, M.; Hayashi, S.; Hidaka, N.; Nakayama, M.; Mitsuda, N. Significance of maternal screening for toxoplasmosis, rubella, cytomegalovirus and herpes simplex virus infection in cases of fetal growth restriction. *J. Obstet. Gynaecol. Res.*, 2013 Mar., 39:653-657.

[47] Blewett, E. L.; White, G.; Saliki, J. T.; Eberle, R. Isolation and characterization of an endogenous cytomegalovirus (BaCMV) from baboons. *Arch. Virol.*, 2001, 146:1723-1738.

[48] Blewett, E. L.; Lewis, J.; Gadsby, E. L.; Neubauer, S. R.; Eberle, R. Isolation of cytomegalovirus and foamy virus from the drill monkey (*Mandrillus leucophaeus*) and prevalence of antibodies to these viruses amongst wild-born and captive-bred individuals. *Arch. Virol.*, 2003 Mar., 148:423-433.

[49] Deitchman, G. I.; Kashkina, L. M.; Slovyova, E. A. Monkey's disease resembling infectious mononucleosis induced by cytomegalovirus. *Vestn. Akad. Med. Nauk. SSSR*, 1972, 27:24-29.

[50] Bommineni, Y. R.; Dick, E. J. Jr.; Malapati, A. R.; Owston, M. A.; Hubbard, G. B. Natural pathology of the Baboon (*Papio* spp.). *J. Med. Primatol.*, 2011 Apr., 40:142-55.

[51] Fox, B.; Owston, M. A.; Kumar, S.; Dick, E. J. Congenital anomalies in the baboon (*Papio* spp.). *J. Med. Primatol.*, 2011 Oct., 40:357-363.

[52] Antonov, P. V.; Vydumkina, S. P.; Tsinzerling, V. A. The role of intrauterine infection of the central nervous system with herpes simplex viruses in development of encephalopathies in infants. *Arkh. Patol.,* 2003 Jul.-Aug., 65:43-46.

[53] Schleiss, M. R. Vertically transmitted herpesvirus infections. *Herpes,* 2003 May,10: 4-11.

[54] Koi, H.; Zhang, J.; Makrigiannakis, A.; Getsios, S.; MacCalman, C. D.; Strauss, J. F. 3rd; Parry, S. Syncytiotrophoblast is a barrier to maternal-fetal transmission of herpes simplex virus. *Biol. Reprod.*, 2002 Nov., 67:1572-1579.

[55] Koi, H.; Zhang, J.; Parry, S. The mechanisms of placental viral infection. *Ann. NY Acad. Sci.*, 2001 Sep., 943:148-156.

[56] Sarmati, L.; Ticconi, C.; Santangelo, R.; Montano, M.; Rezza, G.; Andreoni, M. Does the risk of abortion increase in women with high human herpesvirus-8 antibody titers? *J. Infect. Dis.*, 2003 Jul. 1, 188:173-174; author reply 174-175.

[57] Lanari, M.; Papa, I.; Venturi, V.; Lazzarotto, T.; Faldella, G.; Gabrielli, L.; Guerra, B.; Landini, M. P.; Salvioli, G. P. Congenital infection with human herpesvirus 6 variant B associated with neonatal seizures and poor neurological outcome. *J. Med. Virol.*, 2003 Aug., 70:628-632.

[58] Bigger, J. E.; Martin, D. W. The genome of herpesviruspapio 2 is closely related to the genomes of human herpes simplex viruses. *J. Gen. Virol.*, 2003 Jun., 84:1411-1414.

[59] Haileselassie, M.; Kalayou, S.; Kyule, M.; Asfaha, M.; Belihu, K. Effect of Brucella infection on reproduction conditions of female breeding cattle and its public health significance in Western tigray, northern ethiopia. *Vet. Med. Int.*, 2011, 2011:354943.

[60] O'Callaghan, D. Novel replication profiles of Brucella in human trophoblasts give insights into the pathogenesis of infectious abortion. *J. Infect. Dis.,* 2013 Apr., 207:1034-1036.

[61] Pappas, G.; Papadimitriou, P.; Akritidis, N.; Christou, L.; Tsianos, E. V. The new global map of human brucellosis. *Lancet Infect. Dis.*, 2006 Feb., 6:91-99.

[62] Schlabritz-Loutsevitch, N. E.; Whatmore, A. M.; Quance, C. R.; Koylass, M. S.; Cummins, L. B.; Dick, E. J. Jr.; Snider, C. L.; Cappelli, D.; Ebersole, J. L.; Nathanielsz, P. W.; Hubbard, G. B. A novel Brucella isolate in association with two cases of stillbirth in non-human primates - first report. *J. Med. Primatol.*, 2009 Feb., 38:70-73.

[63] Anderson T. D.; Cheville N. F. Ultrastructural morphometric analysis of Brucellaabortus-infected trophoblasts in experimental placentitis. Bacterial replication occurs in rough endoplasmic reticulum. *Am. J. Pathol.*, 1986 Aug, 124:226-237.

[64] Detilleux, P. G.; Deyoe, B. L.; Cheville, N. F. Penetration and intracellular growth of Brucellaabortus in non-phagocytic cells in vitro. *Infect. Immun.*, 1990 Jul., 58:2320-2328.

[65] Pinkerton, M. E. Bacteremia in Wild Baboons. *Bact. Proc.*, 1967;67:67.

[66] Pinkerton, M. E. Brucella. In: Fiennes RN, editor. Pathology of Simian Primates Part II: Infectious and Parasitic Diseases. London; Karger Publishers; 1972; 289-293.

[67] Celli, J.; de Chastellier, C.; Franchini, D. M.; Pizarro-Cerda, J.; Moreno, E.; Grovel, J. P. Brucella evades macrophage killing via VirB-dependent sustained interactions with the endoplasmic reticulum. *J. Exp. Med.*, 2003 Aug. 18,198:545-556.

[68] Azogue, E. Women and congenital Chagas' disease in Santa Cruz, Bolivia: epidemiological and sociocultural aspects. *Soc. Sci. Med.*, 1993 Aug., 37:503-511.

[69] Blanco, S. B.; Segura, E. L.; Cura, E. N.; Chuit, R.; Tulian, L.; Flores, I.; Garbarino, G.; Villalonga, J. F.; Gurtler, R. E. Congenital transmission of Trypanosoma cruzi: an operational outline for detecting and treating infected infants in north-western Argentina. *Trop. Med. Int. Health*, 2000 Apr., 5:293-301.

[70] Frank, F.; Sartori, M. J.; Asteggiano, C.; Lin, S.; de Fabro, S. P.; Fretes, R. E. The effect of placental subfractions on Trypanosoma cruzi. *Exp. Mol. Pathol.*, 2000 Oct., 69:144-151.

[71] Roellig, D. M.; McMillan, K.; Ellis, A. E.; VandeBerg, J. L.; Champagne, D. E.; Yabsley, M. J. Experimental infection of two South American reservoirs with four distinct strains of *Trypanosoma cruzi*. *Parasitology,* 2010 May, 137:959-966.

[72] Torrico, F.; Alonso-Vega, C.; Suarez, E.; Rodriguez, P.; Torrico, M. C.; Dramaix, M.; Truyens, C.; Carlier, Y. Maternal Trypanosoma cruzi infection, pregnancy outcome, morbidity, and mortality of congenitally infected and non-infected newborns in Bolivia. *Am. J. Trop. Med. Hyg.*, 2004, 70:201-209.

[73] Moretti, E.; Basso, B.; Castro, I.; Carrizo Paez, M.; Chaul, M.; Barbieri, G.; Canal Feijoo, D.; Sartori, M. J.; Carrizo Paez, R. Chagas' disease: study of congenital transmission in cases of acute maternal infection. *Rev. Soc. Bras. Med. Trop.*, 2005 Jan.-Feb., 38:53-55.

[74] Bittencourt, A. L. Congenital Chagas disease. *Am. J. Dis. Child*, 1976 Jan., 130:97-103.

[75] Fretes R. E.; De Fabro S. P. In vivo and in vitro analysis of lysosomes and acid phosphatase activity in human chagasic placentas. *Exp. Mol. Pathol.*,1995 Dec. 63:153-160.

[76] Gorla, N. B.; Ledesma, O. S.; Barbieri, G. P.; Larripa, I. P. Assessment of cytogenetic damage in chagasic children treated with benznidazole. *Mutat. Res.*, 1988 Oct., 206:217-220.

[77] Kirchhoff; L. V., Gam, A. A.; Gilliam, F. C. American trypanosomiasis (Chagas' disease) in Central American immigrants. *Am. J. Med.*, 1987 May, 82:915-920.

[78] Kirchhoff, L. V. American trypanosomiasis (Chagas' disease)--a tropical disease now in the United States. *N. Engl. J. Med.*, 1993 Aug. 26, 329:639-644.

[79] Kirchhoff, L. V. American trypanosomiasis (Chagas' disease). *Gastroenterol. Clin. North Am.*, 1996 Sep., 25:517-533.

[80] Szarfman, A.; Cossio, P. M.; Arana, R. M.; Urman, J.; Kreutzer, E.; Laguens, R. P.; Segal, A.; Coarasa, L. Immunologic and immunopathologic studies in congenital Chagas' disease. *Clin. Immunol. Immunopathol.*, 1975 Nov., 4:489-499.

[81] Gleiser, C. A.; Yaeger, R. G.; Ghidoni, J. J. *Trypanosoma cruzi* infection in a colony-born baboon. *J. Am. Vet. Med. Assoc.*, 1986 Nov. 1,189:1225-1226.

[82] Argañaraz, E. R.; Hubbard, G. B.; Ramos, L. A.; Ford, A. L.; Nitz, N.; Leland, M. M.; Vandeberg, J. L.; Teixeira, A. R. Blood-sucking lice may disseminate Trypanosoma cruzi infection in baboons. *Rev. Inst. Med. Trop. Sao Paulo*, 2001 Sep.-Oct., 43:271-276.

[83] Zabalgoitia, M.; Ventura, J.; Anderson, L.; Carey, K. D.; Williams, J. T.; VandeBerg, J. L. Morphologic and functional characterization of Chagasic heart disease in non-human primates. *Am. J. Trop. Med. Hyg.*, 2003 Feb., 68:248-252.

[84] Zabalgoitia, M.; Ventura, J.; Anderson, L.; Williams, J. T.; Carey, K. D.; VandeBerg, J. L. Electrocardiographic findings in naturally acquired chagasic heart disease in nonhuman primates. *J. Electrocardiol.*, 2003 Apr., 36:155-160.

[85] Grieves, J. L.; Hubbard, G. B.; Williams, J. T.; Vandeberg, J. L.; Dick, E. J. Jr.; López-Alvarenga, J. C.; Schlabritz-Loutsevitch, N. E. Trypanosoma cruzi in non-human primates with a history of stillbirths: a retrospective study (*Papio hamadryas* spp.) and case report (*Macaca fascicularis*). *J. Med. Primatol.*, 2008 Dec., 37:318-328.

[86] Williams, J. T.; Dick, E. J. Jr.; VandeBerg, J. L.; Hubbard, G. B. Natural Chagas disease in four baboons. *J. Med. Primatol.*, 2009 Apr., 38:107-113.

[87] Shikanai-Yashuda, M. A.; Marcondes, C. B.; Guedes, L. A.; Sigueira, G. S.; Barone, A. A.; Dias, J. C.; Amato, Neto, V.; Tolezano, J. E.; Peres, B. A.; Arrudajr, E. R. Possible oral transmission of acute Chagas' disease in Brazil. *Rev. Inst. Med. Trop. Sao Paulo*, 1991 Sep.-Oct., 33:351-357.

[88] Dauby, N.; Goetghebuer, T.; Kollmann, T. R.; Levy, J.; Marchant, A. Uninfected but not unaffected: chronic maternal infections during pregnancy, fetal immunity, and susceptibility to postnatal infections. *Lancet Infect. Dis.*, 2012 Apr.,12:330-340.

[89] Loto, O. M.; Awowole, I. Tuberculosis in pregnancy: a review. *J. Pregnancy*, 2012, 2012:379271.

[90] Martino, M.; Hubbard, G. B.; Schlabritz-Loutsevitch, N. Tuberculosis (Mycobacterium tuberculosis) in a pregnant baboon (*Papio cynocephalus*). *J. Med. Primatol.*, 2007 Apr., 36:108-12.

[91] Chege, G. K.; Warren, R. M.; Gey van Pittius, N. C.; Burgers, W. A.; Wilkinson, R. J.; Shephard, E. G.; Williamson, A. L. Detection of natural infection with Mycobacterium intracellulare in healthy wild-caught Chacma baboons (*Papio ursinus*) by ESAT-6 and CFP-10 IFN-gamma ELISPOT tests following a tuberculosis outbreak. *MC Microbiol.*, 2008 Feb. 7, 8:27.

HORMONAL

In: Monkeys
Editor: Hugo A. Barrera Saldaña
ISBN: 978-1-63117-851-1

Chapter 8

GROWTH HORMONE AND PROLACTIN IN NEW WORLD MONKEYS

Michael Wallis and O. Caryl Wallis
Biochemistry and Molecular Biology Group, School of Life Sciences,
University of Sussex, Brighton, UK

ABSTRACT

In most non-primate mammals pituitary growth hormone (GH) is encoded by a single gene. This is true also for most prosimians, but in higher primates the GH locus comprises a cluster of GH-related genes, including the gene encoding pituitary GH at the 5' end, followed by a number of genes (and/or pseudogenes) at least some of which are expressed in the placenta. Such GH gene clusters arose independently in Old World Monkeys/apes and in New World Monkeys (NWM), as a consequence of multiple, tandem gene duplications followed by divergent evolution. GH-related genes have been investigated in a number of NWM species, but only in marmoset (*Callithrix jacchus*) has the complete structure of the cluster been determined. Here the pituitary-expressed GH gene is followed by 2 GH-like genes, probably expressed in placenta, and five GH-like pseudogenes.

In some NWM the GH gene cluster is more complex than this, comprising up to 40 GH-related genes. Sequence comparison and phylogenetic analysis suggests that the GH-like proteins in NWM can be divided into three groups, designated (1) pituitary GH, (2) GH-like proteins Group A (ghlA) and (3) GH-like proteins Group B (ghlB). Most species have at least one representative in each of these groups. Phylogenetic analysis also shows that rates of evolution of GH in primates have been very variable. An episode of rapid GH evolution preceded the divergence of OWM/apes and NWM. Following the independent duplications that gave rise to GH clusters on each of these lineages the rate of evolution of pituitary GH decreased, but the evolution of the other GH-like proteins remained high, apparently reflecting adaptive evolution. The biological function(s) of the GH-like proteins in NWM remain largely unknown. In contrast to the complexity shown by GH-related genes and proteins in higher primates, the related hormone prolactin is encoded by a single gene in all primates and shows only modest variation between NWM species.

INTRODUCTION

Pituitary growth hormone (GH) and prolactin are structurally related protein hormones produced in, respectively, somatotrophs and lactotrophs in the anterior pituitary gland. They occur as distinct proteins of about 200 amino acid residues in almost all vertebrates, the exception being the primitive jawless fishes (agnathans) in which only a single GH/prolactin-like hormone has been detected. This suggests that a gene duplication occurring early in vertebrate evolution gave rise to two genes which diverged to give GH and prolactin. An additional GH-like protein, somatolactin, is found in many fish, but not tetrapods, while in mammals placental proteins related to GH or prolactin are found in some groups, including man.

The function of GH (also known as somatotropin), in almost all vertebrates, is to regulate growth. Animals lacking GH, due to mutation, genetic manipulation or hypophysectomy, show stunted growth, and this can be reversed by administration of GH. The biological activity of GH shows a rather subtle species specificity. Thus, although human GH is active across a wide range of mammalian, and indeed lower vertebrate species, non-primate GHs are not active in man.

As a result, GH derived from cow, sheep or pig cannot be used to treat human pituitary dwarfism; for clinical purposes human GH obtained from human pituitary glands collected at *post-mortem* or, more recently, produced by recombinant-DNA methods, must be used. Associated with its growth-promoting activity GH shows a range of other actions, including effects on carbohydrate, lipid and protein metabolism.

The function of prolactin in vertebrates is very variable. In mammals its main role is concerned with regulation of lactation and mammary development and growth, though it also has various other roles, including luteotrophic, metabolic and immune system-related actions in some species. In lower vertebrates, regulation of mammary gland/lactation is obviously not involved, and prolactin shows an extremely wide range of actions, varying depending on the group.

Although GH and prolactin have been traditionally viewed as typical pituitary hormones, it is now clear that they are also expressed in a variety of extrapituitary tissues. Such expression is usually at much lower levels than in the anterior pituitary and it is unlikely that extrapituitary expression makes a significant contribution to the level of circulating hormone. However, such expression may produce significant local concentrations, allowing the hormones to have paracrine or autocrine effects in the tissues of origin.

Extrapituitary expression of GH, prolactin or related hormones is particularly notable in the placenta of some mammals, and in some mammalian groups duplications of the GH or prolactin gene have given rise to complex gene families expressed primarily in the placenta. In rodents and ruminants such duplications have involved the prolactin gene. In primates on the other hand, duplications of the GH gene have given rise to complex gene families, mostly expressed in the placenta.

The nature of these GH-related genes and the proteins that they express is considered in detail here, with particular emphasis on the situation in New World Monkeys (NWM). The complex nature of the GH-related genes can be contrasted with that of prolactin, which in primates is encoded by a single gene, showing no sign of recent gene duplication.

MULTIPLE GH-LIKE GENES IN PRIMATES

In most non-primate mammals GH is encoded by a single gene that shows no sign of recent gene duplication. An exception is seen in the sheep and goat, where two GH-like genes are found in tandem on chromosome 11 (Ofir & Gootwine 1997; Valinsky et al. 1990; Wallis et al. 1998). One of these is expressed in the anterior pituitary, and appears to function like normal GH, while the other is expressed in the placenta and has an uncertain role. This duplication appears to have occurred recently; indeed it is not found in all individual sheep - depending on strain some animals have only a single, pituitary-expressed, GH gene.

In some primates the situation is much more complex. The position in human has been investigated most intensively. It was shown by Josimovich & MacLaren (1962) that the human placenta produces a molecule with lactogenic properties, and the nature of this was elucidated by Sherwood (1967) who characterized a placental lactogen from human placenta. Immunological studies suggested that this was related to human GH, and determination of its amino acid sequence (Niall et al. 1971) confirmed this, showing 85% identity between the two proteins. Despite this, the growth-promoting activity of human placental lactogen (hPL; also known as chorionic somatomammotropin hormone, hCSH) is relatively low. It has high lactogenic activity, as does human GH as well as human prolactin. hPL is expressed at high levels during pregnancy, concentration in blood reaching >2 μg/ml (compared with 10-20 ng/ml for GH). The circulating concentration falls following parturition, rapidly becoming undetectable. The physiological role of hPL is still unclear. Despite its name, its role in regulation of lactation must be limited, since it disappears from the circulation at the point when lactation starts. It has been reported that in at least some cases women who lack the gene for hPL can have normal pregnancies and normal babies (Wurzel et al. 1982). It is possible that the role of hPL is to regulate the balance of nutrition between the needs of the mother and those of the foetus, in which case this may be of relatively little importance under the conditions of good regular nutrition that apply normally in Western societies. It has also been suggested that in humans (unlike experimental models) hPL *inhibits* lactation, which could act to delay the initiation of lactation until hPL levels fall following parturition.

hPL clearly arose as a consequence of the duplication of the GH gene, and investigation of the organization of the GH locus in humans revealed that there have in fact been several recent duplications. Initially the exact number of GH-like genes in human was uncertain, but it is now clear that the locus comprises 5 genes organized in tandem, in the same transcriptional orientation, on chromosome 17 (Chen et al. 1989; Figure 1). The GH gene (*GH-N*), expressed primarily in the pituitary, is found at the 5' end of this cluster. It is followed by four genes expressed in the placenta : *PL-L, PL-B, GH-V* and *PL-A*. *PL-A* and *PL-B* are distinct genes but both encode hPL; the mature protein sequences encoded are identical, but there is one difference in the signal peptide (the peptide sequence at the N-terminus of the precursor protein that facilitates passage into the secretory pathway). The expression of *PL-A* appears to be about 2-3 fold greater than that of *PL-B*. *PL-L* (PL-like) could encode a protein similar in sequence to hPL, but the gene has an abnormal splice pattern. It is expressed at very low levels as mRNA, and possibly not at all as protein, and may be in effect a pseudogene. *GH-V* (GH variant) encodes a protein with about 93% sequence identity to pituitary GH.

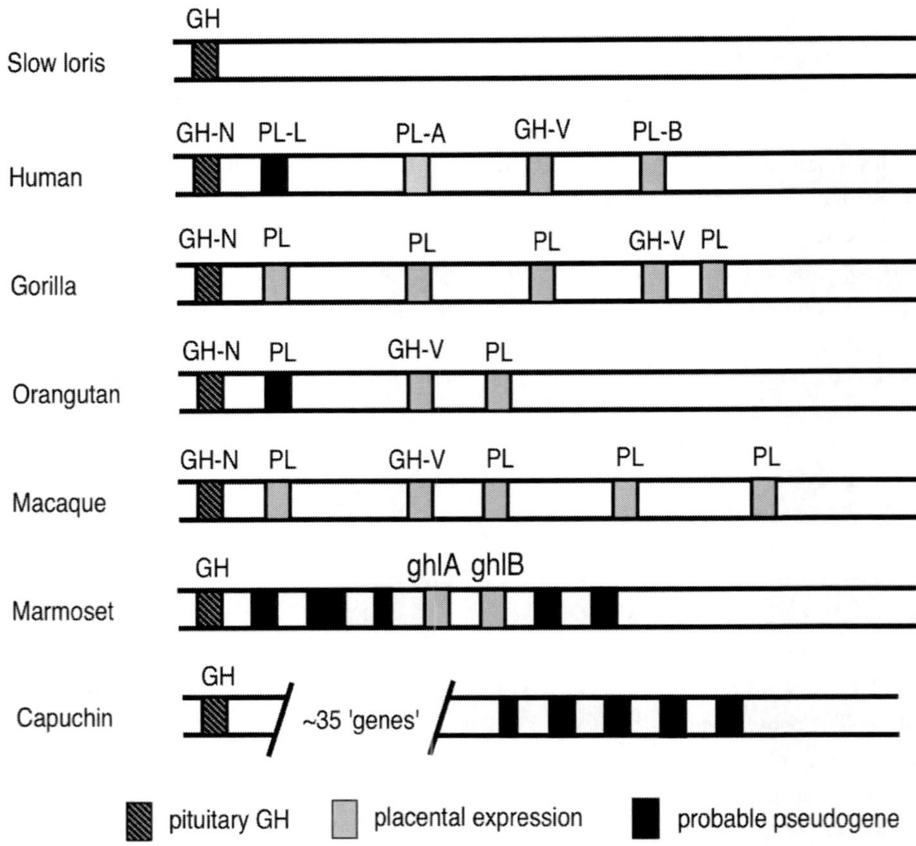

Figure 1. The organization of *GH* gene clusters in primates. In the slow loris, a prosimian, and in most non-primate mammals, there is a single *GH*-like gene, expressed in the pituitary gland. In higher primates there is a complex cluster of *GH*-like genes, with pituitary-expressed *GH* at the 5' end followed by several placentally expressed genes (including those for placental lactogens, PLs), often including one or more pseudogenes. The clusters arose by sequential tandem duplications, apparently independently in Old World monkeys/apes and New World monkeys.

It is expressed at fairly low levels, but during the later stages of pregnancy it appears to replace pituitary GH in the circulation (Frankenne et al. 1988). Its physiological function is not clear, though various actions have been described, including a role in stimulating trophoblast invasion (Lacroix et al. 2005). The *GH-V* and *PL* genes are all expressed on the fetal side of the placenta; it has been suggested that their transfer into the maternal circulation may reflect a maternal-fetal conflict for the control of nutritional resources (Haig 2008).

The GH locus in apes and Old World Monkeys (OWM) is similar to that in man (Figure 1), with a pituitary-expressed gene for GH at the 5' end, and a variable number (3-7) of genes following it that are expressed in the placenta. Some of the latter may be pseudogenes. In NWM the locus is also complex, with again the gene for pituitary GH at the 5' end and a large number of genes/pseudogenes following it. This is discussed in detail below. In most prosimians (lemur, bush baby, slow loris), on the other hand, the situation is similar to that in non-primate mammals, with a single gene for pituitary GH, and no evidence for gene duplication.

The situation in the tarsier (*Tarsius*) is unclear. This group is now considered to be the sister group of NWM/OWM/apes, and thus more 'advanced' than prosimians such as lemur, bush baby or loris. Two GH-related genes have been reported in *Tarsius*, encoding similar sequences that are markedly different from the GH of both other prosimians and of monkeys or apes (Ye et al. 2005). Whether these represent all of the GH-related genes in *Tarsius*, how they are organized in the *Tarsius* genome, and where they are expressed remain unclear.

In all primates, and indeed all mammals, the GH gene comprises 5 exons, split by 4 introns (Figure 2). The mature mRNA is translated to give a precursor molecule comprising a signal peptide of about 26 aas and a mature hormone of about 190 aas. In human, alternative splicing of the initial transcript leads to a shorter protein, 20K hGH, lacking 15 aas normally encoded by exon 3, which makes up 5-10% of the GH in the normal pituitary (Figure 2). The biological properties of this variant are similar, but not identical, to those of the normal 22K GH (Solomon et al. 2006, Takahashi et al. 2001).

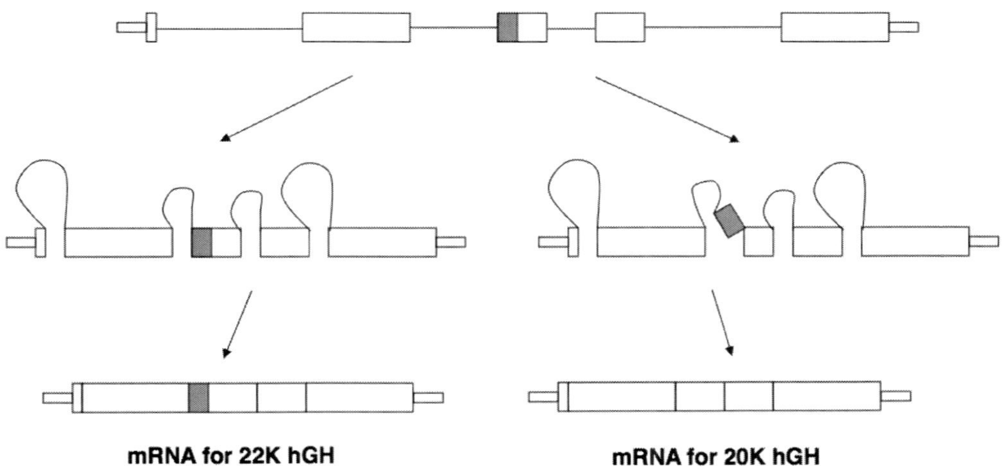

Figure 2. Alternative splicing of the human GH gene. The GH gene comprises 5 exons and 4 introns (top). Normally it is spliced as shown on the bottom left, to give an mRNA encoding 22K hGH. However, 5-10% of initial transcripts are spliced as shown on the bottom right, with 45 nucleotides of what would normally become part of exon 3 being included instead in intron 2. This gives an mRNA that encodes 20K hGH, 15 amino acids shorter than 22K, due to an internal deletion of 45 nucleotides.

THE GH LOCUS IN NEW WORLD MONKEYS

As indicated, NWM, like OWM, apes and man, have a complex GH locus, including the gene for pituitary GH and a number of related genes, at least some of which appear to be expressed in the placenta. The nature of these GH-like genes has been investigated in several NWM species.

GH of marmoset (*Callithrix jacchus*) was shown to be more similar to human GH than to prosimian GH, suggesting that the episode of rapid GH change that occurred during primate evolution (discussed in detail below) was largely complete before divergence of OWM and NWM (Wallis et al. 2001; Wallis & Wallis 2002). The marmoset GH cluster comprises 8

GH-related genes, in the same transcriptional orientation, with the *GH* gene itself being located at the 5' end (Figure 3)

The seven genes following this were clearly derived from the *GH* gene by a series of gene duplications. Five of them are probably pseudogenes; they contain mutations that would prevent normal expression, including stop codons and small insertions/deletions (indels) that would alter the reading frame, as well as in one case a large deletion removing exons 2-4 and in another a large insertion into exon 3 (Figure 3).

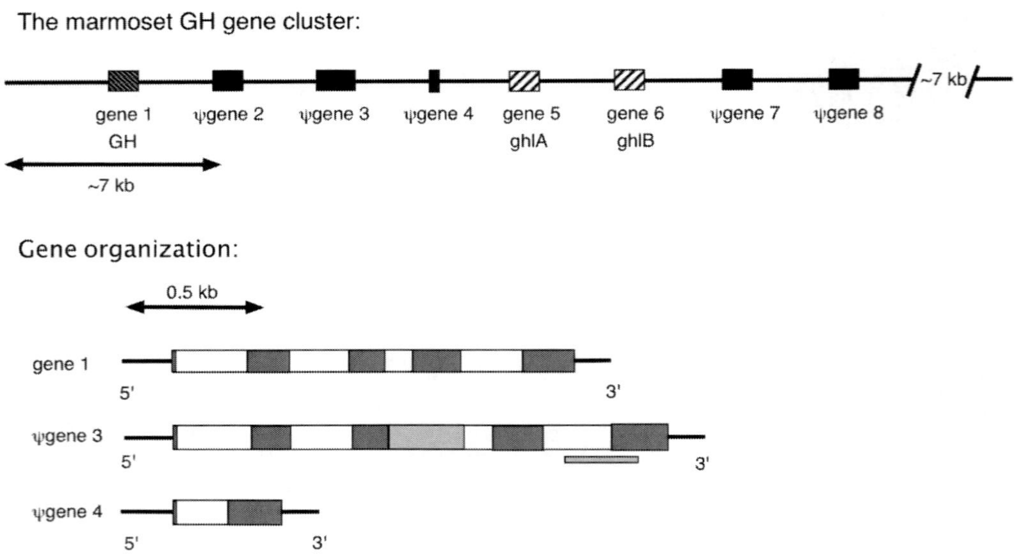

Figure 3. The marmoset GH gene cluster. The cluster (top) comprises 8 genes/pseudogenes, including that at the 5' end (gene 1) encoding pituitary GH, gene 5 (*ghlA*) and gene 6 (*ghlB*) encoding GH-like proteins probably expressed in placenta, and five apparent pseudogenes. The structure shown for gene 1, with 5 exons and 4 introns, is representative of most genes/pseudogenes in the cluster. Pseudogene 3 contains an internal (reverse) repeat of the region marked, spanning intron 4 and exon 5. Pseudogene 4 contains a large deletion, removing exons 2-4 and some flanking sequence. (Based on Wallis & Wallis 2006).

The other two genes have sequences that could be transcribed and translated to give proteins differing substantially from marmoset GH (at 25-30% of residues), and analysis of transcriptome libraries (see below) indicates that these two genes are indeed expressed. Given the expression of equivalent genes in a different NWM, as discussed in the next section, it is likely that they are expressed in placenta. None of the 2 genes or 5 putative pseudogenes in the marmoset *GH* gene cluster is equivalent to the *PL* or *GH-V* genes in man, and this and a number of other considerations suggests that the clusters of GH-like genes in OWM/apes and NWM arose independently, as discussed in detail in a later section.

The marmoset *GH* cluster is notably more regular than the cluster in man, with a rather regular spacing between the 8 genes and similar breakpoints, indicating a regular process of gene duplication. Although it contains more genes, the marmoset cluster is shorter than that in human (about 35kb in the former, about 50 kb in the latter) and contains fewer, more regularly distributed *Alu* repeats. As in the case of the human gene cluster several gene conversion events appear to have occurred in the marmoset GH gene cluster, leading to greater similarity between some regions than would be expected on the basis of phylogenetic divergence (Wallis & Wallis 2006).

Information about the marmoset GH cluster is also now available from the genomic assembly for this species (*http://www.ncbi.nlm.nih.gov*). As has been seen to be the case for chimpanzee (Perez-Maya et al. 2012), automatic assembly of the complex GH locus of primates proves problematical, but detailed analysis of the contigs and sequence reads available confirms the presence of eight marmoset GH-like genes and accords with the organization of the cluster published previously.

The *GH* gene cluster in a second NWM, the white fronted capuchin, *Cebus albifrons*, is much more complex than that in marmoset. In *C. albifrons* there appear to be at least 40 GH-related genes arranged in no more than 2 or 3 clusters, and quite probably a single cluster (Wallis & Wallis 2006).

The complexity of the situation has precluded a complete elucidation of the organization of these genes, but it is clear that again the single gene that appears to correspond to pituitary GH lies at the 5' end of the cluster, and that a large proportion of the 40 detected "genes" are likely to be pseudogenes (Figure 1).

Genes similar to most of the 8 genes detected in marmoset are present, with in most cases several *C. albifrons* genes equivalent to each marmoset gene. It is known that as the number of related genes in a gene cluster increases this makes additional gene duplication events more likely, due to mismatched gene pairing at meiosis (Hancock 2005), and it seems likely that this has occurred in the case of the *C. albifrons* GH locus, leading to an extraordinary expansion of the cluster, probably driven by random processes rather than adaptation.

The sequence of squirrel monkey (*Saimiri boliviensis*) GH was determined by cDNA cloning by Liu et al. (2001). Extensive information about the genome sequence of this species is available (*http://www.ncbi.nlm.nih.gov*). A preliminary examination suggests that this genome includes at least 20 GH-like genes, indicating that *S. boliviensis* may possess a GH locus of complexity approaching that of *C. albifrons*. *Cebus* and *Saimiri* are closely related in the NWM phylogenetic tree (Wildman et al. 2009).

Revol de Mendoza et al. (2001, 2004) examined the GH-like genes in the spider monkey (*Ateles geoffroyi*). A gene for pituitary GH was identified together with 5 additional genes/pseudogenes, three of which were cloned. The cloned genes/pseudogenes were similar to some of those reported for marmoset, including one with a large deletion of exons 2-4. Further analysis of the *A. geoffroyi* GH locus was carried out by Gonzales-Alvarez et al. (2006), who confirmed the presence of six genes with rather similar intragenic regions, as in marmoset, though the complete organization of the locus was not determined.

GH-like genes have also been identified in a second species of spider monkey, *Ateles fusciceps*. Papper et al. (2009) cloned transcripts from at least 3 GH-like genes from placental tissue from this species. Two of these corresponded to genes described by Revol de Mendoza et al. (2004) in *Ateles geoffroyi*, a third was not previously seen in *Ateles*.

GH-like genes from three additional species of NWM, the red howler (*Alouatta seniculus*), the dusky titi (*Callicebus moloch*) and the white-faced saki (*Pithecia pithecia*), were studied by Li et al. (2005). 4-5 genes/pseudogenes were identified in each of these species, corresponding fairly closely to the genes described previously in marmoset and spider monkey. It is not clear whether in these cases the complete set of genes in the corresponding gene cluster was identified. Phylogenetic analysis confirmed the independent origin of GH gene clusters in OWM/apes and NWM.

Table 1 lists the NWM GH-like genes/proteins that are potentially expressed (i.e. excluding probable pseudogenes). The proteins can be divided into three groups on the basis of sequence similarities (see below): (1) pituitary GH, (2) GH-like proteins group A (ghlA), (3) GH-like proteins Group B (ghlB). The nomenclature used in the rest of this article is based on these three groups, and given in the left-hand column of the table.

Table 1. New World Monkey GH-related proteins

Protein	Species	original name (gene and/or protein)	Accession No	Source
Pituitary GH				
Age_GH	*Ateles geoffroyi*	GH-N	AF374234	d,e
Ase_GH	*Alouatta seniculus*	A.sen1, ghlp1	AY744451	f
Cal_GH	*Cebus albifrons*	Capuchin-GH1	AM235213	a
Cja_GH	*Callithrix jacchus*	ghlp1, MarmGH; MarmGene1	AJ297563,	b, c
Cmo_GH	*Callicebus moloch*	C.mol1	AY744456	f
Sbo_GH	*Saimiri boliviensis*	GH	AF339060	h
GH_like A				
Afu_ghlA1	*Ateles fusciceps*	GHC transcript variant 1a	EU935073	g
Afu_ghlA2	*Ateles fusciceps*	GHC transcript variant 1b	EU935073	g
Age_ghlA	*Ateles geoffroyi*	GH-C	AY435434	d, e
Ase_ghlA	*Alouatta seniculus*	A.sen5	AY744455	f
Cja_ghlA	*Callithrix jacchus*	ghlp5, MarmGene5	AJ489810	c
Cmo_ghlA	*Callicebus moloch*	C.mol2	AY744457	f
Ppi_ghlA1	*Pithecia pithecia*	P.pit2	AY744462	f
Ppi_ghlA2	*Pithecia pithecia*	P.pit4	AY744464	f
GH_like B				
Afu_ghlB1	*Ateles fusciceps*	GHB	EU935072	g
Afu_ghlB2	*Ateles fusciceps*	GHD transcript variant 1	EU935080	g
Age_ghlB	*Ateles geoffroyi*	GH-V/ GH-B	AF374235	d, e
Ase_ghlB1	*Alouatta seniculus*	A.sen2	AY744452	f
Ase_ghlB2	*Alouatta seniculus*	A.sen3	AY744453	f
Ase_ghlB3	*Alouatta seniculus*	A.sen4	AY744454	f
Cja_ghlB	*Callithrix jacchus*	MarmGene6, ghlp6	AJ489811	c
Cmo_ghlB	*Callicebus moloch*	C.mol5	AY744460	f
Ppi_ghlB	*Pithecia pithecia*	P.pit1	AY744461	f

The table includes proteins for which expression is likely. Common names of species are: *Ateles fusciceps (Afu)*, brown-headed spider monkey; *Ateles geoffroyi (Age)*, black-handed spider monkey; *Alouatta seniculus (Ase)*, red howler; *Cebus albifrons (Cal)*, white-fronted capuchin; *Callithrix jacchus (Cja)*, common marmoset; *Callicebus moloch (Cmo)*, dusky titi; *Pithecia pithecia (Ppi)*, white-faced saki; *Saimiri boliviensis (Sbo)*, Bolivian squirrel monkey. Age_ghlA is an incomplete sequence. Ase_ghlB1 has a "hybrid" structure, the N-terminal end being B-type, the C-terminal end A-type. Sources were: a) Wallis & Wallis (2006), b) Wallis et al. (2001), c) Wallis & Wallis (2002), d) Revol de Mendoza et al. (2001), e) Revol de Mendoza et al. (2004), f) Li et al. (2005), g) Papper et al. (2009), h) Liu et al. (2001).

EXPRESSION OF GH-LIKE GENES IN NEW WORLD MONKEYS

Most of the previous studies were carried out on genomic material. Relatively little information is available about the transcription of the GH-like genes. That NWM GH is expressed in the pituitary gland was shown by Liu et al. (2001), who cloned cDNA for the GH of *Saimiri boliviensis* from pituitary tissue. Genes encoding a very similar protein, presumed to be pituitary GH, have been identified in six of the species discussed above (*Saimiri boliviensis, Callithrix jacchus, Cebus albifrons, Ateles geoffreyi, Alouatta seniculus* and *Callicebus moloch).*

We have found that transcripts for marmoset (*C. jacchus*) GH are abundant in a transcriptome library made from 15 different tissues, including pituitary gland (Pipes et al. 2013). As indicated in the previous section, evidence that the NWM GH-like genes are transcribed in placenta was obtained by Papper et al. (2009), who studied placental transcripts from a spider monkey (*Ateles fusciceps*). They found expression of 3-4 GH-like genes in the placenta of this species (two transcripts, Afu_ghlA1 and Afu_ghlA2 were very similar - whether they are products of different genes is not clear). Two of these (Afu_ghlA1/2 and Afu_ghlB1) corresponded to genes described in the genomes of *Ateles geoffroyi* and marmoset. The third (Afu_ghlB2) has no described equivalent in *A. geoffroyi*; it may correspond to a pseudogene in marmoset (*ghlp8*). Afu_ghlB1 transcripts were much less abundant than those of Afu_ghlA1/2 or Afu_ghlB2. Expression of several splice variants of Afu_ghlA1/2 and Afu_ghlB2 was detected. The physiological function of these placentally expressed genes and variants is yet to be explored. Pituitary GH gene was not expressed in placenta.

As discussed above, in the marmoset GH gene cluster only two of the 7 genes/pseudogenes, in addition to that for pituitary GH, appear potentially expressible as functional protein. We found transcripts for both these genes expressed at high level in a marmoset transcriptome library (sra study SRP016577, Experiment SRX208981, Washington University Genome Institute). The tissue of origin for this library was not specified, but it does show abundant transcripts of some genes characteristic of placenta. Expression of the five potential pseudogenes, including *ghlp8*, was not detected in this transcriptome library.

EVOLUTION OF GH-LIKE PROTEINS IN NEW WORLD MONKEYS

Figure 4 gives an alignment of protein sequences derived by conceptual translation from the available GH-like gene sequences (excluding likely pseudogenes and incomplete sequences) from NWM. The sequence of human GH is included as an outgroup. Scrutiny of the alignment suggests that the sequences fall into three groups, pituitary GH and two groups of GH-like proteins, designated ghlA and ghlB (see also Table 1).

This division is confirmed by a phylogenetic tree (Figure 5) constructed from these data, using the Neighbor Joining method (Saitou & Nei 1987) in PAUP* (Swofford (1998). Each of the 3 groups receives strong Bootstrap support. Most of the NWM species studied have at least one sequence in each of the three groups. Ase_ghlB1 is an exception - it appears to have a 'hybrid' structure, the N-terminal end being B-type and the C-terminal end A-type; it was omitted for the phylogenetic analysis.

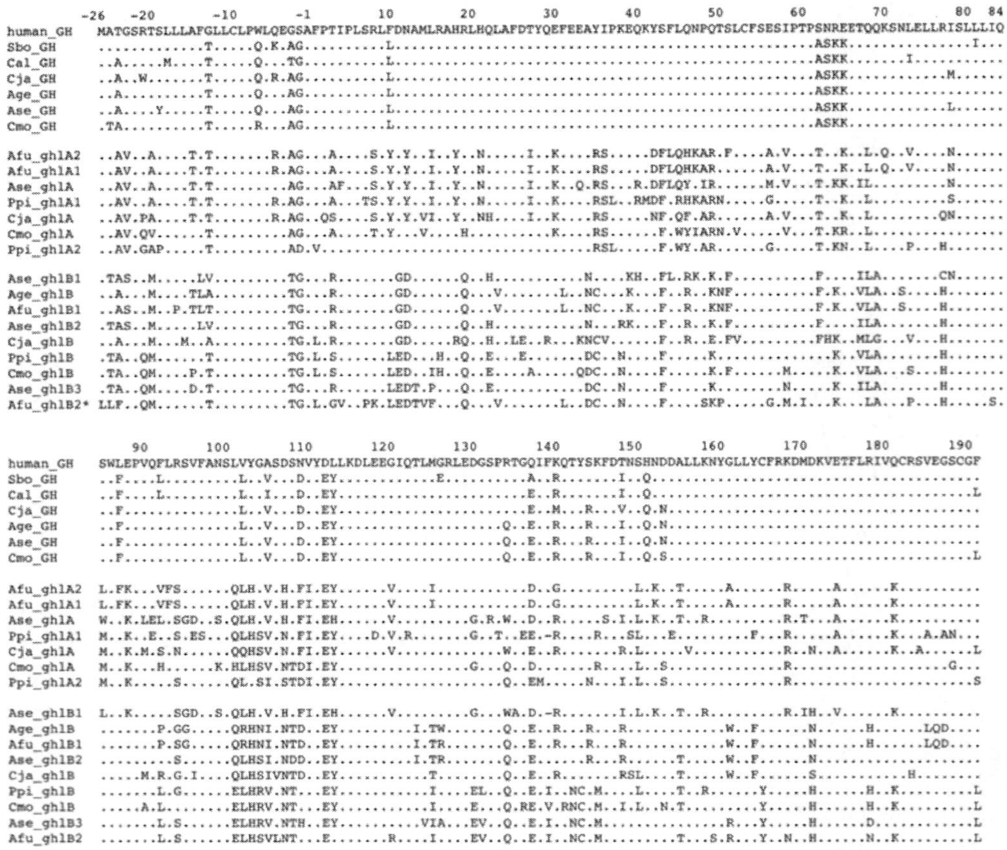

Figure 4. Sequence alignment for New World Monkey GH-like proteins. The sequences of NWM proteins are compared to that of human GH, with . indicating identity and - indicating a gap. The sequences are divided into 3 groups, GH (top), ghlA (middle) and ghlB (bottom). The first 26 residues (-26 - -1) are the signal peptide which is followed by the mature protein (1-192). Pseudogenes and incomplete sequences have been excluded. Afu_ghlB2 has an extended N-terminal sequence (*, MGRSACGPCLFPGF) which has been omitted from the alignment.

It is notable that sequence variation (as indicated by branch lengths in Figure 5), and therefore rate of evolution, is much greater for the two groups of GH-like proteins, than for pituitary GH. A similar tree was presented by Li et al. (2005), based on full gene (and pseudogene) sequences rather than protein sequences, which made the distinction between the ghlA and ghlB groups rather less clear-cut. Groups ghlA and ghlB are also seen in the phylogenetic tree of Papper et al. (2009), though this included fewer species.

Variable Rates of Evolution of Primate GH-Like Genes

In addition to the complex expansion of the GH-gene cluster, GH evolution in primates is also notable for showing variable rates of evolution (Ohta 1993; Wallis 1981, 1994; Figure 6). Most remarkable is the episode of rapid evolution that preceded the divergence of OWM and NWM.

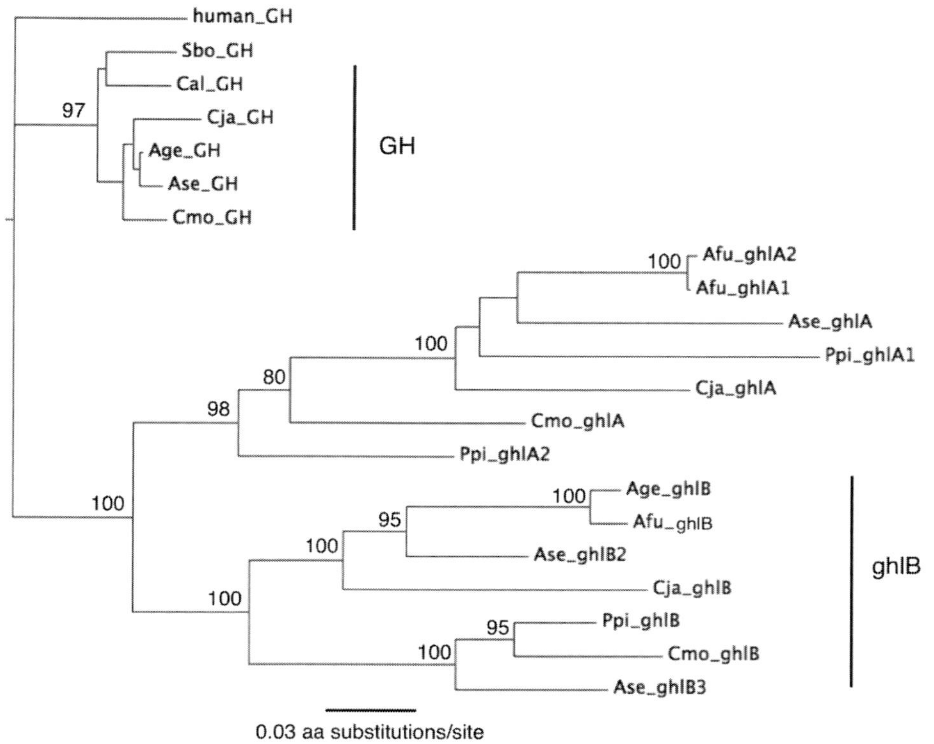

Figure 5. Phylogenetic tree for New World Monkey GH-like proteins. The tree is based on the protein sequences shown in Figure 4, with Ase_ghlB1 excluded (see text). Human GH is included as an outgroup. Numbers at nodes are Bootstrap values (%); values < 80% are omitted. Note that the 3 main groups identified in Figure 4 fall into three well defined clades.

Generally the sequence of GH in mammals is strongly conserved, and this extends to prosimians. Thus, the sequence of slow loris (*Nycticebus pygmaeus*) GH is very similar (~98% identical) to that of GH from pig, dog or elephant (Wallis et al. 2001). However, the sequence of human or marmoset GH differs from that of slow loris at about 35% of all residues. Sequences of human and marmoset GHs are rather similar (differing at about 10% of all residues). Thus there appears to have been a period of very rapid evolution of GH on the lineage leading to higher primates, after separation of prosimians, but before divergence of OWM/apes and NWM, as illustrated in Figure 6. Various lines of evidence support the idea that this was adaptive, involving positive selection, although the nature of the biological changes involved is not known (Forsyth & Wallis 2002). Episodes of rapid evolution of GH can be seen on at least three additional mammalian lineages (Figure 6).

Papper et al. (2009) have suggested that the episodes of rapid evolution of GH seen in mammalian evolution were preceded by gene duplications, with subsequent deletion of paralogues, so that, for example, the GH genes of higher primates and prosimians are not strictly orthologous. They proposed that a total of three duplicate GH genes were present in the last common ancestor of the Boroeutheria (the mammalian group including primates and artiodactyls), with loss of two of these genes along each lineage, without any trace being preserved in the complete genome sequences that are available for most of the groups involved.

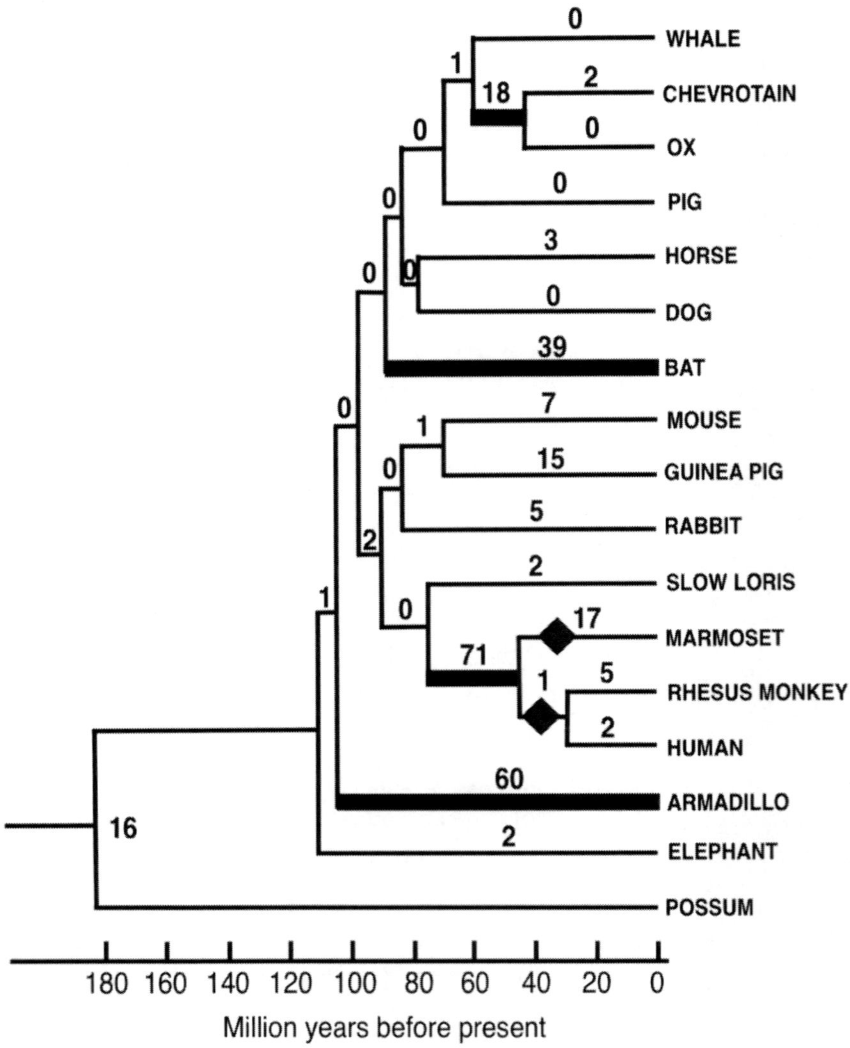

Figure 6. Evolutionary tree for mammalian GHs. Branch lengths were derived from amino acid sequences for mature pituitary GHs from representative species, using the codeml (aaml version) programme (Yang 2007) and a defined tree based on Murphy et al. (2004). Evolutionary rate is markedly variable, with episodes of rapid change occurring on the branches marked with thick lines. Evolutionary time (from Murphy et al. 2004) is represented on the abscissa. Numbers on branches are numbers of amino acid substitutions. The approximate times of duplications that gave rise to the GH-related placental proteins in NWM and OWM are shown by □. GH-like proteins resulting from these duplications are not included.

The proposal, involving gene loss on at least five separate occasions, seems unnecessarily complicated, especially since inclusion of additional mammalian lineages, some associated with further episodes of accelerated evolution (Wallis 2008; Figure 6), would presumably require yet further duplications of the GH gene with subsequent loss of all but one of the paralogues on each lineage.

A more likely explanation of the episodes of rapid GH evolution in mammals, including that in primates, is that they were due to selective forces, possibly associated with subtle and fluctuating changes in specificity. A mechanism for this has been proposed ("Function

Switching") which could arise when a gene has two or more functions, the relative importance of which fluctuates over evolutionary time (Wallis 1997, Forsyth & Wallis 2002). Repeated alternating adaptation of GH to the two functions could lead to accumulation of multiple amino acid substitutions with little overall change in physiological activity. A resolution of the situation could arise with separation of the functions by gene duplication, and it is interesting that the rapid evolution of pituitary GH (but not the other GH-like proteins) slowed after the gene duplication(s) that gave rise to the clusters of GH-like genes in both NWM and OWM/apes. (i.e., in this case rapid evolution preceded rather than followed gene duplication). An important second function for GH in this respect could have involved placental expression, which may well have had an early origin in primate evolution (Papper et al. 2009).

Further periods of rapid evolution occurred in the evolution of the GH-like genes (but not pituitary GH itself) as the GH clusters expanded and diverged in both OWM/apes and NWM. In at least some instances in OWM/apes this has been shown to reflect adaptive evolution (Rodríguez-Sánchez et al. 2010; Wallis 1996). Adaptive evolution also appears to have been involved in the diversification of both the ghlA and ghlB groups in NWM. Notably, the residues varying in these two groups are largely different, suggesting adaptation to different functions. Li et al. (2005) obtained evidence for adaptive evolution in their Cluster 2 genes (approximately equivalent to the ghlA group here) but not in Cluster 3 (similar to the ghlB group).

Independent Origins of the GH Clusters in NWM and OWM/Apes

As already indicated, various lines of evidence suggest that the gene duplications that gave rise to the clusters of GH-related genes in OWM/apes and NWM occurred independently (Li et al. 2005; Revol de Mendoza et al. 2004; Wallis & Wallis 2002, 2006). It seems a remarkable coincidence that such similar events should have occurred on each of the two lineages (rather than a single one before divergence). The evidence supporting independent origins is:

1. The positions of break-points between unique DNA sequence and repeated sequence are quite different for the clusters in OWM/apes and NWM.
2. The overall organization of the clusters is very different for the two groups, with very regular, relatively short repeats in NWM and more variable, longer repeats in OWM/apes.
3. The distribution of *Alu* elements in NWM is very even, with just 2 such repeats in every gene. In OWM/apes there are more *Alu* elements, showing a rather uneven distribution.
4. Phylogenetic analysis shows the GH-like genes of NWM clustering with the gene for pituitary GH, excluding the pituitary GH gene and GH-like genes of OWM/apes.

The independent evolution of complex GH gene clusters on these two lineages may be a consequence of "pressure" generated by the function-switching process referred to previously.

If this pressure was not released before splitting of the two lineages it would remain on each. This would eventually lead to resolution by gene duplication, divergent evolution and separation of pituitary and placental expression - essentially the same process on each lineage, but differing in details.

GH-Receptor Interactions

GH acts by binding to a specific receptor on the plasma membrane of its target cells. The marked difference between the GHs of higher primates and non-primate mammals is associated with species specificity for receptor binding. Human GH binds to the receptors of non-primate mammals (and shows growth-promoting activity in the rat for example), but non-primate GHs do not bind to the human GH receptor, and have little growth-promoting activity in man. A major determinant of this species specificity was identified by Souza et al. (1995) and Behncken et al. (1997) who showed that it reflects replacement on the primate lineage of Leu by Arg at position 43 in the receptor and His by Asp at position 171 in GH. Repulsion between positively charged Arg 43 (receptor) and His 171 (GH) prevents binding of non-primate GH to human GH receptor, while formation of a salt bridge between Arg 43 (receptor) and Asp 171 (GH) enhances binding of human GH.

The evolution of this species specificity has been explored by Liu et al. (2001) who showed that in NWM (squirrel monkey, *Saimiri* and spider monkey, *Ateles*) there is an intermediate situation, in which GH has Asp 171 (as in human GH) but the GH receptor has Leu 43, as in prosimians and non-primates. Yi et al. (2002) examined the binding properties of the squirrel monkey GH receptor and showed that it is able to bind rat GH, unlike the human receptor. That the species specificity of the receptor is largely due to the residue at position 43 was shown by altering this position by site-directed mutagenesis, to Leu in the human receptor and Arg in the squirrel monkey receptor, which was accompanied by the anticipated changes in binding properties.

The change of His-->Asp at position 171 of GH is seen for all available NWM GH sequences (Figure 4). It clearly preceded the corresponding change of Leu-->Arg in the receptor. This militates against the idea of coevolution of hormone and receptor as far as the development of species specificity is concerned, since change in hormone was complete well before change in receptor began. However, although the change in species specificity is the most marked biological change seen in the evolution of primate GHs it is probably misleading to put emphasis on this as a driver of evolution, since in nature the GH receptor from one species will never encounter GH from another. The development of species specificity is more likely to be a side effect of structural changes driven by changes in other biological properties involving the coevolution of GH and its receptor. It is known that the episode of rapid change seen during primate GH evolution was paralleled by a corresponding burst of change in the receptor. It is notable that NWM GH-like proteins show considerable variation at position 171 (Figure 4). In all NWM GHs this residue is Asp, as in human GH. In group A GH-like proteins it is also Asp, except in the marmoset protein (Cja_ghlA) where it is Asn. In Group B GH-like proteins it is His (as in non-primate GHs), Asn or Ser. In the light of the work of Yi et al. (2002) these changes are likely to reflect variation in binding properties for the various GH-like proteins.

```
                    -29      -20       -10       -1        10        20        30        40        50        60        70        80         90
humanPRL            MNIKGSPWKG-SLLLLLVSNLLLCQSVAPLPICPGGAARCQVTLRDLFDRAVVLSHYIHNLSSEMFSEFDKRYTHGRGFITKAINSCHTSSLATPEDKEQAQQMNQKDFLSLIVSILRS
Callithrix_PRL      ...EA....AG............W...T...V.....V....................................Q.....S..L...................IS..............
Saguinas_PRL        .......C.AG................T...V.....V......Q.............................Q.....S......................IS..............
Saimiri_PRL         ..........-................TS..V.....V.............I...................Q..Q.....S......................IS........I.....
Nycticebus_PRL      --------SLLL...I.L......K...S.........N...S.........I.............N......AQ.................S..........IHHE.L.N.VLRV...

                          100       110       120       130       140       150       160       170       180       190      199
humanPRL            WNEPLYHLVTEVRGMQEAPEAILSKAVEIEEQTKRLLEGMELIVSQVHPETKENEIYPVWSGLPSLQMADEESRLSAYYNLLHCLRRDSHKIDNYLKLLKCRIIHNNNC
Callithrix_PRL      ..........................I.....S..........L.......R..AV.S.................F...........................V.D...
Saguinas_PRL        ..........................I................L...........V.S.................F...........................V.D...
Saimiri_PRL         ...........................................L........G..V.S.................F...........................V.D...
Nycticebus_PRL      ..D................D....R.I.....N........K..G....RIR...V.S......A......D...F.F..........................YDS..
```

Figure 7. Sequence alignment for New World Monkey prolactins. The sequences of prolactins from NWM and a prosimian (slow loris) are compared to that of human prolactin, with indicating identity. The first 29 residues (-29 - -1) are the signal peptide which is followed by the mature protein (1-199). Incomplete sequences have been excluded.

PROLACTIN IN NEW WORLD MONKEYS

Pituitary prolactin is encoded by a single gene in most mammals. The structure of this gene is similar to that of GH, in that it comprises 5 exons split by 4 introns, but the prolactin gene is much larger than that for GH, about 10 kb compared with 2kb for GH. The difference in size is accounted for mainly by larger introns in the prolactin gene. In addition, the prolactin gene in those mammals where it has been studied is subject to alternative splicing in many extrapituitary tissues, with an alternative exon 1a, about 5 kb upstream of the rest of the gene, being spliced onto part of the normal exon 1b and the remaining 4 exons. This means that the regulation of expression of the gene in such tissues is markedly different from that in the pituitary.

A single prolactin gene has been identified and characterized in marmoset (Wallis et al. 2005). In overall structure it is similar to the human prolactin gene. The rate of evolution of prolactin in primates is elevated compared with that in most other mammals (Wallis 2000) and the marmoset prolactin sequence confirmed this. Examination of the marmoset genomic sequence confirms the existence of a single prolactin gene and allows the completion of its sequence (exon 1 was missing). A single prolactin gene was also identified and partially sequenced in the white-fronted capuchin (Wallis et al. 2005). This was very similar to the corresponding gene for marmoset. The single prolactin gene contrasts markedly with the very complex GH gene cluster in this species discussed above.

Examination of genomic data available for two additional NWM species (*Saimiri boliviensis* and *Saguinas midas*) shows that in each of these there is also a single prolactin gene. Derived sequences for prolactins of these species are compared with that of marmoset in Figure 7. The NWM prolactin mature protein sequences are similar, differ from human prolactin at 15-18 residues and from a prosimian (slow loris, *Nycticebus pygmaeus*) at about 38 residues.

We have found that transcripts for marmoset prolactin are abundant in the transcriptome library made from 15 different tissues, including pituitary gland (Pipes et al. 2013). Almost all of these transcripts appear to be transcribed from the proximal promoter, without involvement of the upstream exon 1a.

CONCLUSION

GH-like and prolactin genes in NWM show similarities to those seen in OWM/apes. In the case of prolactin the genes are clearly orthologous and equivalent in each group. In the case of GH, orthology presumably applies to the genes expressing pituitary GH, but the GH-like genes, although expressed in the placenta in each case are not strictly orthologous, having originated independently in each group. The NWM GH-like proteins appear to fall into three main groups, pituitary GH, GH-like A and GH-like B. The rate of evolution for the GHs is fairly slow, having decreased markedly following an episode of rapid evolution of GH that occurred before divergence of OWM and NWM. On the other hand, the NWM GH-like proteins, for both groups ghlA and ghlB, show very high evolutionary rates, apparently reflecting ongoing adaptive evolution. Although much information about GH-like genes and proteins in NWM is now available, there are still large gaps in our understanding. In only one case, the marmoset, is the overall organization of the GH locus known, and it is clear that this organization varies considerably among NWM species, showing remarkable complexity in some cases. Expression of some ghlA- and ghlB-type genes in placenta has been demonstrated, but whether expression occurs also in in other tissues is not known, nor is the full extent and significance of the alternative transcript forms that have been detected in the case of spider monkey. The biggest gap in our knowledge is perhaps in understanding the biological functions of the GH-like genes expressed in the placenta, and this will probably be the most difficult area in which to make progress, given how little is understood about the real physiological functions of the much more studied placental GH and lactogens in human.

The evolution of GH-like and prolactin-like proteins/genes in various mammalian groups, including primates, shows a number of remarkable features of evolutionary interest, including episodes of very rapid molecular change, rapid expansion of gene families by multiple gene duplications, adaptive evolution of recently evolved proteins and maternal-fetal competition. Unraveling these will continue to be a challenge for many years to come.

REFERENCES

Behncken, S. N., Rowlinson, S. W., Rowland, J. E., Conway-Campbell, B. L., Monks, T.A. & Waters, M. J. (1997). Aspartate 171 is the major primate-specific determinant of human growth hormone. Engineering porcine growth hormone to activate the human receptor. *Journal of Biological Chemistry* 272, 27077–27083.

Chen, E.Y., Liao, Y.-C., Smith, D.H. Barrera-Saldaña, H.A. Gelinas, R.E. & Seeburg, P.H. (1989). The human growth hormone locus: nucleotide sequence, biology, and evolution. *Genomics* 4, 479-497.

Forsyth, I.A. & Wallis, M. (2002). Growth hormone and prolactin - molecular and functional evolution. *Journal of Mammary Gland Biology and Neoplasia* 7, 291-312.

Frankenne, F., Closset, J., Gomez, F., Scippo, M.L., Smal, J. & Hennen, G. (1988). The physiology of growth hormones (GHs) in pregnant women and partial characterization of the placental GH variant. *Journal of Clinical Endocrinology and Metabolism* 66, 1171-1180.

González Alvarez, R., Revol de Mendoza, A., Esquivel Escobedo, D., Corrales Félix, G., Rodríguez Sánchez, I., González, V., Dávila, G., Cao, Q., de Jong, P., Fu, Y.-X. & Barrera Saldaña, H.A. (2006). Growth hormone locus expands and diverges after the separation of New and Old World Monkeys. *Gene* 380. 38-45.

Haig, D. (2008). Placental growth hormone-related proteins and prolactin-related proteins. *Placenta* 29 (Suppl), S36–S41.

Hancock, J.M. (2005). Gene factories, microfunctionalization and the evolution of gene families. *Trends in Genetics* 21, 591–595.

Josimovisch, J.B. & MacLaren, J.A. (1962). Presence in the human placenta and term serum of a highly lactogenic substance immunologically related to pituitary growth hormone. *Endocrinology* 71, 209-220.

Lacroix, M.-C., Guibourdenche, J., Fournier, T., Laurendeau, I., Igout, A., Goffin, V., Pantel, J., Tsatsaris, V. & Evain-Brion, D. (2005). Stimulation of human trophoblast invasion by placental growth hormone. *Endocrinology* 146, 2434-2444.

Li, Y., Ye, C., Shi, P., Zou, X.-J., Xiao, R., Gong, Y.-Y. & Zhang, Y.-P. (2005). Independent origin of the growth hormone gene family in New World monkeys and Old World monkeys/hominoids. *Journal of Molecular Endocrinology* 35, 399-409.

Liu, J-C., Makova, K.D., Adkins, R.M., Gibson, S. & Li, W.-H. (2001). Episodic evolution of growth hormone in primates and emergence of the species specificity of human growth hormone receptor. *Molecular Biology and Evolution* 18, 945–953.

Murphy, W.J., Pevzner, P.A. & O'Brien, S.J. (2004). Mammalian phylogenomics comes of age. *Trends in Genetics* 20, 631-639.

Niall, H.D., Hogan, M.L., Sauer, R., Rosenblum, I.Y. & Greenwood, F.C. (1971) Sequences of pituitary and placental lactogenic and growth hormones: evolution from a primordial peptide by gene reduplication. *Proceedings of the National Academy of Sciences U.S.A.* 68, 866-869.

Ofir, R. & Gootwine, E. (1997). Ovine growth hormone gene duplication - structural and evolutionary implications. *Mammalian Genome* 8, 770–772.

Ohta, T. (1993). Pattern of nucleotide substitutions in growth hormone-prolactin gene family: a paradigm for evolution by gene duplication. *Genetics*, 134, 1271 -1276

Papper, Z., Jameson, N.M., Romero, R. Weckle, A.L., Mittal, P., Benirschke, K., Santolaya-Forgas, J., Uddin, M., Haig, D., Goodman, M. & Wildman, D.E. (2009). Ancient origin of placental expression in the growth hormone genes of anthropoid primates. *Proceedings of the National Academy of Sciences U.S.A.* 106, 17083–17088.

Pérez-Maya, A.A., Rodríguez-Sánchez, I.P., de Jong, P., Wallis, M. & Barrera-Saldaña, H.A. (2012) The chimpanzee *GH* locus: composition, organization, and evolution. *Mammalian Genome* 23, 387-398.

Pipes, L., Li.S., Bozinoski, M., Palermo, R., Peng, X. et al. (2013). The non-human primate reference transcriptome resource (NHPRTR) for comparative functional genomics. *Nucleic Acids Research* 41, D906–D914.

Revol de Mendoza, A., Esquivel Escobedo, D., Santiago Alarcón, D. & Barrera Saldaña, H. (2001). Independent duplication of the growth hormone gene in three Anthropoidean lineages. *Journal of Endocrine Genetics* 2, 151–159.

Revol de Mendoza, A., Esquivel Escobedo, D., Martínez Dávila, I., Barrera Saldaña, H. (2004). Expansion and divergence of the GH locus between spider monkey and chimpanzee. *Gene* 336, 185–193.

Rodríguez-Sánchez, I.P., Tejero, M.E., Cole, S.A., Comuzzie, A.G., Nathanielsz, P.W., Wallis, M. & Barrera-Saldaña, H.A. (2010). Growth hormone-related genes from baboon (*Papio hamadryas*): characterization, placental expression and evolutionary aspects. *Gene* 450, 1–7.

Saitou, N. & Nei, M. (1987). The Neighbor-Joining method: a new method for reconstructing phylogenetic trees. *Molecular Biology and Evolution* 4, 406-425.

Sherwood, L.M. (1967). Similarities in the chemical structure of human placental lactogen and pituitary growth hormone. *Proceedings of the National Academy of Sciences U.S.A.* 58, 2307-2314.

Solomon, G., Reicher, S., Gussakovsky, E.E., Jomain, J.-B., & Gertler, A. (2006). Large-scale preparation and *in vitro* characterization of biologically active human placental (20 and 22K) and pituitary (20K) growth hormones: Placental growth hormones have no lactogenic activity in humans. *Growth Hormone & IGF Research* 16, 297–307.

Souza, S. C., Frick, G. P., Wang, X., Kopchick, J. J., Lobo, R. B. & Goodman, H. M. (1995). A single arginine residue determines species specificity of the human growth hormone receptor. *Proceedings of the National Academy of Sciences U.S.A.* 92, 959–963.

Swofford, D.L. (1998). PAUP*. Phylogenetic Analysis Using Parsimony (*and other methods), version 4. Sinauer, Sunderland, MA.

Takahashi, S., Shiga, Y., Satozawa, N. & Hayakawa, M. (2001). Diabetogenic activity of 20 kDa human growth hormone (20K-hGH) and 22K-hGH in rats. *Growth Hormone & IGF Research*, 11, 110–116.

Valinsky, A., Shani, M. & Gootwine, E. (1990). Restriction fragment length polymorphism in sheep at the growth hormone locus is the result of variation in gene number. *Animal Biotechnology* 1 135–144.

Wallis, M. (1981). The molecular evolution of pituitary growth hormone, prolactin and placental lactogen: a protein family showing variable rates of evolution. *Journal of Molecular Evolution* 17, 10-18.

Wallis, M. (1994). Variable evolutionary rates in the molecular evolution of mammalian growth hormones. *Journal of Molecular Evolution* 38, 619-627.

Wallis, M. (1996). The molecular evolution of vertebrate growth hormones: a pattern of near-stasis interrupted by sustained bursts of rapid change. *Journal of Molecular Evolution* 43, 93-100.

Wallis, M. (1997). Function switching as a basis for bursts of rapid change during the evolution of pituitary growth hormone. *Journal of Molecular Evolution* 44, 348-350.

Wallis, M. (2000). Episodic evolution of protein hormones: molecular evolution of pituitary prolactin. *Journal of Molecular Evolution* 50, 465-473.

Wallis, M. (2008) Mammalian genome projects reveal new growth hormone (GH) sequences. Characterization of the GH-encoding genes of armadillo (*Dasypus novemcinctus*), hedgehog (*Erinaceus europaeus*), bat (*Myotis lucifugus*), hyrax (*Procavia capensis*), shrew (*Sorex araneus*), ground squirrel (*Spermophilus tridecemlineatus*), elephant (*Loxodonta africana*), cat (*Felis catus*) and opossum (*Monodelphis domestica*). *General and Comparative Endocrinology* 155, 271-279.

Wallis, M., Lioupis, A. & Wallis, O.C. (1998). Duplicate growth hormone genes in sheep and goat. *Journal of Molecular Endocrinology* 21, 1-5.

Wallis, O. C. & Wallis, M (2002). Characterisation of the GH gene cluster in a new-world monkey, the marmoset (*Callithrix jacchus*). *Journal of Molecular Endocrinology* 29, 89-97.

Wallis, O.C. & Wallis, M. (2006). Evolution of growth hormone in primates: the GH gene clusters of the New World monkeys marmoset (*Callithrix jacchus*) and white-fronted capuchin (*Cebus albifrons*). *Journal of Molecular Evolution* 63, 591–601.

Wallis, O.C., Zhang, Y.-P. & Wallis, M. (2001). Molecular evolution of growth GH in primates. Characterisation of the GH genes from slow loris and marmoset defines an episode of rapid evolutionary change. *Journal of Molecular Endocrinology* 26, 249-258.

Wallis, O.C., Mac-Kwashie, A.O., Makri, G. & Wallis, M. (2005). Molecular evolution of prolactin in primates. *Journal of Molecular Evolution* 60, 606-614.

Wildman, D.E., Jameson, N.M., Opazo, J.C. & Yi, S.V. (2009). A fully resolved genus level phylogeny of neotropical primates (Platyrrhini). *Molecular Phylogenetics and Evolution* 53, 694-702.

Wurzel, J.M., Parks, J.S., Herd, J.E. & Nielsen, P.V. (1982). A gene deletion is responsible for absence of human chorionic somatomammotropin. *DNA* 1, 251–257.

Yang, Z. (2007). PAML 4: Phylogenetic analysis by maximum likelihood. *Molecular Biology and Evolution* 24, 1586-1591.

Ye, C., Li, Y., Shi, P. & Zhang, Y.-P. (2005). Molecular evolution of growth hormone gene family in old world monkeys and hominoids. *Gene* 350, 183-192.

Yi, S., Bernat, B., Pál, G., Kossiakoff, A. & Li, W.-H. (2002). Functional Promiscuity of squirrel monkey growth hormone receptor toward both primate and nonprimate growth hormones. *Molecular Biology and Evolution* 19, 1083-1092.

In: Monkeys
Editor: Hugo A. Barrera Saldaña

ISBN: 978-1-63117-851-1

Chapter 9

REVELATIONS OF THE PRIMATE GENOME PROJECTS: THE CASE OF THE GROWTH HORMONE *LOCUS*

Antonio Alí Pérez-Maya* and Hugo Alberto Barrera-Saldaña

Genomic and Bioinformatics Laboratory at the Department of Biochemistry and Molecular Medicine of Medicine School of Autonomous University of Nuevo León. Monterrey, México

ABSTRACT

Comparative genomic analyses have revealed unexpected and complex patterns of changes during the evolution of primates. Phylogenetic analysis is given us the unprecedented opportunity to reconstruct the evolutionary history of the human genome. It has been possible to identify genetic characteristics that distinguish humans from other organisms, including our evolutionary close relatives and some extinct hominid species. The unique attributes of hominids can be explained as a consequence of changes in key amino acids within the coding sequences of critical genes or as a result of dramatic changes in the temporal and spatial regulation of gene expression. Comparative genomic analyses of the growth hormone (GH) *locus* provide a framework for understanding the evolution of its structure at the DNA level and the corresponding impact on gene expression regulation and function of the encoded hormones. In primates, the GH *locus* evolved from a unigenic (non-primate mammals and prosimians) to a multigenic unit (anthropoids). Interestingly, while the ancestral gene continued to be expressed in the pituitary, acquired novel genes became expressed in placenta and other tissues. Tissue specificity is attributed to the presence of defined regulatory elements at the DNA level (transcriptional initiation control) as well as to mechanisms of epigenetic control. The Growth Hormone in humans affects many organs and systems, acting on various metabolic and homeostatic processes: brain development, neurogenesis, neuroprotection, learning and memory. Novel genes within the GH *locus* (CSHs and GH-variant) are associated with fetal growth and the production of insulin growth factor I (IGF-I) during pregnancy. In our laboratory, we have been investigating the evolutionary path followed

* Corresponding author: Antonio Alí Pérez-Maya. E-mail: bioquimicomty@gmail.com.

by this gene family in primates, by studying the sequences of their genomic *locus* in the genome of different species of this, referencing the human GH *locus* whose anatomy and expression one of our laboratories (Chen et al., 1989) also helped to reveal years over a ago. Our groups have solved the organization of the *loci* of lemur (*Lemur catta*), chimpanzee (*Pan troglodytes*), gorilla (*Gorilla gorilla*), rhesus monkey (*Macaca mulatta*) and vervet (*Cercopithecus aethiops*). From their analysis we have gained insights into structure of some representative *loci* of primates along the evolutionary scale.

The inspection of the sequences of genes and intergenic regions, allowed us to begin understanding the dramatic evolution of this *locus*.

I. Introduction to Comparative Genomics: The Case of Hominids

Primate genome projects in progress are not held to the same standard of quality as that of the human genome. This increases the frustration in differentiating artifacts from real sequence differences in low coverage draft assemblies (Marques-Bonet et al., 2005). So far, more than ten primate genomes have been sequenced and about twenty new sequencing projects are underway (GenBank Database, November, 2012).

What have we learned so far from the genomes of primates? That they are highly dynamic entities that have evolved rapidly but unevenly over time. With the exception of gibbons, there are relatively few differences between the karyotypes of primates. Much of this has been well-characterized with respect to the human karyotype. However, most of the smaller structural nucleotide changes cannot be detected by standard cytogenetic techniques, but rather by sequencing the genomes of non-human primates (Marques-Bonet et al., 2005).

Due to the extensive genomic information available and new bioinformatics tools developed, structural variation and differences in the number of copies have emerged as another important aspect of the genetic variation of primates (Marques-Bonet et al., 2005).

By obtaining the sequence of a genome, access to the information it contains is achieved, but this information lacks interpretation. Thus, the analysis of genomes refers to the task of understanding what the sequence says; basically what genes it contains, where they are, and what functions the proteins that are encoded carry out (Rouze et al., 1999; Stein, 2001; Rust et al., 2002). Throughout evolution, the changes that genomes and their genes undergo reflect the action of natural selection. The advantages or disadvantages of these changes will determine if the following generations consolidate or not. The comparison of genomes, genes, and proteins of different species is the most intuitive and straightforward approach to interpret "the instruction manual" in genomes (Koonin, 2001; Salzberg, 2003).

When amino acid sequences of proteins that perform the same function in different organisms are compared, they are found to be similar and a correspondence between the amino acid residues at certain positions of one and another can be established with some being identical and others different. In the context of molecular evolution, these proteins in different species are similar because they have a common origin (they are said to be orthologous). These differences occur because over time the genes have diverged by the accumulation of changes or mutations in their sequences (Alberts, 2009).

Comparison of orthologous proteins provide much information about how evolution has occurred and therefore about the organisms that produce them, since the fact that amino acid positions are conserved is related to how they are key to the structure or protein function in question (Miller et al., 2004) as well as the importance for the survival of the organism. In this regard, residues more directly involved in the performance of function, generally have less variation, since these changes can result in loss of function (Alberts, 2009; Khaitovich et al., 2006), and of the ability of the organism to compete for survival.

The importance of these comparative studies of homologous proteins was highlighted for the first time half a century ago (Zuckerkandl and Pauling, 1965) and has since this been demonstrated again and again in numerous works. It is not always easy to obtain information comparing sequences because many times what we see is not there because it is optimal, but because it is the result of a long evolutionary process. This can be expressed more precisely by using the terms positive, negative and neutral selection, which refer to a change to an organism (Abascal, 2003).

Gene duplication events have been instrumental in shaping genomes and gene families. In these, the generation of new ancestral gene copies allows copies to be free to mutate and specialize in order to perform different functions, representing an advantageous shift for the individual carrier (Khaitovich et al., 2006). Specialization occurs gradually, allowing repeated series of duplication and divergence over millions of years, leading to a gene that originates from a family of genes within the same genome.

The relevant nucleotide differences between members of the same family are those that focus on their coding and regulatory regions. There are examples of families in which differentiation appears in both regions, with those occurring in genes with different functions being more pronounced, such as tissue-specific genes with varied expression (Barsh et al., 1983). Such is the case of the GH family, especially in primates, which also includes genes of chorionic somatomammotropins or CSHs. Our laboratory has been using this multigenic family as the model for studying the molecular mechanisms by which very closely related genes express similar proteins, but being active in different tissues at very different levels—more than a thousand times—without mentioning their apparent functional ambiguity (Barrera-Saldaña, 1998).

Gene duplication is considered one of the most relevant phenomena in the evolution of proteins and their new functions (Ohno et al., 1968; Watts and Watts, 1968; Ohta, 1989; Lang et al., 2000). It has been postulated that when two copies of a gene exist, the selective pressure of changes in important residues is less because if one of the genes loses function, the other remains. This facilitates the exploration of new functional niches for protein family members that begin to change.

There are families of proteins in which this phenomenon is especially noticeable; where a variety of functions among its members can be seen (Alberts, 2009). As mentioned, proteins with sufficiently similar sequences tend to have a common origin (Zuckerkandl and Pauling, 1965) and present a similar three-dimensional architecture (Chothia and Lesk, 1986), but it is not uncommon for these to perform different functions (Henikoff et al., 1997), although others may display overlapping functions.

In many situations, families of proteins that have a common origin and a similar three-dimensional structure, exhibit similar enzymatic activities. Thus it could be considered that these proteins perform the same function. But the fact is that this activity may contribute to different cellular processes (Abascal, 2003).

Researchers have learned much about the evolution and function of human genes by examining their counterparts in animal models.

Now, comparison of their genomes to determine similarities and differences in their sequences, the genes´ location, length and number of exons, the highly conserved genomic regions and the amount of non-coding DNA have been added (Perez-Maya et al., 2012; Birney et al., 2007). Sequencing of the genomes of primates provides a means to study structural and functional aspects of human genome evolution, from their divergence from a common ancestor (Leakey et al., 1998; Glazko and Nei, 2003; Kumar et al., 2005).

Structural variation, which encompasses a group of numerical submicroscopic genomic alterations of DNA fragments whose sizes are greater than 1000 bp (Iafrate et al., 2004; Sebat et al., 2004), is a recurring phenomenon that shapes and creates new variants of the genome. These changes that include deletions, insertions, and duplications with a population frequency greater than 1%, are known as variations in the number of copies (VNCs). The VNCs represent a major source of genomic variation with about 1,500 variable regions described, covering approximately 12% of the genome and thousands of genes; this seems to have an important role in the evolution of the human genome (Redon et al., 2006).

One obvious way by which VNCs result in phenotypic diversity in humans is by altering the expression levels of genes that are in a variable number of copies. This correlation has been demonstrated for VNCs of certain genes at the transcriptional (Hollox et al., 2003; Aldred et al., 2005; McCarroll et al., 2006) and translational levels (Linzmeier and Ganz, 2005). Studies correlating the levels of mRNA and protein with the number of genomic copies of gene VNCs, need to always consider that some VNCs may result in phenotypic effects that seem to appear only in certain tissues and/or stages of development.

VNCs reinforce the idea that the human genome is not a static entity, but a constantly changing and evolving one. Another aspect of this type of variation that must be addressed is whether the levels and patterns of variation among humans are similar to those of non-human primates and other organisms. Gene duplication is known to be an important long-term evolutionary force (Freeman et al., 2006)

Genomic Comparisons in Hominids

What is the usefulness of sequencing and comparing the genomes of the great apes, our closest relatives on the evolutionary scale? The comparison of their genomes with ours may help reveal the molecular basis and the evolutionary forces that have shaped the genome of our species, including the underlying processes of mutation and selective constraint, among other issues. Sequencing the genome of a species close to ours provides a vantage point from which to contemplate our genome.

Because of their evolutionary proximity to humans, in 2005 the chimpanzee genome was sequenced. The analysis of the genomic sequence obtained showed that this primate shares 98.8% of its genetic material with our species. The remaining 1.2% of the nucleotide positions where both genomes differ (approximately 35 million), include about five million insertions, deletions, duplications and inversions. Comparative studies of both genomes show a highly correlated evolution pattern in the genes that encode proteins (Mikkelsen et al., 2005; Khaitovich et al., 2006). Given the small difference between the sequences of the genomes of humans and chimpanzees, it can be presumed that these contain crucial molecular changes that led to the evolutionary course that marked the emergence of our species (Khaitovich et al., 2006; Kehrer-Sawatzki and Cooper, 2007).

Most phenotypic differences are the result of specific genomic changes that have resulted in differences in gene expression, causing a gain, loss, and changes in function (Mikkelsen et al., 2005; King and Wilson, 1975; Carroll, 2003; Olson and Varki, 2003; Enard and Paabo, 2004; Carroll, 2005; Goodman et al., 2005; Varki and Altheide, 2005; Khaitovich et al., 2006).

Recent progress in the comparisons of human and chimpanzee genomes has revealed that insertions and deletions have significantly altered the genomic landscape of these species, after separation as independent lineages (Khaitovich et al., 2006; Kehrer-Sawatzki and Cooper, 2007). However, the low resolution of the chimpanzee genome assembly challenges the ability of current methods to identify many of these structural changes (Mikkelsen et al., 2005; Kehrer-Sawatzki and Cooper, 2007). Thus, comparative genomics of specific gene regions, from which lessons can be drawn, are necessary.

Six years after the chimpanzee, the sequencing of the orangutan genome was done, with this being the most different in the family of hominids. The orangutan is situated at a greater evolutionary distance than the chimpanzee; therefore, its genome provides new details on evolution and new knowledge for scientists about the unique aspects that separate man from the great apes at the genomic level. Research shows that the human and the orangutan genome are 97% similar. From an evolutionary point of view, the orangutan genome is quite special among great apes because it has remained remarkably stable over the past 15 million years. This makes it different from the human and chimpanzee, which have experienced large-scale structural rearrangements, which might have accelerated their evolution (Locke et al., 2011).

When looking at the genomes of humans and chimpanzees, there is an acceleration of structural changes during the course of evolutionary history. However, for some unknown reason, the orangutan genome has not experienced this acceleration. One possible key that can explain this lack of dynamism could be its low content of repetitive elements of the Alu family type. These account for about 10% of the human genome and can appear on "unexpected" sites where they generate mutations or genetic rearrangements. Of the approximately half a million copies, the human genome has about 5,000 human specific Alu copies, while the chimpanzee has about 2,000 specific Alu copies. But in the genome of the orangutan, only 250 copies of new Alu elements have been found over 15 million years of evolution. This is the closest thing to irrefutable evidence that explains the structural stability of the orangutan genome (Locke et al., 2011).

In March 2012 the gorilla genome, which shares 98.2% DNA with humans, and who had a common ancestor about eight to twelve million years ago, was deciphered. This great ape shares more genetic material with humans than previously thought. Over 11,000 genes in humans, chimpanzees, and gorillas were examined, in search of evolutionary clues. Initial results revealed that although the overall human genome is more similar to the chimpanzee, almost one third of it is evolutionarily closer to the gorilla. Fifteen percent of the gorilla genome is closer evolutionarily to the human or chimpanzee, that the latter two to each other (Gibbs and Rogers, 2012; Scally et al., 2012).

Genes related to sensory perception, hearing and brain development showed "accelerated evolution" in the three groups (humans, chimpanzees and gorillas), but particularly in humans and gorillas. The study of the gorilla genome may help develop new treatments for human diseases caused by protein variants associated with a normal gorilla phenotype (Scally et al., 2012).

For example, a variant of the protein that causes hereditary disease in humans is the only version found in gorillas [For example: the variant Arg432Cys of the progranulin (PGRN) growth factor is associated with dementia, a disease that gorillas do not suffer].

The two species of the Pan genus differ in their genomes by only 0.4% and yet, their social and sexual behaviors differ greatly. While chimpanzees are widespread across equatorial Africa, bonobos live only south of the Congo River in the Democratic Republic of the Congo. The territories of the Bonobos and chimpanzees in Central Africa are close to each other and separated only by the Congo River.

It has been hypothesized that the formation of the Congo River 1.5-2.5 million years ago, separated the ancestors of both apes, creating a barrier to gene flow that allowed different phenotypes to evolve over a relatively short period of time. Examination of the relationship between bonobos and chimpanzees show that a clear divergence seems to have occurred without subsequent crosslinking, which supports this hypothesis (Prufer et al., 2012).

Although extinct, Neanderthals are other close evolutionary relatives to modern humans. This group arose about 400,000 years ago, spread throughout Europe and Western Asia, and then disappeared approximately 30,000 years ago.

The first draft of the genome of this extinct hominid, published in 2010, shows that during these migrations, the ancestors of modern humans interbred after their departure from Africa with Neanderthals, less than 100,000 years ago. This gene flow between Neanderthals and the common core of all non-African human populations resulted in non-African modern humans sharing with the extinct Neanderthals, 1 to 4% of their genes.

Almost all our DNA is very similar to Neanderthals, as we are evolutionarily very close. What non-Africa humans especially share with them, are variations of non-fundamental traits of our biology. The "relatively low" percentage of Neanderthal DNA in the genome of our species, might suggest that interbreeding was "quite limited" (Green et al., 2010; Burbano et al., 2010).

Despite this small genetic mixing, when the Neanderthal genome is compared with the genomes of five modern humans from different parts of the world (a South African San group, an African Yoruba group, a Chinese Han, a Frenchman and a native of Papua-New Guinea), the results reveal a variety of genes that are unique to modern humans, including a small group of genes that spread rapidly among the latter, after humans and Neanderthals diverged from a common ancestor (Green et al., 2010).

For the first time it was possible to identify genetic characteristics that distinguish humans from other organisms, including our extinct closest evolutionary relatives. These findings also provide a list of genomic regions and key genes for human identification.

Such investigations have identified 83 specific *Homo sapiens* genes, which are related with cognitive function, the anatomy and physiology of the skin, and skeletal development of the skull.

For now, the genes identified correspond to physiological, metabolic, morphological and cognitive—these involved in Down syndrome, schizophrenia and autism—aspects, which differ between *sapiens* and Neanderthals; however, functions related to specific genetic changes are still unknown (Green et al., 2010; Burbano et al., 2010).

One of these genes, whose rapid development has attracted particular interest, is FOXP2. This gene is involved in language and in important brain functions. All vertebrates possess it, but particularly in humans, there are two unique variants.

Researchers suggest that language ability would be present in at least three *homo* species (sapiens, Neanderthals and Denisovans) and obviously in their common ancestors between 600,000 and 800,000 years ago. However, having this gene is not enough; for there to be communication between individuals a neurological and cognitive structure, a vocal apparatus for emitting a range of sounds, a hearing apparatus that allows understanding, and finally, a social environment in which language can develop are required (Konopa et al. 2009; Meyer et al., 2012).

After disappearing, *Homo neanderthalensis* left a genetic, cultural, and social legacy that is not always recognized, but that may have contributed in part to what is now our species (Green et al., 2010; Burbano et al., 2010).

Complete genome reconstruction of the Denisovan hominid discovered in Asia in 2010, has shed light on the relationship between Denisovans and modern humans, and on a list of genetic changes associated with the development of brain function and the nervous system, which allowed differentiation from archaic hominids (Meyer et al., 2012).

Genome sequencing of the Denisova man has revealed unexpected aspects that indicate that the evolutionary process of human populations in the last hundred thousand years was far more complex than previously thought, demonstrating that the ancestor of modern humans and Denisovans may have lived 800,000 years ago (Reich et al., 2010; Reich et al., 2011; Meyer et al., 2012). In this scenario, human populations seem to be divided into two main evolutionary lines. One of these, which lived on the African continent, eventually originated our species, while the other lineage corresponded to the common trunk from which the lineages of Neanderthals in Europe and the Denisovans in Asia arose.

Variant structural regions of the genome are typically difficult to characterize and are related to human diseases that indicate at the same time that the genome of the Denisovans, at least some regions, are more archaic than those of other hominids because they share some characteristics with the chimpanzee genome (Reich et al., 2010; Reich et al., 2011; Zhang et al., 2011; Meyer et al., 2012). Combined with the Neanderthal genome sequence, the Denisovan genome suggests a complex picture of genetic interactions between our ancestors and other ancient hominid groups with whom they lived at the time (Reich et al., 2010; Reich et al., 2011; Meyer et al., 2012).

A recent comparison of the Denisovan genome with that of Neanderthals and eleven modern human populations worldwide confirmed that aboriginal human groups of the islands of Southeast Asia (where Denisovans once dwelt) share more genetic variants with the Denisovans than with other human populations (Reich et al., 2010; Reich et al., 2011; Meyer et al., 2012). The Denisovans and the Neanderthals descended from the same ancestral population that previously separated from the ancestors of modern humans (Figure 1).

However, unlike Neanderthals, Denisovans did not contribute genes to all current Eurasian populations. Similar to how Neanderthals interbred with the ancestors of the current African humans, current Melanesians have between 4 and 6% of the genetic material of extinct Denisovans (Reich et al., 2010; Reich et al., 2011; Meyer et al., 2012).

The genome sequence has generated a list of about 100,000 recent changes in the human genome after separation of the Denisovans. Some of these changes affect genes associated with the development of brain function and the nervous system.

Others affect the skin, eyes, and dental morphology. The analysis of all of these provide new insight to understand what makes our species so unique (Reich et al., 2010; Reich et al., 2011; Meyer et al., 2012).

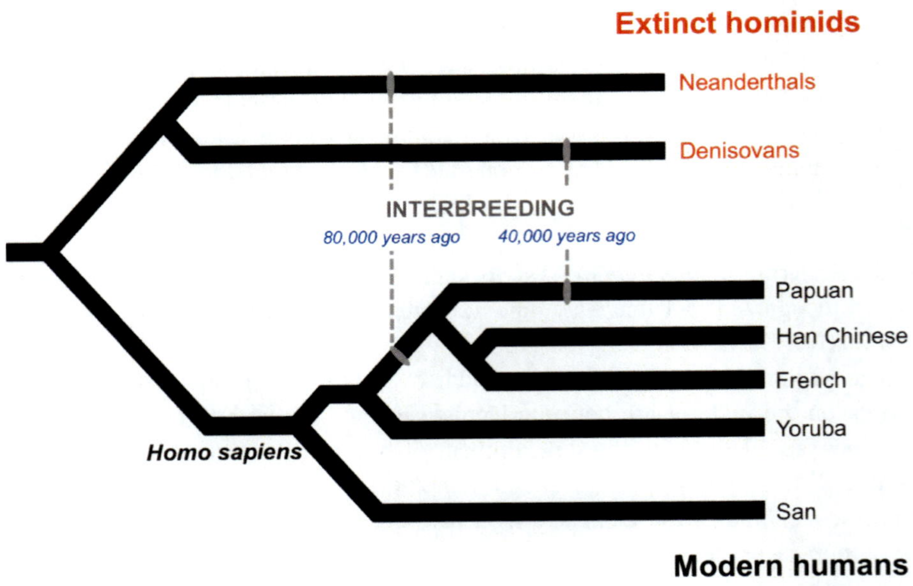

Figure 1. A new member of the human family. The DNA obtained from a tooth showed that Denisovans were closely related to Neanderthals and consequently related to humans.

II. GH *Locus*: Our Model to Study Molecular Evolution of the Primate Genome

The growth hormone (GH) *locus* in humans which is located in band q24.2 of the long arm of chromosome 17, has been our "window" to study the human genome (Barrera-Saldaña, 1998). It covers an area of approximately 50,000 base pairs (bp) containing five highly similar genes with the same transcriptional orientation (Figure 2) (Chen et al., 1989).

At its 5' end it is flanked by the CD79b gene, which is active in B lymphocytes (Cajiao et al., 2004; Yoo et al., 2006) and in its 3' end by the TCAM gene, a testicular adhesion molecule that has become a pseudogene in the human genome (Sakatani et al., 2000; Kimura et al., 2004). The *locus* starts with the pituitary GH gene or the hGH-N gene (referred to as normal), whose expression is specific to the anterior pituitary somatotrophs.

Followed by the remaining four genes with specific expression in the syncytiotrophoblast of the placenta. Of these, two codify for an identical mature CSH (hCSH-A y hCSH-B) (Barrera-Saldana et al., 1983); one (also a CSH, referred to as hCSH-L) was originally considered a pseudogene [because it possessed a mutation in the second intron that would prevent proper processing of its transcripts (Resendez-Perez et al., 1990), although this is in doubt, since evidence of expression in placenta has been reported (Misra-Press et al., 1994; MacLeod et al., 1992; Hu et al., 1999; Männik et al., 2010)], and the remaining is the hGH-V gene, which encodes a GH variant that has been postulated to replace pituitary GH during pregnancy (Chen et al., 1989; Alsat et al., 1997). In our efforts to understand the evolution of the GH *locus* by ourselves and in collaboration with Dr. Michael Wallis, a leading expert in GH family evolution, we have undertaken the challenge of determining the composition and gene organization of this gene family in representative species of different groups of primates.

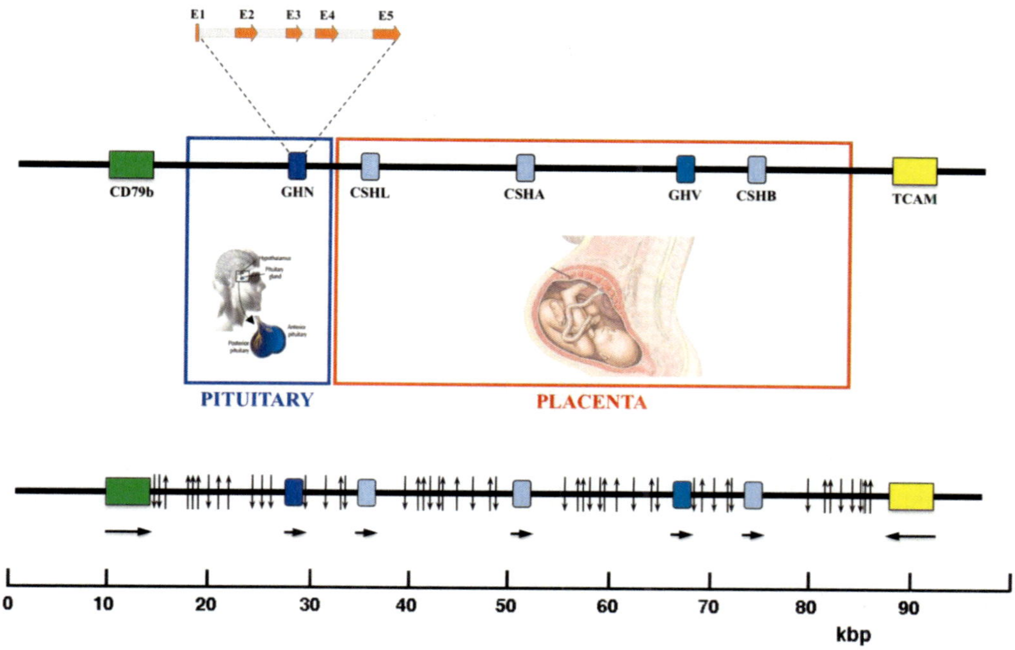

Figure 2. Representation of the human GH *locus*. At top, canonical anatomy the GH/CSH genes; in middle, tissues where gene members are expressed; at bottom, arrangement and orientation (horizontal arrows) of genes, including distribution of repetitive elements of the Alu family (vertical arrows). Distances can be deduced from scale shown.

In a painstaking process we have performed *loci* annotations, to decipher the evolutionary pathway and the mechanisms by which primates have acquired new genes to transit from a unigenic to a multigenic *locus*, diverged some of the new copies to give rise to GH-V type and CSH type genes, and acquired tissue specificity of their expression in placenta and other non-pituitary tissues. The key to our progress has been our collection of BACs carrying GH *loci* of eleven primate species (Figure 3).

Function of Encoded Hormones

The main function of HGH-N is to promote postnatal linear growth and regulate diverse metabolic activities, such as glucose homeostasis, protein synthesis in muscle tissue, lipolysis in adipose tissue, tissue repair, and bone regeneration (Farreras-Rozma, 2000). More recently, it has been involved in neurogenesis and neuroprotection, learning and memory (Sonntag et al., 2005).

Meanwhile, the levels of GH-V are associated with intrauterine growth rates, birth weight, and fetal levels of insulin-like growth factor I (IGF-I) (Verhaeghe et al., 2002; Chellakooty et al., 2004; Rygaard et al., 1998). This isoform has somatogenic (MacLeod et al., 1991) and lactogenic (Nickel et al., 1990) activity and, as mentioned before, replaces GH-N during the second half of pregnancy (Frankenne et al., 1987-A; Frankenne et al., 1987-B); it has also been implicated in increasing the levels of IGF-I in maternal serum (Caufriez et al., 1990).

A similar role is awarded to CSH, whose levels have been associated with placental functioning and also with fetal levels of IGF-1 (Hardwerger et al., 2000; Karabulut et al., 1972). This hormone, which also has 22 kDa, is thought to play an important role as an antagonist of insulin in the regulation of maternal metabolism during pregnancy (Josimovich et al., 1964).

Potential anti-insulin effects of human CSH comprise inhibiting glycogen synthesis, glycolysis, lipogenesis and glucose entry into cells, in addition to activating glycogenolysis, muscle proteolysis, and gluconeogenesis from amino acids.

These actions are intended to increase the plasma concentration of nutrients and energy to the fetus (Ruiz-Parra et al., 1999; Felber et al., 1972; Williams and Coltart, 1978; Walker et al., 1991; Kim and Felig, 1971).

It also exerts direct somatotropic activity on fetal tissues (Felber et al., 1972; Coltart Williams, 1978). During the third trimester of pregnancy there is an increase in the synthesis of this hormone, accounting for almost 10% of the total production of placental proteins (Barrera-Saldana et al., 1982).

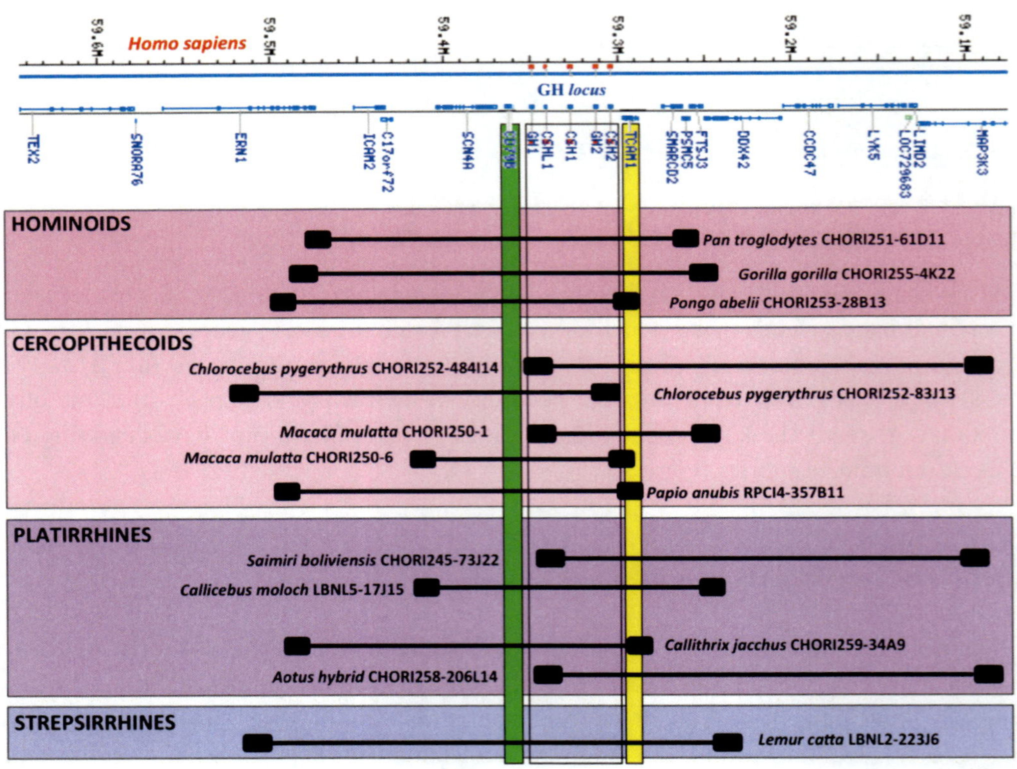

Figure 3. Genomic regions in our collection of BACs. Mapping of BACs of eleven species of primates that are part of our collection and that were obtained screening the Bacterial Artificial Chromosome Resources kindly made available to us by Dr. Pieter de Jong from the Children's Hospital Oakland Research Institute in Oakland, CA, USA.

III. The Unusual Evolution of the Human GH *Locus* in Primates

First with molecular biology tools and now with powerful sequencing methods, our laboratory has been investigating the evolution of the GH family in primates. At the bottom of the primates evolutionary tree, the *locus* originated from an evolutionary branch initially unigenic, which continued largely unchanged until the appearance of platyrhine (*Ceboidea*).

Then the *locus* experienced a dramatic evolutionary acceleration.

At first, *locus* expansion events were privileged because of duplication caused by the presence of repetitive Alu elements surrounding the ancestral *locus* (Chen et al., 1989). But after separation of the tarsiers and anthropoids, but before divergence of platyrrhines and catarrhines, multiple gene duplication events led to an increase in the number of GH genes, which show the remarkable expansion that this *locus* has suffered in primates (Revol de Mendoza et al., 2004; Gonzalez Alvarez et al., 2006; Wallis et al., 2006). Besides the GH type gene whose expression remained restricted to the pituitary, CSHs genes appeared, as well as GH-V, both with placental expression (Perez-Maya et al., 2012; Chen et al., 1989; Golos et al., 1993; Ye et al., 2005). Molecular evidence suggests that the GH *locus* not only underwent an expansion in the number of genes and but also in the size of its intergenics regions (IGs) (Figure 4).

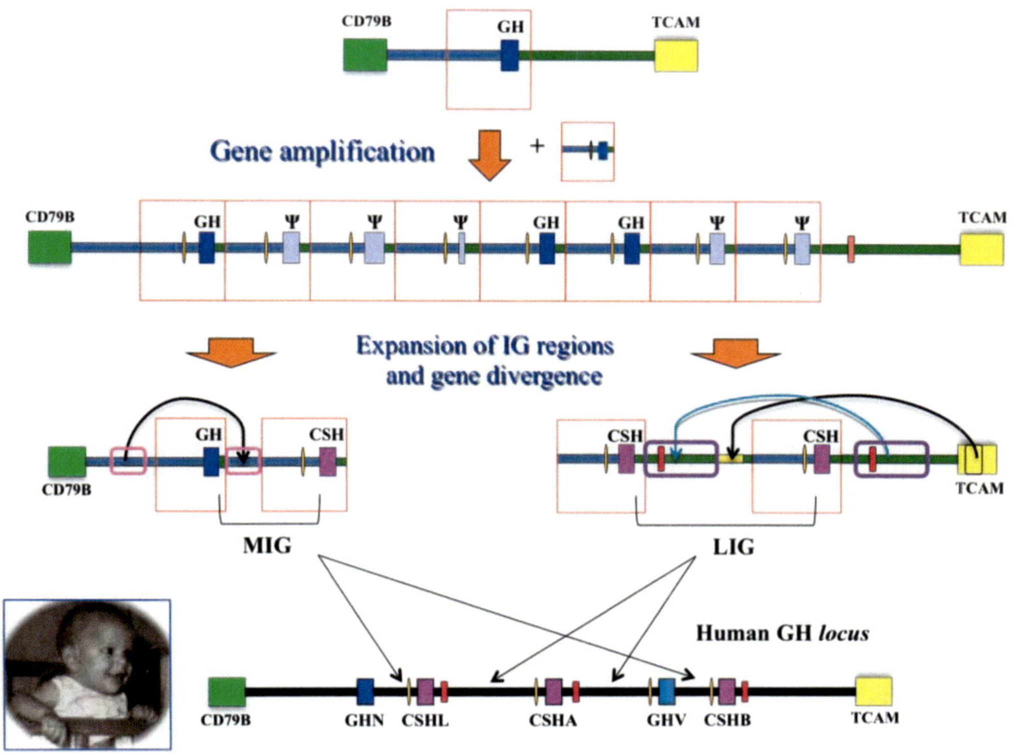

Figure 4. Evolution of the GH *locus* in primates. Likely mechanisms that shaped the anatomy of the GH *locus* in its evolution from a unigenic *locus* in prosimians to a multigenic arrangement in humans are represented. This expansion of the GH *locus* occurred due to an increase in the number of genes and in the length of the intergenic regions, accompanied by gene divergence events.

Our evidence suggests that during these events key genomic regions were incorporated for the acquisition of tissue specific expression and that gene divergence processes occurred that lead to the appearance of new hormones. In anthropoids, the GH *loci* are composed of multiple related non-allelic genes and, in general, GHs in this group of primates are well preserved. So far it is not clear how some GH genes are expressed in placenta in platyrrhines. However, since Cercopithicoidea, members of this family of genes accumulated amino acid substitutions that predict the acquisition of new distinctive features of placental hormones from the human *locus*. One of the key points of gene expression regulation is transcriptional control (Figure 5).

Included in this category is the presence of ancillary regulatory sequences (enhancers and repressors), and the interaction of multiple inhibitory and activating proteins that bind to these sequences.

In Old World monkeys (OWM) and Great Apes (GA), placental genes are subject to transcriptional repression mediated by a conserved region, called P element or pituitary inhibitor situated approximately 2 kb opposite of the initiation site of transcription (Norquay et al., 2001), to which proteins in the hypophysis bind to impede expression in this gland (Nachtigal et al., 1993; Barrera-Saldana, 1998). Approximately 2.3 kbp after each CSH gene an enhancer is located, which increases the expression of these in the placenta (Jacquemin et al., 1994; Barrera-Saldana, 1998; Rogers et al., 1986). It has been reported that the function of this enhancer is independent of its orientation and position (Rogers et al., 1986).

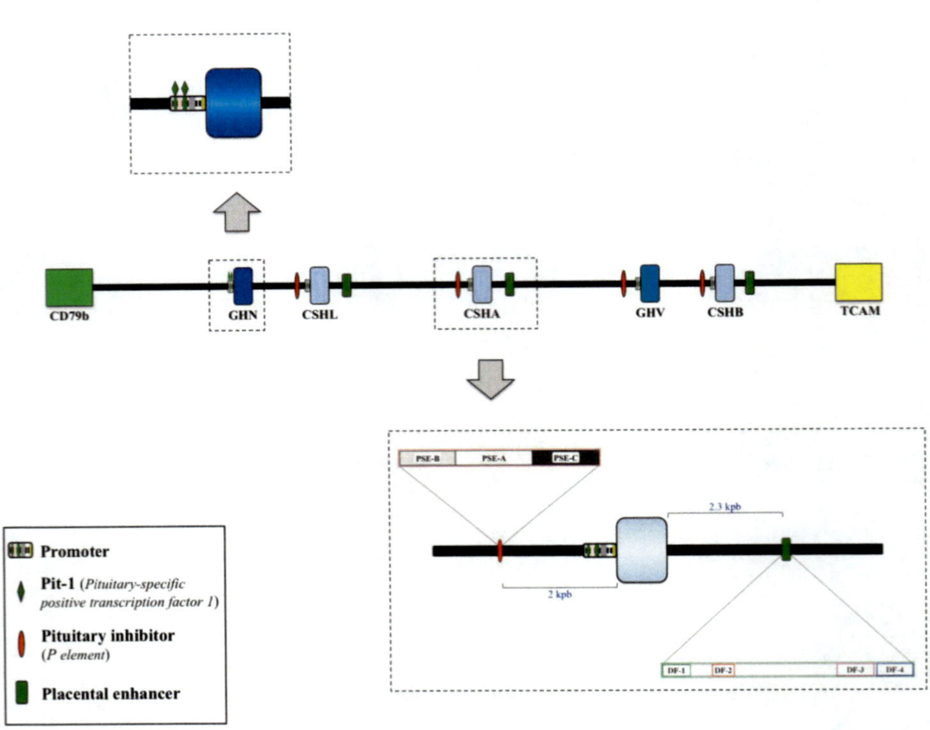

Figure 5. Key DNA elements regulating gene expression at the human GH *locus*. The two Pit-1 sites present in the hGH-N promoter are outlined before the gene. The site of the pituitary inhibitor or P element is represented before each placental gene and amplified are their PSE-A, PSE-B and PSE-C domains. The placental enhancer is shown after the CSHS genes and amplified are the four binding sites for the nuclear placenta proteins it presents: DF-1, DF-2, DF-3, and DF-4.

IV. THE ORIGIN OF DIFFERENTIAL EXPRESSION

Some of the questions that come to light in the scenarios where GH *locus* evolution took place in primates: How did the expansion phase take place in the GH *locus*?, Was a same or different route followed by the different primate suborders? What changes have occurred in the coding of their genes? How did the genes´ tissue specificity originated?

Regarding the rise of differential expression of genes that are *locus* members, activity in the placenta of GH-related genes has been documented in apes (Chen et al., 1989; Golos et al., 1993; Papper et al., 2009). However, so far no reports on GH gene expression in placenta in prosimians have appeared.

Control of tissue specificity in the GH *locus* of the human genome is carried out largely by the pituitary inhibitor and the placental enhancer. To date, the functionality of the repressor element (P element) of placental genes in the pituitary in the GH *loci* New World monkeys (NWM) is not entirely clear; it is assumed that there is no differentiation of expression of tissue specificity, thus making it imperative to further analyze in detail the origin and function of the regulatory element and the enhancer of placental gene expression in this extraembryonic organ in primates, to subsequently determine when regulation of tissue specificity of the GH-related genes originated.

The origin of the enhancer that is present in the GH *locus* in prosimians and NWM, is another of the questions in this evolutionary history. Our bioinformatic analysis has identified a region after the last gene in the marmoset GH *locus* with similarity to the enhancer identified in OWM and GA, whose function in these has been shown to be the enhancement of expression of CSHs genes in placenta. Likewise, we identified a similar region in the lemur (a prosimian) whose GH *locus* is characterized by a single gene, before which we also identified a region similar to the P element in NWM, OWM and GA and which in the case of OWM and GA is awarded the ability to suppress placental gene expression in the pituitary.

It has been seen in the case of OWM and GA that similar sequences are present in the end region of the TCAM gene in the long intergenic regions (LIGs) of the GH *locus* (Figure 6).

These sequences were probably incorporated during duplication or transposition events that have occurred in this *locus*. Study of the GH *loci* in primates suggests that repetitive elements have played an important role in these events.

Finally, it has been known for decades that GH in humans affects many organ and systems, acting on different metabolic and homeostatic processes. But more recently, evidence has accumulated of their involvement in brain development, neurogenesis, neuro-protection, learning, and memory. For its part, maternal levels of CSH have been associated with fetal growth and the production of IGF-I during pregnancy (Sonntag, 2005).

Given the impact of GH on health, even during embryogenesis and fetal development, there is no doubt that as a result of this project, in the medium term, clinical applications and healthcare biotechnology will arise.

CONCLUSION

The comparison of genomes, genes and proteins of different species, is the most intuitive and straightforward approach to interpret "the instruction manual" of genomes.

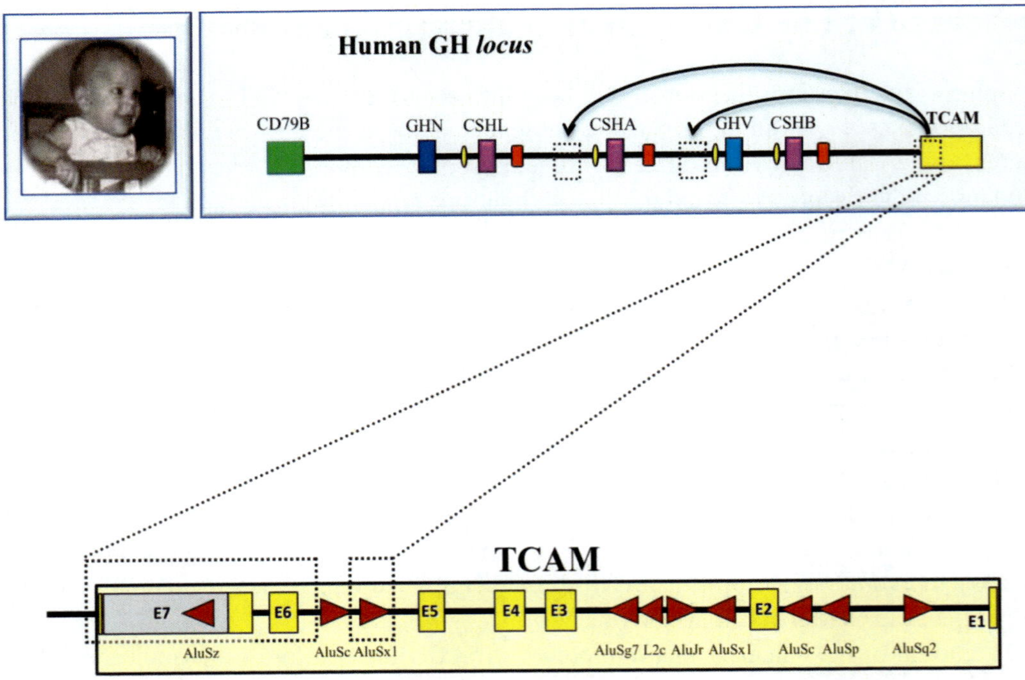

Figure 6. Detail of the evolution of the human GH *locus*. An extension of the TCAM gene sequence is represented that allows visualization of the presence of repetitive elements within this gene, which probably played a decisive role in the variation of the genomic landscape of the region. The dotted lines represent sequences during the evolutionary process which were incorporated into the LIGs of the GH *locus*.

In the case of hominids, their genomes provide new details on the evolution and new insights to scientists about the unique aspects that separate man from great apes at the genomic level. Completing the encyclopedia of hominid genes will allow tracing the evolutionary relationship of hominids. The genome sequence has generated a list of recent changes in the human genome after its separation from other hominids and contributes new insights to understanding what makes our species so unique.

References

Abascal, F. (2003). Análisis de genomas. Métodos para la predicción y anotación de la función de las proteínas. *Tesis doctoral*, Universidad Autónoma de Madrid.

Alberts, B. How Genes and Genomes Evolve. In: *Essential cell biology,* 3rd edition New York: Garland Science; 2009; 297-326.

Aldred, P. M., E. J. Hollox and J. A. Armour, Copy number polymorphism and expression level variation of the human alpha-defensin genes DEFA1 and DEFA3. *Hum. Mol. Genet.,* 2005. 14(14): p. 2045-52.

Alsat, E., J. Guibourdenche, et al., 1997. Human placental growth hormone. *Am. J. Obstet. Gynecol*. 177(6): 1526-34.

Barrera-Saldana, H. A., Growth hormone and placental lactogen: biology, medicine and biotechnology. *Gene*, 1998. 211(1): p. 11-8.

Barrera-Saldana, H. A., Robberson, D. L. and Saunders, G. F. Transcriptional products of the human placental lactogen gene. *J. Biol. Chem.* 1982;257:12399-12404.

Barsh, G. S., P. H. Seeburg and R. E. Gelinas, The human growth hormone gene family: structure and evolution of the chromosomal *locus*. *Nucleic Acids Res.*, 1983. 11(12): p. 3939-58.

Birney, E., et al., 2007. Identification and analysis of functional elements in 1% of the human genome by the ENCODE pilot project. *Nature* 447(7146): 799-816.

Cajiao, I., A. Zhang, et al., 2004. Bystander gene activation by a *locus* control region. *EMBO J.* 23(19): 3854-63.

Caufriez, A., F. Frankenne, et al., 1990. Placental growth hormone as a potential regulator of maternal IGF-I during human pregnancy. *Am. J. Physiol.* 258(6 Pt 1): E1014-9.

Chellakooty, M., et al., A longitudinal study of intrauterine growth and the placental growth hormone (GH)-insulin-like growth factor I axis in maternal circulation: association between placental GH and fetal growth. *J. Clin. Endocrinol. Metab.*, 2004. 89(1): p. 384-91.

Chen, E. Y., et al., 1989. The human growth hormone *locus*: Nucleotide sequence, biology, and evolution. *Genomics*, 4(4):479-497.

Chothia, C. and A. M. Lesk, The relation between the divergence of sequence and structure in proteins. *EMBO J.*, 1986. 5(4): p. 823-6.

Farreras-Rozman, 2000. Medicina Interna. Capitulo 16. *Endocrinología*. Decimocuarta Edición. Ediciones Harcourt. ISBN Obra Completa: 84-8174-357-7. Publicación Harcourt: www.harcourt.es.

Felber, J. P., N. Zaragoza, et al., 1972. The double effect of human chorionic somato-mammotropin (HCS) and pregnancy on lipogenesis and on lipolysis in the isolated rat epididymal fat pad and fat pad cells. *Horm. Metab. Res.* 4(4): 293-6.

Frankenne, F., et al., Discovery of a placental variant of human growth hormone: biochemistry, physiology and implication in the secretion of hypophyseal forms. *Reprod. Nutr. Dev.*, 1987. 27(2B):523-4.

Frankenne, F., F. Rentier-Delrue, M. L., Scippo, J. Martical, G. Hennen 1987. Expression of the growth normone variant gene in human placenta. *J. Clin. Endocrin. Metab.* 64:635-637.

Freeman, J. L., et al., Copy number variation: new insights in genome diversity. *Genome Res.*, 2006. 16(8): p. 949-61.

Golos, T. G., Durning, M., Fisher, J. M., y Fowler, P. D., 1993. Cloning of four growth hormone/chorionic somatomammotropin-related complementary deoxyribonucleic acids differentially expressed during pregnancy in the rhesus monkey placenta. *Endocrinology*, 133: 1744.

Gonzalez Alvarez, R., et al., Growth hormone *locus* expands and diverges after the separation of New and Old World Monkeys. *Gene*, 2006. 380(1):38-45.

Handwerger, S., Freemark, M., 2000. The roles of placental growth hormone and placental lactogen in the regulation of human fetal growth and development. *J. Pediatr. Endocrinol. Metab.* 13, 343-356.

Henikoff, Steven, Elizabeth A. Greene, Shmuel Pietrokovski, Peer Bork, Teresa K. Attwood, y Leroy Hood, 1997. Gene Families: The Taxonomy of Protein Paralogs and Chimeras. *Science* 278:609.

Hollox, E. J., J. A. Armour and J. C. Barber, Extensive normal copy number variation of a beta-defensin antimicrobial-gene cluster. *Am. J. Hum. Genet.*, 2003. 73(3): p. 591-600.

Iafrate, A. J., et al., Detection of large-scale variation in the human genome. *Nat. Genet.*, 2004. 36(9): p. 949-51.

Josimovich, J. B. y Atwood, B. L., 1964. Human Placental Lactogen (Hpl), a Trophoblastic Hormone Synergizing with Chorionic Gonasotropin and Potentiating the Anabolic Effects of Pituitary Growth Hormone. *Am. J. Obstet. Gynecol.* 88:867-79.

Karabulut, A. K., Layfield, R., Pratten, M. K., 2001. Growth promoting effects of human placental lactogen during early organogenesis: a link to insulin-like growth factors. *J. Anat.* 198, 651-662.

Khaitovich, P., Enard, W., Lachmann, M., y Pääbo, S., 2006. Evolution of primate gene expression. Review. *Nature Reviews Genetics* 7, 693-702.

Kim, Y. J. y Felig, P., 1971. Plasma chorionic somatomammotropin levels during starvation in midpregnancy. *J. Clin. Endocrinol. Metab.* 32(6): 864-7.

Kimura, M., 1968. Evolutionary rate at the molecular level. *Nature* 217(5129): 624-6.

Koonin, E. V., 2001. Computational genomics. *Curr. Biol.* 11:R155-158.

Lang, D., Thoma, R., Henn-Sax, M., Sterner, R., Wilmanns, M., 2000. Structural evidence for evolution of the beta/alphabarrel scaffold by gene duplication and fusion. *Science*. 289: 1546-1550.

Linzmeier, R. M. and T. Ganz, Human defensin gene copy number polymorphisms: comprehensive analysis of independent variation in alpha- and beta-defensin regions at 8p22-p23. *Genomics*, 2005. 86(4): p. 423-30.

MacLeod, J. N., Lee, A. K., Liebhaber, S. A., Cooke, N. E. 1992. Developmental control and alternative splicing of the placentally expressed transcripts from the human growth hormone gene cluster. *J. Biol. Chem.* 267:14219–14226.

MacLeod, J. N., Worsley, I., et al., 1991. Human growth hormone-variant is a biologically active somatogen and lactogen. *Endocrinology* 128(3): 1298-302.

Männik, J., Vaas, P., Rull, K., Teesalu, P., Rebane, T., and Laan, M. 2010. Differential Expression Profile of Growth Hormone/Chorionic Somatomammotropin Genes in Placenta of Small- and Large-for-Gestational-Age Newborns. *J. Clin. Endocrinol. Metab.* 95 (5): 2433–2442.

Marques-Bonet, T., O. A. Ryder and E. E. Eichler, Sequencing primate genomes: what have we learned? *Annu. Rev. Genomics Hum. Genet.*, 2009. 10: p. 355-86.

McCarroll, S. A., et al., Common deletion polymorphisms in the human genome. *Nat. Genet.*, 2006. 38(1): p. 86-92.

Miller, W., Makova, K. D., Nekrutenko, A., and Hardison, R. C. Comparative genomics. *Annu. Rev. Genomics Hum. Genet.*, 2004. 5: p. 15-56.

Misra-Press, A., Cooke, N. E., Liebhaber, S. A. 1994. Complex alternative splicing partially inactivates the human chorionic somatomammotropin-like (hCS-L) gene. *J. Biol. Chem.* 269:23220–23229.

Nickel, B. E., E. Kardami y P. A. Cattini, 1990. Differential expression of human placental growth-hormone variant and chorionic somatomammotropin in culture. *Biochem. J.*, 267 (3):653-8.

Ohno, S., U. Wolf and N. B. Atkin, Evolution from fish to mammals by gene duplication. *Hereditas*, 1968. 59(1): p. 169-87.

Ohta, T., 1989. Role of gene duplication in evolution. *Genome*. 31:304-310.

Papper, Z., et al., Ancient origin of placental expression in the growth hormone genes of anthropoid primates. *Proc. Natl. Acad. Sci. US*, 2009. 106(40): p. 17083-8.

Perez-Maya, A. A., et al., The chimpanzee GH *locus*: composition, organization, and evolution. *Mamm. Genome*, 2012. 23(5-6): p. 387-98.

Redon, R., et al., Global variation in copy number in the human genome. *Nature*, 2006. 444 (7118): p. 444-54.

Reséndez-Pérez, D., Ramírez-Solis, R., Varela-Echavarria, A., Martínez- Rodríguez, H. G., Barrera-Saldaña, H. A. 1990. Coding potential of transfected human placental genes. *Nucleic Acids Res.*, 18: 4665-4670.

Revol de Mendoza, A., et al., 2004. Expansion and divergence of the GH *locus* between spider monkey and chimpanzee. *Gene*, 336(2): p. 185-193.

Rouze, P., Pavy, N., Rombauts, S., 1999. Genome annotation: which tools do we have for it? *Curr. Opin. Plant Biol.* 2:90-95.

Ruiz Parra, A. I., Metabolismo en el embarazo. Entre lo clásico y lo actual. *Rev. Col. Obstet. Ginecol.*, 1999. 50(3): 125-126.

Rust, A. G., E. Mongin and E. Birney, Genome annotation techniques: new approaches and challenges. *Drug Discov. Today*, 2002. 7(11): p. S70-6.

Rygaard, K., A. Revol, et al., 1998. Absence of human placental lactogen and placental growth hormone (HGH-V) during pregnancy: PCR analysis of the deletion. *Hum. Genet.* 102 (1): 87-92.

Salzberg, S. L., 2003. Genomics: Yeast rises again. *Nature*. 423:233-234.

Sakatani, S., et al., 2000. Structure, expression, and conserved physical linkage of mouse testicular cell adhesion molecule-1 (TCAM-1) gene. *Genome* 43(6): 957-62.

Sebat, J., et al., Large-scale copy number polymorphism in the human genome. *Science*, 2004. 305(5683): p. 525-8.

Sonntag, W. E., M. Ramsey, et al., 2005. Growth hormone and insulin-like growth factor-1 (IGF-1) and their influence on cognitive aging. *Ageing Res. Rev.* 4(2): 195-212.

Stein, L., 2001. Genome annotation: from sequence to biology. *Nat. Rev. Genet.* 2:493-503.

Verhaeghe, J., Pintiaux, A., Van Herck, E., Hennen, G., Foidart, J. M., Igout, A., 2002. Placental GH, IGF-I, IGF-binding protein-1, and leptin during a glucose challenge test in pregnant women: relation with maternal body weight, glucosa tolerance, and birth weight. *J. Clin. Endocrinol. Metab.* 87, 2875-2882.

Walker, W. H., Fitzpatrick, S. L., Barrera-Saldaña, H. A., Reséndez-Perez, D., Saunders, G. F., 1991. The human placental lactogen genes: Structure, function, evolution and transcriptional regulation. *Endocrine Reviews* 12, 316-328.

Wallis, O. C. y W. Michael, 2006. Evolution of Growth Hormone in Primates: The GH Gene Clusters of the New World Monkeys Marmoset (Callithrix jacchus) and White-Fronted Capuchin (Cebus albifrons). *Journal of Molecular Evolution* V63(5): 591-601.

Watts, R. L., Watts, D. C., 1968. Gene duplication and the evolution of enzymes. *Nature*. 217: 1125-1130.

Williams, C. y T. M. Coltart, 1978. Adipose tissue metabolism in pregnancy: the lipolytic effect of human placental lactogen. *Br. J. Obstet. Gynaecol.* 85(1): 43-6.

Ye, Chun, e.a., 2005. Molecular evolution of growth hormone gene family in old world monkeys and hominoids. *Gene*, 350: p. 183-192.

Yoo, E. J., I. Cajiao, et al., 2006. Tissue-specific chromatin modifications at a multigene *locus* generate asymmetric transcriptional interactions. *Mol. Cell. Biol.* 26(15): 5569-79.

Zuckerkandl, Emile, Linus Pauling., 1965. The Molecular Evolutionary Clock. *Journal of the History of Biology.* Springer Nerherlands. Vol. 31.

In: Monkeys
Editor: Hugo A. Barrera Saldaña

ISBN: 978-1-63117-851-1

Chapter 10

NONHUMAN PRIMATES USED FOR RESEARCH OF OBESITY AND RELATED METABOLIC DISEASES: THE BABOON AS A MODEL

***M. Elizabeth Tejero*[1,*]*, Iram P. Rodríguez-Sánchez*[2] *and Hugo A. Barrera-Saldaña*[2]**

[1]Laboratorio de Nutrigenómica y Nutrigenética, Insitituto Nacional de Medicina Genómica, Mexico City, Mexico
[2]Genomic and Bioinformatics Laboratory at the Department of Biochemistry and Molecular Medicine of Medicine School of Autonomous University of Nuevo León. Monterrey, México

ABSTRACT

Given the increase in the rates of obesity and comorbidities around the world, scientific research in these areas has grown very rapidly. Although rodents have been used for investigation of these diseases, their limitations as animal models are well known. Many other mammal species such as dogs, cats and pigs develop spontaneous obesity with some of the characteristic metabolic abnormalities; however, the genetic and physiologic differences between these species and humans are considerable, due to their phylogenic distance. It is also clear that animal models that closely resemble the mechanisms underlying the health risks observed in humans are required. Nonhuman primates (NHP) have been extensively used in the study of obesity and related diseases. Numerous investigations have shown variation in body weight, body composition, endophenotypes and metabolic response to nutrients that resemble human metabolism.

Primates such as baboons, vervets, marmosettes, rhesus monkeys, tamarins, and macaques show augmentation of body fat in response to aging and excessive caloric intake, and these changes are associated with metabolic abnormalities such as insulin resistance and dyslipidemia. The present chapter integrates information regarding the use of the most common NHP in the study of obesity and related diseases, with emphasis in

* Corresponding author: ME Tejero. Laboratorio de Nutrigenómica y Nutrigenética, Insitituto Nacional de Medicina Genómica. Periférico sur 4809, Col Arenal Tepepan. Mexico DF. 52(55)53501900 ext 1145, Mexico City. E-mail: etejero@inmegen.gob.mx.

investigations conducted in the baboon (*Papio* sp), which has been used as an animal model in a large number of metabolic, feeding, pharmacological, genetic and epigenetic studies on obesity, cardiovascular disease, type 2 diabetes mellitus (T2DM) and related traits.

1. Animal Models in Research on Obesity and Related Diseases

Obesity is one of the most prevalent diseases in the world, with an alarming increase in the last decades. It is characterized by excess adipose tissue in an organism with an adverse effect on health [1]. The increase in adiposity is the consequence of an imbalance between energy intake and expenditure. The excess energy is stored as body fat triggering a variety of biologic responses that involve many cell types and organs that have been related to metabolic abnormalities and diseases such as insulin resistance, subclinical inflammation, T2DM, cardiovascular disease, and some types of cancer. Human obesity is considered a complex disease influenced by the interaction between genetic and environmental factors [2]. The research of obesity and related diseases requires appropriate animal models for in vivo studies [3].

Experimental animal models have been widely used in laboratory research to investigate the physiopathology mechanisms by conducting experiments that are not feasible to be carried out in humans due to ethical or practical reasons. The choice of an animal model depends, to a great extent, on the proposed research question, the nature of the studied phenomenon, the study design, resources, and other factors that contribute to answer that specific question.

The most commonly used species in research are murine models (rats and mice) [4]. These species have been used in a large variety of experiments, and have well-known benefits and limitations. Although the use of these models have significant advantages derived from their size, life span, rapid reproduction and adaptation to captivity, they are not considered optimal for obesity research since rodent genomes diverge substantially from the human genome, and therefore their metabolism and etiology of diseases; making it difficult to apply findings in these models to human health. In addition, rodent models do not typically develop spontaneous T2DM, one of the most important comorbidities of obesity [6].

Although they may show other metabolic abnormalities derived from diet-induced obesity [5, 7]. The genetic and physiologic differences between human and rodents derive from the ancient divergence of the primate and rodent lineages [6]. It is of interest to point out that rodents (particularly mice) have been widely used in a large variety of investigations that involve genetic manipulation, such as transgene overexpression, gene knockout and other genetically modified models. These models have been of great value for the study of molecular mechanisms; however single gene modifications do not resemble the polygenic nature of human common obesity [7].

Other species used in obesity research are hamsters, dogs, cats and pigs [7-9].

According to Page and Goodman (2001), the separation of human and other great apes lineages occurred approximately 25 million years ago [10, 11] (Figure 1). Thus, the genetic similarity between human and Old World monkeys [12], such as macaques [13] and chimpanzees [14] is higher than with any other species.

Many non-human primates (NHP) in captivity develop spontaneous obesity, including rhesus monkeys (*Maccaca mulatta*), macaques (*Macaca* sp), vervets (*Chlorocebus aethiops*), bonnet macaques (*Macaca radiata*), baboons (*Papio* sp), chimpanzees (*Pan troglydytes*), mandrills (*Mandrillus sphinx*), marmosetes (*Callithrix jacchus*), squirrel monkeys (*Saimiri sciureus*), capuchins (*Cebys apella*), Tamarins (*Saquinus sp*), and cynomolgus monkeys (*Macaca fascicularis*). As they age, these species show changes in body composition with the metabolic abnormalities that typically couple with obesity [15], such as insulin resistance, dyslipidemia, and pancreatic abnormalities [13, 16].

Among the observed changes in pancreas, the deposition of amyloid is a trait that is observed mainly in primates. Amylin is a protein that is co-secreted with insulin in the pancreas, and may form amyloid that replaces islet mass [17]. Islet amyloidosis has been found in 90% of human with T2DM, and in macaques, vervets and baboons. In primates, the amino acid sequence of amylin allows for deposition and formation of amyloid.

As observed in humans, T2DM has been associated with age and body mass in NHP4,6, 13,20 [18].

Gestational diabetes (GDM) has also been identified among cynomolgus and rhesus monkeys. This condition is considered a risk factor for macrosomy and development of T2 DM after pregnancy 21.

2. Metabolic and Morphometric Traits in Nonhuman Primates

Traits related to obesity and the metabolic syndrome show variation across different studies conducted in non-human primates (NHP). The associations between metabolic and body composition traits in NHP are very similar to those found in humans.

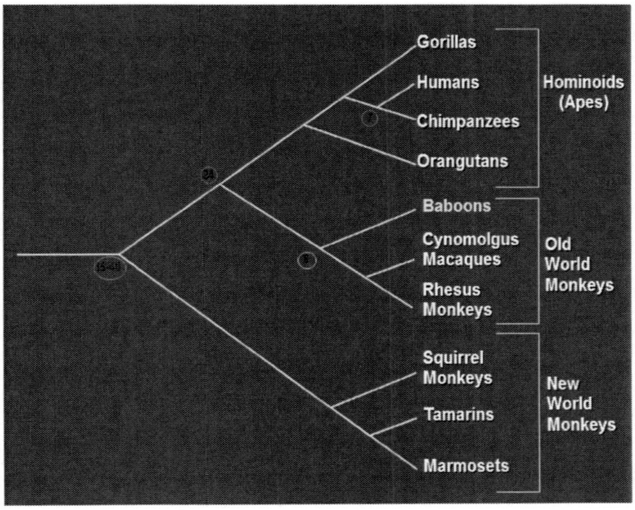

Figure 1. Timeline of primate evolution. Circled numbers represent millions of year since the species diverge from their common ancestor.

Obesity and features of the metabolic syndrome (dyslipidemia, insulin resistance, high blood pressure, central adiposity, glucose intolerance, low grade inflammation, among other traits) may be induced by dietary manipulation in these species (see section for Feeding studies). Morphometric measures such as segment length, circumferences and skin folds have been conducted in NHP, [19, 22] Methods for body composition analysis have been applied to study the amount and distribution of body fat and non-fat tissues in NHP [20, 22, 23].

Findings of these investigations have shown that most of these species have less subcutaneous fat than humans, with a significant amount of visceral fat, and association between these measures and insulin resistance. A study by Chavez et al. (2008) tested three models using simple morphometric measurements of adiposity, such as abdominal circumference and body weight, plus measures of glucose and insulin parameters. The best predictive value was found for the model that included abdominal circumference and fasting insulin levels [23]. The development of this method allows for a fast and accurate identification of overweight, obese, and insulin resistant baboons for investigation of human metabolic diseases. As observed in Table 1, variation in body weight is associated with the circulating levels of hormones and metabolites involved in insulin sensitivity.

2.1. Biochemical Parameters

Phenotypes such as blood metabolites, proteins and hormones are routinely measured for research in NHP. In many cases, the similarity between the human and nonhuman primate proteins is very high, and antibodies developed for human samples may be used for accurate analysis.

Given the characteristics of lipid transport and metabolism, inflammation and glucose utilization, NHP such as macaques, baboons and other species have been used in the investigation of atherosclerosis and mechanisms related to its complications [23-27].

Table 1. Variation in metabolic traits according to weight in baboons

	Male			**Female**		
Phenotype	**Obese n= 5**	**Non-obese n=5**	**p**	**Obese n= 5**	**Non- obese n=5**	**p**
Age (y)	**12.1 (2.1)**	**10.4 (1.3)**	**NS**	**17.1 (2.2)**	**16.4 (3.3)**	**NS**
Weight (kg)	**32.8 (5.6)**	**24.8 (3.6)**	**0.03**	**20.8 (5.6)**	**16.3 (0.6)**	**0.001**
Waist (cm)	**71.3 (9.3)**	**53.9 (7.2)**	**0.01**	**51.3 (5.3)**	**43.9 (4.2)**	**0.001**
Glucose (mg/dl)	**86.7 (19.0)**	**83.5 (12.4)**	**NS**	**82.7 (10.1)**	**79.5 (8.4)**	**NS**
Insulin (mU/dl)	**47.3(35.9)**	**17.3 (15.4)**	**0.01**	**31.3(20.9)**	**11.3 (14.4)**	**0.01**

2.2. Studies of Pharmacological Treatment of Obesity

Considering the significance of an effective and safe medication for weight loss, a large amount of research in this area has been carried out during the last decades 28 (Foltin).

Baboons, rhesus monkeys, and other NHPs have been used to investigate the effects of anorexigenic drugs such as amphetamines, fenfluramine, compounds that affect dopaminergic and serotonergic functions, diazepam, and dexfenfluramine, among others. These studies have contributed to the understanding of the mechanisms involved in feeding behavior [28-31].

3. The Baboon As a Model for the Study of Obesity-Related Abnormalities

3.1. Baboons

The baboon (*Papio* sp) is a primate that lives in semi-deserts, dry savanna and rocky areas of Egypt, Sudan, Eritrea, Ethiopia and Somalia. As in the rest of the *Papio* species, the baboon has a profound sexual dimorphism.

Males are much larger and stout (60 to 85 cm/22-37 kg) than females (53-70 cm/12-15 kg), fitted with larger tusks they use in fights and a mane on the head and shoulder that makes them appear larger and dangerous. [32] The baboon fur is gray in the case of males and light brown in females, and in both sexes longer than other *Papio* species (five species) (Figure 2): *Papio hamadryas* (*P hamadryas*), *Chacma baboon* (*P ursinus*), Guinea baboon, olive (*P anubis*) and yellow (*P cynocephallus*), although it is not clear whether they are truly species or subspecies. The face, hands and buttocks are red. Species of baboons live in drier environments than other *Papio* species (collectively "savanna *Ph*"), but need to drink quite often, so they never stray from a source of water. They sleep in caves or take shelter among rocks and have an omnivorous diet, although most of its diet consists of roots and plants. As for their social life, they form groups or tribes of an adult male and one to three adult females, most adolescents and young.

Several tribes move together during the day in a small herd or band. The bands come together, especially at bedtime, to form troops or hordes. Troops can get to have up to 800, although usually there are 40-140. The estimated lifespan for this species is of about 25 to 40 years [32]. Baboons are considered adults around 7 years of age, and have a menopause. [27]

3.2. In Vivo Metabolic Studies

Baboons have been used for *in vivo* metabolic studies, providing highly valuable information on the physiological and molecular mechanisms involved in energy metabolism. This model has significant advantages due to the homogenous conditions, such as diet and other environmental factors, throughout their life.

Taken from: WIKIPEDIA [71].

Figure 2. Male and female *Papio hamadryas*.

3.3. Studies Using Euglycemic, Hyperinsulinemic Clamp

The euglycemic hyperinsulinemic clamp method measures the amount of insulin required to maintain a stable glucose concentration in blood and is considered the gold standard for the study of insulin resistance. The procedure is time consuming, and is considered invasive and costly. Some simple and inexpensive methods to screen for insulin resistance in human are available, and require measurements of glucose and insulin [20, 33].

Given the need for a simple method to identify obese, insulin resistant baboons for research, Chávez et al. investigated a group of obese, non-diabetic, insulin resistant baboons with the euglycemic clamp method to develop predictive models for the analysis of insulin sensitivity. Findings of this investigation indicate that insulin resistance in this species is highly predicted by adiposity, metabolic markers of obesity and impaired glucose metabolism [21, 34]. A study was conducted using the euglycemic clamp in lean and obese nondiabetic baboons, with collection of skeletal muscle and adipose tissue biopsies for the analyses of the expression of molecules involved in insulin metabolism at baseline, and 30 and 120 min after the influx of insulin. Findings of this investigation suggested a wide range of insulin sensitivity among the studied baboons, with a significant inverse correlation between indicators of adiposity and insulin sensitivity. The genes encoding for the analyzed proteins were sequenced, and were found to be approximately 98% identical to the human nucleotide and amino acid sequence.

Obese and insulin resistant baboons showed abnormal insulin-mediated signaling in key molecules in muscle and adipose tissue as compared with observations in lean baboons.

A study conducted by Kamath et al. (2011) analyzed the hepatic and peripheric insulin resistance using the euglycemic clamp, and the fatty acid composition of the liver. Obese baboons had higher triglyceride content in the liver as compared with insulin-sensitive animals. The levels of fatty acids showed significant correlations with traits such as waist circumference, BMI, plasma insulin and lipids. According to this investigation, insulin resistance was associated with histological findings of non-alcoholic fatty acid disease [35].

This species has also been utilized for the adaptation and optimization of clinical protocols, such as the measurement of fatty acid in blood and glycerol fluxes. Such studies allow the assessment of *in vivo* mechanisms involved in lipid metabolism. Thus, the final protocol demonstrated maintenance of food intake and body weight [36], limited the stress on the animal, and allowed for blood sampling to perform the analyses of a group of metabolites and hormones such as glucose, insulin, cortisol, plasma triglycerides, apolipoproteins, free fatty acids, and glycerol concentrations and the turnover of fatty acid metabolites. Overall, the described data add support for the use of NHP to investigate different aspects of energy metabolism and related abnormalities.

3.4. Adipose Tissue

Excess adipose tissue is caused by an imbalance between energy intake and expenditure. As observed in other mammals, NHP store energy as triglycerides in adipocytes, which is the characteristic cell type in adipose tissue, among other types of cells that are present in smaller numbers, and have relevant biological functions. During the last decades the adipose tissue has been thoroughly investigated across many different animal models and human studies, and has been identified as the largest endocrine organ in the organism [22, 23].

This tissue produces a wide variety of proteins with complex biological activities. Studies in baboons have shown that this tissue expresses, as observed in human, leptin [24, 39], adiponectin [25, 40], resistin [26, 41], and *Glut4* transporter transcripts [27, 42], among many other transcripts. Other features of adipocytes such as size, triglyceride content and number [39, 40], show variation and are related to functional traits and inflammatory markers. Leptin mRNA and circulating protein are heritable, with their levels being higher in female than male baboons, as observed in humans, and expression levels correlate with adipocyte size [39]. The presence of inflammation in this tissue is consistent with the increase in adiposity and adipocyte size [43]. Transcripts for pro-inflammatory cytokines are expressed in adipose tissue, and their concentration is related to metabolic abnormalities. Resistin mRNA was used as a quantitative phenotype in a linkage study, which showed pleiotropic effects influencing its levels and traits related to the metabolic syndrome [41].

A similar study was conducted using *Glut4* mRNA, which encodes for an insulin-dependent glucose transporter, as a quantitative phenotype. Findings suggest that insulin, body weight, and this transporter may share genetic effect [42]. Adipose tissue studies in baboons have been conducted mainly in omental adipose tissue, although other fat depots have been investigated, and differences in gene expression have been documented in cells derived from specific adipose tissue depots [44]. Apolipoprotein E (apoE) expression was compared across different fat depots in five species.

ApoE expression is higher in subcutaneous fat compared with visceral fat in humans, mice, rats and baboons. This apolipoprotein is a key modulator of adipose tissue triglyceride and lipoprotein metabolism [44, 45].

Because of its role in modulating adipocyte triglyceride and lipoprotein metabolism, depot-specific differences in endogenous adipocyte apoE may affect the lipid accumulation pattern. Compared with subcutaneous abdominal fat, subcutaneous gluteal fat had a similar expression level of apoE, being lower in visceral and pericardial and higher in epicardial adipocytes. Inflammation in adipose tissue was associated with suppression of apoE expression [45].

A study by Bose et al. used the homeostatic model assessment method (HOMA-IR), a surrogate measure of insulin resistance, to estimate systemic insulin resistance in a group of unrelated baboons.

Significant correlations were found between plasma monocyte chemoattractant protein 1 (MCP-1) and traits in adipocytes (number and volume), HOMA –IR index, and levels of liver enzymes, after adjustment for age and sex [46]. Inflammation in baboon adipose tissue is also related to macrophage infiltration and IL-6 expression [43]. These findings suggest that MCP-1 may be involved in insulin resistance, and may participate in the development of NAFLD. Overall, these investigations indicate that biological activity of the baboon adipose tissue shows very similar characteristics to human tissue, with heritable variation in number and size of adipocytes, infiltration of macrophages and expression of adipokines. The expression of these proteins has been related to phenotypes of insulin resistance, reinforcing the concept that adipose tissue cross talks with peripheral tissues [39-43]. Expression studies have been conducted in other baboon tissues associated with obesity and comorbidities such as T2DM and cardiovascular disease including skeletal muscle, liver, endothelial cells, pancreas, hypothalamus, kidney, etc. [47-49], allowing the opportunity to analyze tissues that are not available for in vivo human studies. Significant differences were found in the transcriptome of the analyzed tissues between obese and non-obese animals, and validated by qPCR assays: The baboon was found to be a suitable model for research of the metabolic pathways involved in the pathogenesis of multifactorial diseases such as obesity [48]. These results describe for the first time 43 genes differentially expressed by gender and the presence of obesity. Of all the discovered genes, 38% of them participate directly or indirectly in the endocrine system, 22% in the regulation of immunity and inflammation, 8% in lipid metabolism, 4% in glycolysis and glycogenesis and the remaining 28% they did not relate experimentally or theoretically with any condition related to obesity.

Pathway analysis showed differences in genes involved in energy expenditure, insulin signaling, cell cycle regulation, neurotransmission, and cancer, among others. The identifycation of genes that are differentially expressed between obese and non-obese animals across different tissues may allow for the identification of therapeutic targets and biomarkers related with obesity and its comorbidities.

3.5. Feeding Studies

A large number of feeding studies have been conducted in NHP, particularly in baboons and rhesus monkeys. These studies have addressed the effect of different types of diets and ingredients on body composition and metabolic indicators, across the life span.

During the last decades some investigations were conducted to compare the long-term effect of breast feeding compared to formula, and overfeeding on the risk for metabolic diseases in adult life.

According to the results of these experiments, long-term deferred effects of infant nutrition affect serum HDL-cholesterol concentrations, adiposity, and atherosclerosis in the baboon. Adiposity, low HDL-C levels, and obesity are associated with accelerated progression of atherosclerosis and with an increased risk of coronary heart disease in humans.

This experiment identified effects of infant nutrition on phenotypes related to metabolic abnormalities such as bile acid metabolism, enzyme activities, and water and electrolyte balance. These findings suggest that infant nutrition may program metabolic systems for life, and that these effects may contribute to other chronic diseases of adults [51].

A series of studies has been conducted using a high fat diet to induce atherosclerosis in baboons. As observed in humans, differences between responders and non-responder have been identified, and genetic variation has been considered a possible contributor to variation in the metabolic effect of dietary interventions. The changes in the lipoprotein sub-fractions were used to identify differences between subjects and develop lines of baboons [52]. It has been observed that baboons living in the wild at National parks in Africa, where they have contact with human food, may increase their body weight and develop characteristics of the metabolic syndrome [53].

An interventional dietary study using a high sugar, high fat diet was carried out in adult male baboons to induce weight gain and an abnormal metabolic profile. The diet was composed by solid (a modified Chow-like pellet) and liquid energy sources (a sugary beverage), and water. The animals were assigned to three groups with ad libitum access to food. The control group was fed a regular low fat Chow diet and water, a second group had the experimental diet consisting of solid high fat-high sugar pellets, and water, and the third group received the experimental diet and a sugary drink, although they had full access to water. The experimental diets were provided for 8 weeks. Phenotypes were measured at baseline and at week 8 of treatment. Body composition was determined by dual-energy X-ray absorptiometry (DEXA) and metabolic biomarkers were analyzed using standard methods. No changes in body composition and blood biomarkers were observed in the control group. The intervention group with the solid and liquid sources of energy showed the most significant changes, with increase in fat mass in the absence of weight gain, plasma triglyceride, leptin and percentage of HbA1c, and decrease in adiponectin. These findings indicate that the consumption of a diet with a high sugar and fat content, increased body fat and triglyceride concentrations, induced abnormal adipokine concentrations and glucose metabolism in baboons; these findings are concordant with observations in human fed high fructose diets [54]. Although the focus of this section is the use of baboons as an animal model for feeding studies on obesity and metabolic disease, a brief comment on the use of rhesus monkeys in studies of the effects of caloric restriction (CR) is provided. This model has been used for long term CR, feeding approximately 70% of the energy requirement to adult monkeys. Results of these studies consistently show that aging, and the onset of metabolic diseases is significantly delayed in these animals. Features such as insulin resistance, dyslipidemia, and inflammation are significantly lower across lifetime in the treated animals as compared with ad libitum fed peers [55]; similar findings have been found in other species, such as yeast, worms, flies and rodents [56].

4. Genetics of Obesity in Non-Human Primates. Studies in Baboons

4.1. Genetic Sequence

The application of new technologies to research of genomics of multiple species has allowed a rapid evolution of this field [57]. The whole genome sequence of several NHP species is available, as well as information on their transcriptome profiles and sequences, thus, the analysis and comparison of DNA sequences coding and non-coding region across NHP and human species is available for the study of inter and intra-species variation, exploration of the biomedical value of currently used species, and investigation on evolution and conservation. Some of the currently available sequences are baboon, rhesus monkeys, chimpanzee, cynomolgus monkeys; information is available at public gene databases for the comparison and analyses of these sequences [58]. A high percentage of identity in ORFs and proteins (90-99%) has been reported among primates. The high percentage of identity between baboon and human genomes allows the use of tools designed for the analysis of human tissues, such as antibodies, RNA and microRNA transcript microarrays [34, 48, 49, 59] (Table 2). The bioinformatic analyses of these data sets may be challenging due to the sequence differences between the two species, although highly productive, given that the information may be used for multiple purposes. The sequence and expression pattern of numerous genes that encode for proteins that participate in insulin sensitivity, appetite regulation and adipose tissue biology has been studied in baboons, showing remarkable similarity with the human homologues (Table 2).

4.2. Heritability of Metabolic Traits

The investigation of heritability of traits is required to establish the genetic contribution of genetic factors to the variation of the phenotype of interest, and to conduct further studies to identify the effect of specific genes and their variation to a given phenotype. The calculation of heritability (h^2) requires the analysis of traits in genetically related individuals. The observed heritability of many obesity-related traits has been estimated in baboons.

Phenotypes such as glucose, insulin, lipoproteins, liver enzymes, metabolites, hormones, and even mRNA transcripts have shown h^2 values between 0.14 to 0.8 (Table 3), suggesting that they are influenced by genetic factors that contribute to their variation.

Table 2. Percentage of identity between baboon and human ORF sequences of selected genes

Molecule	% of identity mRNA	% of identity amino acid
GLUT4 transporter	95	97
Ghrelin	97	96
Lipoprotein lipase	95	96
Adiponectin Receptor 1	99	99
Adiponectin Receptor 2	97	98
Resistin	95	95

Table 3. Heritabilities of obesity-related traits in baboons

Trait	h^2	p-value
Weight (kg)	0.8	< 1 × 10-7
Weight adjusted for length (BMI)	0.46	< 1 x 10-5
Fat Mass (kg)	0.41	0.00017
Fat Free Mass (kg)	0.32	6 × 10 -6
Plasma insulin	0.46	0.006
Plasma glucose	0.19	0.02
HOMA-IR	0.47	0.006
Plasma alanine aminotransferase (ALT)	0.24	< 0.001
Plasma aspartate aminotransferase (AST)	0.32	< 0.0001
Adipose cell volume	0.36	9 x10 -5
TNF-a in plasma	0.21	0.009
GLUT4 mRNA in adipose tissue	0.24	0.005
Resistin mRNA in adipose tissue	0.23	0.003
Adiponectin mRNA in adipose tissue	0.21	0.005
Plasma ghrelin	0.25	0.0004
Plasma CCK	0.14	0.03
Plasma IL-6	0.27	0.003
Plasma CRP (C reactive protein)	0.2	0.07

Some of these values are higher than those found in humans, probably due to the smaller variation in non- genetic (environmental) factors found among NHPs. The heritability data may be used for the estimation of genetic correlations. These data identify the presence of pleiotropy, or genetic factors that are influencing two different traits. Several investigations have demonstrated the presence of significant pleiotropy between traits in NHPs [40-43].

4.3. Linkage Studies

A linkage map of 7.2 cM, built with 325 human and 6 species-specific microsatellites was developed for the genetic study of the baboon [60] to allow for the identification of chromosomal regions that harbor genes that contribute to the variation of traits of interest, and localize them in the corresponding region on human chromosomes. The position of the analyzed microsatellites on the baboon map showed, in some cases, correspondence with the position in the human chromosome (chromosome 1). In other cases, the microsatellites position suggested the presence of rearrangements of chromosome segments (chromosome 7).

The identification of the peaks flanked by microsatellites, with significant LOD scores, defines a sequence for further analysis. This sequence will be used for the selection of positional candidates and fine mapping of the genes of interest. The identified genetic variation in the sequenced genes will be statistically and functionally tested to confirm the contribution to the variation on phenotypes of interest. Many phenotypes have been analyzed in NHP, and significant LOD scores (logarithm of odds >3) have been found for them. Cross species replication of a QTL has been reported for mRNA levels of baboon and human resistin on chromosome 19.

Significant QTLs have been found for plasma resistin in both species, but the region was not replicated [61]. One of the baboon traits that has been found to have a significant QTL is adipocyte volume (LOD score 3.2) on the baboon homologue of human chromosome 6, near marker D6S1028. Other QTLs identified after genome scan analyses have been found for gamma glutamyl transpeptidase (GGT) plasma levels located on the baboon homologue of human chromosome 22, between markers D22S304 and D22S280, and for plasma albumin levels on marker D10S1432 on the baboon homologue of human chromosome 10 [62]. A genome-wide scan was conducted using plasma ghrelin as a quantitative trait, and evidence of suggestive linkage (LOD score 2.3) for a locus on chromosome 9q22 was found (between markers D9S910 and D9S261) [63]. Cholecystokinin (CCK) is a hormone produced in the gastrointestinal tract that inhibits food intake. This molecule is involved in satiety, and is a signal for controlling meal volume. A genome-wide scan of plasma CCK detected a significant signal on chromosome 17p12-13 (LOD = 3.1), near marker D17S804. A second signal was suggestive of linkage with a LOD score of 2.3, near marker D4S2374, on chromosome 4q34-35. According to previous studies, chromosome 17p seems to harbor genes that influence several traits related to obesity, such as BMI, leptin, adiponectin, and acylation stimulating protein [64].

5. Studies on Genetics of Obesity

5.1. The Baboon As a Model for Obesity during Pregnancy

Maternal obesity is associated with a higher risk for gestational diabetes, incidence of stillbirths, fetal macrosomy, fetal intrauterine growth restriction and pre-eclampsia [65]. The placenta is a very unique organ that secures the supply of nutrients to the fetus, among other functions during pregnancy. Information on the effects of maternal obesity on placental performance is scarce in human or animals. Given that the baboon placenta closely resembles the morphology of the human placenta, this animal model is ideal for investigation of the structure and function of this organ. Maternal obesity is present in 20-34% of pregnant women and has been linked to abnormalities in fetal development such as intrauterine growth restriction and large-for-gestational age products. An investigation conducted by Farley and cols. studied the effects of maternal obesity in a baboon model for pregnancy. They hypothesized that obesity during pregnancy would induce an inflammatory state, and this condition may be related with structural and functional changes in the placenta [65]. Findings of this study showed that obese mothers had decreased placental syncytiotrophoblast amplification factor, and decreased system A placental amino acid transport activity. Fetal serum amino acid composition and mononuclear cells (PBMC) transcriptome were different between fetuses from obese and non-obese mothers. In addition, increased placental and adipose tissue macrophage infiltration, and increased CD14 expression in maternal PBMC and maternal hyperleptinemia were found in obese baboons. These data suggested that placental and fetal traits resemble findings in human obesity during pregnancy, even in the absence of increased fetal weight. An investigation by Samson et al. analyzed structural changes in the placental villous membrane (VM) during obesity. The study compared 4 obese to 4 non-obese pregnant baboons [66].

The analysis was performed by transmission electron microscopy to estimate VM oxygen diffusing capacities and morphometric in placental stereology. This study found that specific placental oxygen diffusing capacities per unit of fetal weight were similar between baboons and humans.

A significant, negative correlation between leptin and placental basement membrane thickness was found, while fetal leptin levels correlated negatively with endothe-lial thickness of fetal capillaries.

No differences were observed in total and specific villous membrane oxygen diffusing capacities between the compared groups. According to these results, obesity is associated with placental oxygen diffusing capacities and placental ultra-structural changes.

A very similar model was used for the comparison of endocanabinoid metabolism during pregnancy in obese and non-obese pregnant baboons. Significant differences were found in the placenta and adipose tissue parameters for endocanabinoid metabolism [67]

An investigation of the effects of betamethasone administered to pregnant baboons at doses equivalent to those given to pregnant women during and after 3 weekly courses was conducted. Glucocorticoids are known to promote obesity in adults. According to this study, betamethasone administration during the second half of pregnancy had an effect on maternal weight, but not on maternal food intake nor on fetal weight, as assessed at the end of gestation. Betamethasone induced metabolic changes including an increase in maternal serum glucose levels, the ratio of insulin-like growth factor-I and insulin-like growth factor binding protein-3, and serum leptin during treatment.

Maternal and fetal serum leptin concentrations did not differ between groups at the end of gestation [68]. Results of this investigation suggest that prolonged maternal hyperleptinemia caused by betamethasone administration in the second half of gestation did not have an effect on fetal metabolic parameters and placental leptin distribution at the end of gestation.

Overall, these studies show very significant similarities between baboon and human physiology during pregnancy, and support the use of baboons as a model for human diseases during this stage of life.

6. PHARMACOGENOMICS

Pharmacogenomics investigates the interaction between the genome and biologically active substances used for the prevention, and mainly treatment of disease. The metabolism of drugs and other xenobiotics is performed by a variety of enzymes. Some of these enzymes are clustered in families with structural and functional similarities. The baboon has been explored as a model for pharmacogenomics studies, and some of the major groups of enzymes involved in drug metabolism have been investigated.

Some examples are UDP glucuronosyltransferase 1A enzymes (UGT1As), carboxyleste-rase (CES). The UDP glucuronosyltransferase 1A enzymes participate in major pathways for elimination of drugs and endogenous substances. CES enzymes exert the detoxification of a wide range of drugs and narcotics, and catalyze some reactions in cholesterol and fatty acid metabolism. Baboon CES1 and CES2 were sequenced; the protein secondary and tertiary structure were predicted using *in silico* methods. The evolutionary relationships for these enzymes with other primate CES orthologues were analyzed.

The baboon CES1 and CES2 proteins showed extensive percentage of identity with human CES1 and CES2, conserving the key structural features reported for the human enzyme, and showed family specific sequences that allow for the characteristic structure [70]. Studies of the genetic and biochemical properties of primate CES may contribute to understand the mechanisms underlying human disease, including atherosclerosis, obesity and drug addiction, for which non-human primates serve as useful animal models [28, 69, 70] Although many other mechanisms involved in drug metabolism remain to be investigated, findings of these studies suggest that the baboon may be an excellent model for research on genome-drug interactions.

In some areas of Ancient Egypt, the *Ph* was domesticated for recollection of fruits of palm trees and even to guard herds of ruminants. The relationship between men and monkeys became so close that the *Ph* was worshiped as a local god minor, Babi (from whose name comes the word *Ph*), one of the many forms taken by Dyehuty (for Egyptians) or Tot (for the Greeks), the god of wisdom, writing, music, spells and magic spells and symbol of the Moon [71, 74]. It is for this reason that sometimes the name "Sacred baboon" is used (Figure 3).

Currently, the baboon is not considered an endangered species (IUCN: LC). Human pressure (weak in the regions where they live) has expelled them from some areas, but the extermination of natural predators (lion, eagle Kaffir, leopard, striped hyena, spotted hyena, common jackal and black-backed jackal) by man has helped to reduce mortality of baboons to a minimum.

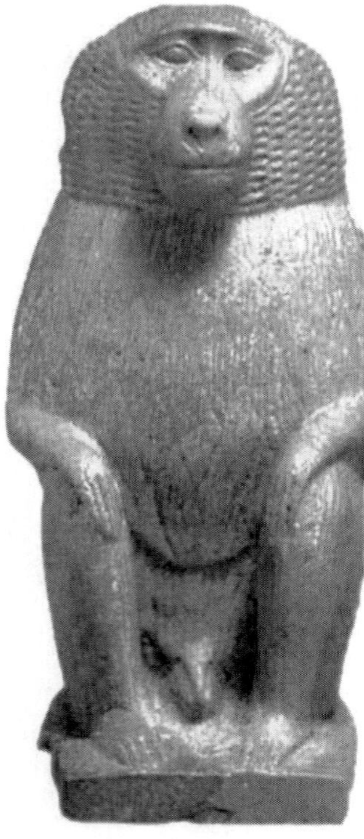

Taken from: WIKIPEDIA [71].

Figure 3. Dyehuty baboon depicted as ([illegible]). This is considered one of the many shapes that this deity could acquire.

CONCLUSION

NHP are the group of species with the closest phylogenetic relationship with humans, with a very high physiological and genetic similarity. These characteristics make NHPs valuable models for research of complex human diseases, including obesity and the variety of metabolic abnormalities that are triggered by the excess of adipose tissue. Although there is a high similarity between Old World Monkeys and humans, some species have been more used for specific research questions. A vast body of evidence has demonstrated that the baboon is an excellent animal model for the study of genetic and metabolic factors underlying obesity and related diseases.

REFERENCES

[1] Obesity as a medical problem. Kopelman, P. G. *Nature*. 2000 Apr. 6;404(6778):635-43.

[2] Genetics of nonsyndromic obesity. Lee, Y. S. *Curr. Opin. Pediatr*. 2013 Dec.;25(6): 666-73.

[3] Obesity in rhesus and cynomolgus macaques: a comparative review of the condition and its implications for research. Bauer, S. A., Arndt, T. P., Leslie, K. E., Pearl, D. L., Turner, P. V. *Comp. Med.* 2011 Dec.;61(6):514-26.

[4] The use of animal models in the study of diabetes mellitus. Chatzigeorgiou, A., Halapas, A., Kalafatakis, K., Kamper, E. *In Vivo*. 2009 Mar.-Apr.;23(2):245-58.

[6] Old world nonhuman primate models of type 2 diabetes mellitus. Wagner, J. E., Kavanagh, K., Ward, G. M., Auerbach, B. J., Harwood, H. J. Jr, Kaplan, J. R. *ILAR J.* 2006;47(3):259-71.

[7] The contribution of animal models to the study of obesity. Speakman, J., Hambly, C., Mitchell, S., Król, E. *Lab. Anim*. 2008 Oct.;42(4):413-32.

[8] Type 2 diabetes-an introduction to the development and use of animal models. Kaplan, J. R., Wagner, J. D. *ILAR J.* 2006;47(3):181-5.

[9] Old world nonhuman primate models of type 2 diabetes mellitus. Wagner, J. E., Kavanagh, K., Ward, G. M., Auerbach, B. J., Harwood, H. J. Jr, Kaplan, J. R. *ILAR J.* 2006;47(3):259-71.

[10] The contribution of animal models to the study of obesity. Speakman, J., Hambly, C., Mitchell, S., Król, E. *Lab. Anim*. 2008 Oct.;42(4):413-32.

[11] Catarrhine phylogeny: noncoding DNA evidence for a diphyletic origin of the mangabeys and for a human-chimpanzee clade. Page, S. L., Goodman, M. *Mol. Phylogenet. Evol.* 2001 Jan.;18(1):14-25.

[12] Molecular dating and biogeography of the early placental mammal radiation. Eizirik, E., Murphy, W. J., O'Brien, S. J. *J. Hered*. 2001 Mar.-Apr.;92(2):212-9.

[13] Nonhuman primates and other animal models in diabetes research. Harwood, H. J. Jr, Listrani, P., Wagner, J. D. *J. Diabetes Sci. Technol*. 2012 May 1;6(3):503-14.

[14] Growth hormone locus expands and diverges after the separation of New and Old World Monkeys. González Alvarez, R., Revol de Mendoza, A., Esquivel Escobedo, D., Corrales Félix, G., Rodríguez Sánchez, I., González, V., Dávila, G., Cao, Q., de Jong, P., Fu, Y. X., Barrera Saldaña, H. A. *Gene*. 2006 Sep. 15;380(1):38-45.

[15] The chimpanzee GH locus: composition, organization, and evolution. Pérez-Maya, A. A., Rodríguez-Sánchez, I. P., de Jong, P., Wallis, M., Barrera-Saldaña, H. A. *Mamm. Genome*. 2012 Jun.;23(5-6):387-98.

[16] Obesity in rhesus and cynomolgus macaques: a comparative review of the condition and its implications for research. Bauer, S. A., Arndt, T. P., Leslie, K. E., Pearl, D. L., Turner, P. V. *Comp. Med.* 2011 Dec.;61(6):514-26.

[17] Nonhuman primates and other animal models in diabetes research. Harwood, H. J. Jr, Listrani, P., Wagner, J. D. *J. Diabetes Sci. Technol.* 2012 May 1;6(3):503-14.

[18] Accumulation of intrahepatic islet amyloid in a nonhuman primate transplant model. Liu, C., Koeberlein, B., Feldman, M. D., Mueller, R., Wang, Z., Li, Y., Lane, K., Hoyt, C. C., Tomaszewski, J. E., Naji, A., Rickels, M. R. *Endocrinology*. 2012 Apr.;153(4): 1673-83.

[19] Obesity, body fat distribution, insulin sensitivity, and islet β-cell function as explanations for metabolic diversity. Kahn, S. E., Prigeon, R. L., Schwartz, R. S., Fujimoto, W. Y., Knopp, R. H., Brunzell, J. D., Porte, D. 2001. *J. Nutr.* 131:354S–360S.

[20] Metabolic consequences of the early onset of obesity in common marmoset monkeys. Power, M. L., Ross, C. N., Schulkin, J., Ziegler, T. E., Tardif, S. D. *Obesity* (Silver Spring). 2013 Mar. 20.

[21] Gestational diabetes mellitus in a cynomolgus monkey with group A streptococcal metritis and hemolytic uremic syndrome. Wagner, J. D., Jayo, M. J., Bullock, B. C., Washburn, S. A. *J. Med. Primatol.* 1992;21(7-8):371–4.

[22] Bauer, S. A., Arndt, T., Leslie, K. E., Pearl, D. L., Turner, P. V. Obesity in Rhesus and Cynomolgus Macaques: A Comparative Review of the Condition and Its Implications for Research. *Comparative medicine* 2011, 61: 514-526.

[23] Predictive models of insulin resistance derived from simple morphometric and biochemical indices related to obesity and the metabolic syndrome in baboons. Chavez, A. O., Gastaldelli, A., Guardado-Mendoza, R., Lopez-Alvarenga, J. C., Leland, M. M., Tejero, M. E., Sorice, G., Casiraghi, F., Davalli, A., Bastarrachea, R. A., Comuzzie, A. G., DeFronzo, R. A., Folli, F. *Cardiovasc. Diabetol.* 2008 23;8:22.

[24] Ratio of leptin to adiponectin as an obesity index of cynomolgus monkeys (Macaca fascicularis). Chen, Y., Ogawa, H., Narita, H., Ohtoh, K., Yoshida, T., Yoshikawa, Y. *Exp. Anim.* 2003 Apr.;52(2):137-43.

[25] Characterization and heritability of obesity and associated risk factors in vervet monkeys. Kavanagh, K., Fairbanks, L. A., Bailey, J. N., Jorgensen, M. J., Wilson, M., Zhang, L., Rudel, L. L., Wagner, J. D. 2007. *Obesity* (Silver Spring) 15:1666–1674.

[26] Adiponectin but not resistin is associated with insulin resistance-related phenotypes in baboons. Tejero, M. E., Freeland-Graves, J. H., Proffitt, J. M., Peebles, K. W., Cai, G., Cole, S. A., Comuzzie, A. G. *Obes. Res.* 2004 May;12(5):871-7.

[27] The baboon as a nonhuman primate model for the study of the genetics of obesity. Comuzzie, A. G., Cole, S. A., Martin, L., Carey, K. D., Mahaney, M. C., Blangero, J., VandeBerg, J. L. *Obes. Res.* 2003 Jan.;11(1):75-80.

[28] The behavioral pharmacology of anorexigenic drugs in nonhuman primates: 30 years of progress. Foltin, R. W. *Behav. Pharmacol.* 2012, 23: 461-477).

[29] Antagonism of glutamatergic NMDA and mGluR5 receptors decreases consumption of food in baboon model of binge-eating disorder. Bisaga, A., Danysz, W., Foltin, R. W. *Eur. Neuropsychopharmacol.* 2008 Nov.;18(11):794-802.

[30] A novel procedure for assessing the effects of drugs on satiation in baboons: effects of memantine and dexfenfluramine. Foltin, R. W., Danysz, W., Bisaga, A. *Psychopharmacology* (Berl.). 2008 Sep.;199(4):583-92.
[31] The behavioral pharmacology of anorexigenic drugs in nonhuman primates: 30 years of progress. Foltin, R. W. *Behav. Pharmacol.* 2012 Sep.;23(5-6):461-77.
[32] Lifetime reproductive success, longevity, and reproductive life history of female yellow baboons (Papio cynocephalus) of Mikumi National Park, Tanzania. Rhine, R. J., Norton, G. W., Wasser, S. K. *Am. J. Primatol.* 2000 Aug.;51(4):229-41.
[33] Selection of the appropriate method for the assessment of insulin resistance. Borai, A., Livingstone, C., Kaddam, I., Ferns, G. *BMC Med. Res. Methodol.* 2011 Nov. 23;11:158.
[34] Physiological and molecular determinants of insulin action in the baboon. Chavez, A. O., Lopez-Alvarenga, J. C., Tejero, M. E., Triplitt, C., Bastarrachea, R. A., Sriwijitkamol, A., Tantiwong, P., Voruganti, V. S., Musi, N., Comuzzie, A. G., DeFronzo, R. A., Folli, F. *Diabetes.* 2008 Apr.;57(4):899-908.
[35] Coordinated defects in hepatic long chain fatty acid metabolism and triglyceride accumulation contribute to insulin resistance in non-human primates. Kamath, S., Chavez, A. O., Gastaldelli, A., Casiraghi, F., Halff, G. A., Abrahamian, G. A., Davalli, A. M., Bastarrachea, R. A., Comuzzie, A. G., Guardado-Mendoza, R., Jimenez-Ceja, L. M., Mattern, V., Paez, A. M., Ricotti, A., Tejero, M. E., Higgins, P. B., Rodriguez-Sanchez, I. P., Tripathy, D., DeFronzo, R. A., Dick, E. J. Jr, Cline, G. W., Folli, F. *PLoS One.* 2011;6(11):e27617.
[36] Protocol for the measurement of fatty acid and glycerol turnover in vivo in baboons. Bastarrachea, R. A., Veron, S. M., Vaidyanathan, V., Garcia-Forey, M., Voruganti, V. S., Higgins, P. B., Parks, E. J. *J. Lipid Res.* 2011 Jun.;52(6):1272-80.
[37] Adipose tissue dysfunction in nascent metabolic syndrome. Bremer, A. A., Jialal, I. *J. Obes.* 2013;2013:393192.
[38] Adipose tissue: friend or foe? Hassan, M., Latif, N., Yacoub, M. *Nat. Rev. Cardiol.* 2012 Dec.;9(12):689-702.
[39] Genetics of leptin expression in baboons. Cole, S. A., Martin, L. J., Peebles, K. W., Leland, M. M., Rice, K., VandeBerg, J. L., Blangero, J., Comuzzie, A. G. *Int. J. Obes. Relat. Metab. Disord.* 2003 Jul.;27(7):778-83.
[40] Genetics of variation in adiponectin in pedigreed baboons: evidence for pleiotropic effects on adipocyte volume and serum adiponectin. Tejero, M. E., Voruganti, V. S., Rodríguez-Sánchez, I. P., Proffitt, J. M., Blangero, J., Cox, L. A., Mahaney, M. C., Rogers, J., VandeBerg, J. L., Cole, S. A., Comuzzie, A. G. *Heredity.* 2008 Apr.;100 (4): 382-9.
[41] Genome-wide scan of resistin mRNA expression in omental adipose tissue of baboons. Tejero, M. E., Cole, S. A., Cai, G., Peebles, K. W., Freeland-Graves, J. H., Cox, L. A., Mahaney, M. C., Rogers, J., VandeBerg, J. L., Blangero, J., Comuzzie, A. G. *Int. J. Obes.* (Lond.). 2005 Apr.;29(4):406-12.
[42] Quantitative genetic analysis of glucose transporter 4 mRNA levels in baboon adipose. Tejero, M. E., Proffitt, J. M., Cole, S. A., Freeland-Graves, J. H., Cai, G., Peebles, K. W., Cox, L. A., Mahaney, M. C., Rogers, J., Vandeberg, J. L., Blangero, J., Comuzzie, A. G. *Obes. Res.* 2004 Oct.;12(10):1652-7.

[43] Tejero, M. E., Proffitt, J. M., Rodríguez, I. P., Cole, S. A., Freeland-Graves, J. H., Peebles, K. W., Comuzzie, A. G. Association between adipokines expression, adipocyte number and volume in adult baboons. *Cytokine* 2008, 41: 150-54.

[44] Mechanisms of obesity and related pathologies: role of apolipoprotein E in the development of obesity. Kypreos, K. E., Karagiannides, I., Fotiadou, E. H., Karavia, E. A., Brinkmeier, M. S., et al. *FEBS J.* 2009, 276: 5720–5728.

[45] Adipose tissue depot-specific differences in adipocyte apolipoprotein E expression. Huang, Z. H., Espiritu, D. J., Uy, A., Holterman, A. X., Vitello, J., Mazzone, T. *Metabolism*. 2011 Dec.;60(12):1692-701.

[46] Association of monocyte chemoattractant protein-1 with adipocyte number, insulin resistance and liver function markers. Bose, T., Alvarenga, J. C., Tejero, M. E., Voruganti, V. S., Proffitt, J. M., Freeland-Graves, J. H., Cole, S. A., Comuzzie, A. G. *J. Med. Primatol.* 2009 Dec.;38(6):418-24.

[47] Endogenous ApoE expression modulates adipocyte triglyceride content and turnover. Huang, Z. H., Reardon, C. A., Mazzone, T. *Diabetes* 2006, 55: 3394–3402.

[48] Perfiles transcriptómicos de tejidos de papiones con y sin obesidad. Iram Pablo Rodríguez Sánchez. *Tesis de Licenciatura*. Facultad de Ciencias Biológicas. UANL. 2012.

[49] The baboon kidney transcriptome: analysis of transcript sequence, splice variants, and abundance. Spradling, K. D., Glenn, J. P., Garcia, R., Shade, R. E., Cox. *PLoS One*. 2013 Apr. 23;8(4):e57563. doi: 10.1371/journal.pone.0057563.

[50] Early determinants of adult metabolic regulation: effects of infant nutrition on adult lipid lipoprotein metabolism. Mc Gill, H. C. Jr, Mott, G. E., Lewis, D. S., Mc Mahan, C. A., Jackson, E. M. *Nutr. Rev.* 1996 Feb.; 54: S31-40.

[51] Localization of multiple pleiotropic genes for lipoprotein metabolism in baboons. Rainwater, D. L., Cox, L. A., Rogers, J., VandeBerg, J. L., Mahaney, M. C. *J. Lipid Res.* 2009 Jul.;50(7):1420-8.

[52] Serum leptin levels as a marker for a syndrome X-like condition in wild baboons. Banks, W. A., Altmann, J., Sapolsky, R. M., Phillips-Conroy, J. E., Morley, J. E. *J. Clin. Endocrinol. Metab.* 2003 Mar.;88(3):1234-40.

[53] Eight week exposure to a high sugar high fat diet results in adiposity gain and alterations in metabolic biomarkers in baboons (Papio hamadryas sp.). Higgins, P. B., Bastarrachea, R. A., Lopez-Alvarenga, J. C., Garcia-Forey, M., Proffitt, J. M., Voruganti, V. S., Tejero, M. E., Mattern, V., Haack, K., Shade, R. E., Cole, S. A., Comuzzie, A. G. *Cardiovasc. Diabetol.* 2010 Oct 29;9:71.

[54] Fructose-containing sugars, blood pressure, and cardiometabolic risk: a critical review. Ha, V., Jayalath, V. H., Cozma, A. I., Mirrahimi, A., de Souza, R. J., Sievenpiper, J. L. *Curr. Hypertens. Rep.* 2013 Aug.;15(4):281-97. doi: 10.1007/s11906-013-0364-1.

[55] Caloric restriction and aging in nonhuman primates. Kemnitz, J. W. *ILAR* 2011, 52: 66-77.

[56] 2010. Honoring Clive McCay and 75 years of calorie restriction research. McDonald, R. M., Ramsey, J. J. *J. Nutr.* 140:1205-1210.

[57] In transition: primate genomics at a time of rapid change. Rogers, J. *ILAR J.* 2013 Jan.; 54(2):224-33. doi: 10.1093/ilar/ilt042.

[58] http://ncbi.genomes/primates.

[59] Identification and comparative analyses of myocardial miRNAs involved in the fetal response to maternal obesity. Maloyan, A., Muralimanoharan, S., Huffman, S., Cox., Nathanielsz, P. W., Myatt, L., Nijland, M. J. *Physiol. Genomics*. 2013 Oct. 1;45 (19):889-900. doi: 10.1152/physiolgenomics.00050.2013. Epub. 2013 Aug.

[60] Linkage mapping in Papio baboons: conservation of a syntenic group of six markers on human chromosome 1. Rogers, J., Witte, S. M., Kammerer, C. M., Hixson, J. E., MacCluer, J. W. *Genomics*. 1995 Jul. 20;28(2):251-4.

[61] Cross-species replication of a QTL for resistin on chromosome 19p13. Tejero, M. E., Voruganti, V. S., Curran, J. E., Göring, H. H. H., Johnson, M. P., Dyer, T. D., Jowett, J. B., Collier, G. R., Moses, E. K., MacCluer, J. W., Mahaney, M. C., Blangero, J., Comuzzie, A. G., Cole, S. A. *Heredity* 2008, 101: 60-6.

[62] Examining priorities for a primate genome project. VandeBerg, J. L., Williams-Blangero, S., Dyke, B., Rogers, J. *Science*. 2000 Nov. 24;290(5496):1504-5.

[63] Characterization of ghrelin in pedigreed baboons: evidence for heritability and pleiotropy. Voruganti, V. S., Tejero, M. E., Proffitt, J. M., Cole, S. A., Cox, L. A., Mahaney, M. C., Rogers, J. A., Freeland-Graves, J. H., Comuzzie, A. G. *Obesity* (Silver Spring). 2008 Apr.;16(4):804-10.

[64] Genome-wide scan of plasma cholecystokinin in baboons shows linkage to human chromosome 17. Voruganti, V. S., Tejero, M. E., Proffitt, J. M., Cole, S. A., Freeland-Graves, J. H., Comuzzie, A. G. *Obesity* (Silver Spring). 2007 Aug.;15(8):2043-50.

[65] Feto-placental adaptations to maternal obesity in the baboon. Farley, D., Tejero, M. E., Comuzzie, A. G., Higgins, P. B., Cox, L., Werner, S. L., Jenkins, S. L., Li, C., Choi, J., Dick, E. J. Jr, Hubbard, G. B., Frost, P., Dudley, D. J., Ballesteros, B., Wu, G., Nathanielsz, P. W., Schlabritz-Loutsevitch, N. E. *Placenta*. 2009 Sep.;30(9):752-60.

[66] The morphometry of materno-fetal oxygen exchange barrier in a baboon model of obesity. Samson, J. E., Mari, G., Dick, E. J. Jr, Hubbard, G. B., Ferry, R. J. Jr, Schlabritz-Loutsevitch, N. E. *Placenta*. 2011 Nov.;32(11):845-51.

[67] Endocannabinoid crosstalk between placenta and maternal fat in a baboon model (Papio spp.) of obesity. Brocato, B., Zoerner, A. A., Janjetovic, Z., Skobowiat, C., Gupta, S., Moore, B. M. 2nd, Slominski, A., Zhang, J., Schenone, M., Phinehas, R., Ferry, R. J. Jr, Dick, E. Jr, Hubbard, G. B., Mari, G., Schlabritz-Loutsevitch, N. *Placenta*. 2013 Nov.; 34(11):983-9.

[68] The prolonged effect of repeated maternal glucocorticoid exposure on the maternal and fetal leptin/insulin-like growth factor axis in Papio species. Schlabritz-Loutsevitch, N. E., Lopez-Alvarenga, J. C., Comuzzie, A. G., Miller, M. M., Ford, S. P., Li, C., Hubbard, G. B., Ferry, R. J. Jr, Nathanielsz, P. W. *Reprod. Sci*. 2009 Mar.;16 (3):308-19.

[69] Molecular cloning of the baboon UDP-glucuronosyltransferase 1A gene family: evolution of the primate UGT1 locus and relevance for models of human drug metabolism. Caspersen, C. S., Reznik, B., Weldy, P. L., Abildskov, K. M., Stark, R. I., Garland, M. *Pharmacogenet. Genomics*. 2007 Jan.;17(1):11-24.

[70] Baboon carboxylesterases 1 and 2: sequences, structures and phylogenetic relationships with human and other primate carboxylesterases. Holmes, R. S., Glenn, J. P., VandeBerg, J. L., Cox, L. A. *J. Med. Primatol*. 2009 Feb.;38(1):27-38.

[71] http://es.wikipedia.org/wiki/Papio_hamadryas.

[72] Pathologically changed skeletal remains of baboons from ancient Egyptian times. von den Driesch, A., Boessneck, J. *Tierarztl. Prax.* 1985;13(3):367-72.

[73] The raising and worship of monkeys in the late period of ancient Egypt. von den Driesch, A. *Tierarztl. Prax.* 1993 Apr.;21(2):95-101.

[74] http://es.wikipedia.org/wiki/Thot.

INDEX

A

B

C

D

E

F

G

H

I

J

K

L

M

N

O

P

Q

R

S

T

U

V

W

X

Y

Z

α

γ